recent advances in phytochemistry

volume 11

The Structure, Biosynthesis, and Degradation of Wood

RECENT ADVANCES IN PHYTOCHEMISTRY

A Continuation Order Plan is available for this series. A continuation order will bring delivery of each new volume immediately upon publication. Volumes are billed only upon actual shipment. For further information please contact the publisher.

recent advances in phytochemistry

volume 11

The Structure, Biosynthesis, and Degradation of Wood

Edited by

Frank A. Loewus
Washington State University
Pullman, Washington

and

V. C. Runeckles
The University of British Columbia
Vancouver, British Columbia, Canada

PLENUM PRESS • NEW YORK AND LONDON

Library of Congress Cataloging in Publication Data

Phytochemical Society of North America.
  The structure, biosynthesis, and degradation of wood.

  (Recent advances in phytochemistry; v. 11)
  "Proceedings of the sixteenth annual meeting of the Phytochemical Society of North
America held at the University of British Columbia, Vancouver . . . August 1976."
  Includes index.
  1. Wood—Chemistry—Congresses. I. Loewus, Frank Abel, 1919-   II. Runeckles,
Victor C. III. Title. IV. Series.
TS921.P46 1977   582'.15'041   77-8275
ISBN 0-306-34711-3

Proceedings of the Sixteenth Annual Meeting of the Phytochemical Society of North
America held at the University of British Columbia, Vancouver, British Columbia,
Canada, August, 1976

© 1977 Plenum Press, New York
A Division of Plenum Publishing Corporation
227 West 17th Street, New York, N.Y. 10011

Printed in the United States of America

Douglas fir (*Pseudotsuga taxifolia*) and redwood (*Sequoia sempervirens*) were the main timber sources for this sawmill scene which was recorded in Humbolt County, California near the town of Carlotta. The mature Douglas fir log being toppled into the mill pond dwarfs the truck driver as he stands alongside his rig. Wood scrap, which included unmarketable lumber, sawdust, bark and slash, was burned in the large "teepee" burner as seen in the background. Such practice is rapidly receding into history as the forest products industry recognizes and utilizes recent advances in the chemistry and biochemistry of wood.

F. A. Loewus, 1956.

Contributors

W. J. CONNORS, Forest Products Laboratory, USDA, Forest Service, Madison, WI 53705.

WILFRED A. CÔTÉ, N. C. Brown Center for Ultrastructure Studies, State University of New York, College of Environmental Science and Forestry, Syracuse, NY 13210.

DEBORAH P. DELMER, MSU/ERDA Plant Research Laboratory, Michigan State University, East Lansing, MI 48824.

GEORG G. GROSS, Lehrstuhl für Pflanzenphysiologie, Ruhr-Universität, 4630 Bochum, FRG.

HERBERT L. HERGERT, ITT Rayonier Incorporated, 605 Third Avenue, New York, NY 10016.

FRANKLIN W. HERRICK, Olympic Research Division, ITT Rayonier Incorporated, Shelton, WA 98584.

W. E. HILLIS, Forest Products Laboratory, Division of Building Research, CSIRO, Highett, Victoria, 3190, Australia.

T. KENT KIRK, Forest Products Laboratory, USDA, Forest Service, Madison, WI 53705.

PAPPACHAN E. KOLATTUKUDY, Department of Agricultural Chemistry, Washington State University, Pullman, WA 99164.

DEREK T. A. LAMPORT, MSU/ERDA Plant Research Laboratory, Michigan State University, East Lansing, MI 48824.

D. BIR MULLICK, Pacific Forest Research Centre, Canadian Forestry Service, Department of the Environment, 506 West Burnside Road, Victoria, British Columbia, Canada.

E. T. REESE, Food Science Laboratory, U.S. Army Research and Development Command, Natick, MA 01760.

AKIRA SAKAKIBARA, Department of Forest Products, Faculty of
    Agriculture, Hokkaido University, Sapporo, Japan.

J. G. ZEIKUS, Department of Bacteriology, University of
    Wisconsin, Madison, WI   53706.

Preface

     Forest trees constitute one of the major resources of
the world and their utilization, either for structural
purposes or for the materials which they yield, dates back
to antiquity. Over the centuries, the exploitation of
this resource has become progressively more sophisticated,
and, in many parts of the world has led to the development
of highly complex forest-based industries. The research
and development work which led to these industrial uses
fostered the formation of numerous technical societies
and associations, which, through their meetings and publications, have facilitated communication and the exchange of
ideas.

     Over the years, there have been numerous symposia
devoted to wood and the many facets of its properties and
utilization. However, rarely has the emphasis in such
symposia been placed upon the living tree and the changes
which it undergoes in relation to its ultimate utilization.
Hence the Phytochemical Society of North America arranged
the symposium, "The Structure, Biosynthesis, and Degradation
of Wood", held at the University of British Columbia in
August, 1976, the contributions to which form the basis of
the present volume.

     The first chapter (W. E. Côté) reviews the ultrastruc-
ture of woody tissues and provides a frame of reference
for many of the subsequent contributions. In the chapters
which follow, several deal with the chemical structure and
biosynthesis of carbohydrate polymers, glycoproteins, lignin,
and lipid polymers, the major components of wood. D. Delmer
reviews the formation of cell wall polysaccharides with
particular emphasis on recent developments in our understanding of the biosynthesis of cellulose. D. T. A. Lamport
examines cell wall glycoprotein. His chapter includes a
summary of recent progress and some thought-provoking
suggestions as to the functional and evolutionary significance of these structures. A. Sakakibara discusses the

ix

evidence for the "fine" structure of lignin, based on mild
hydrolysis procedures. His chapter is followed by a comprehensive review by G. Gross that deals with lignin biosynthesis
and the formation of related monomers. P. E. Kolattukudy's
detailed account of his research on the structure, biosynthesis, and degradation of lipid polymers, cutin, and
suberin provides a fitting climax to these chapters on the
major components of wood.

Now the focus shifts to secondary changes in wood
with W. E. Hillis' extensive review of changes which occur
within the tree during aging (e.g. the transition to heartwood) as a result of various environmental conditions
encountered during growth. Other biological agents also
influence wood structure and chemistry. Thus, E. T.
Reese's chapter deals with the degradation of cell wall
carbohydrates by microorganisms while the following chapter
by T. K. Kirk, W. J. Conners, and J. G. Zeikus reviews the
microbial degradation of lignin. In the next chapter,
D. B. Mullick presents a detailed review of recent work on
the cytological changes in both bark and wood which result
from wounding or attack by insects and pathogens.

The final chapter by Herrick and Hergert leads the
reader back to the potentials and practicalities, successes
and failures in the development of commercial wood products,
emphasizing those materials which can be obtained directly
as extractives or by treatment of wood.

We wish to thank the contributors for their role in
making the original symposium a success, and for their
cooperation in bringing this volume to completion. Special
notes of acknowledgment go to the National Research Council
of Canada, and to the Deans of the Faculties of Agricultural
Sciences and Forestry at the University of British Columbia
for financial support of the symposium, and to the Canadian
Forestry Service for supporting the costs of the color
plates included in the chapter by Dr. D. B. Mullick.

Special appreciation goes to Mrs. Donna Verstrate and
Mrs. Paula Jenkins who contributed much of their time to
retyping this volume for publication and to Washington
State University for the use of its facilities in preparing
final copy.

Contents

Chapter One

WOOD ULTRASTRUCTURE IN RELATION TO CHEMICAL COMPOSITION

WILFRED A. CÔTÉ

*N. C. Brown Center for Ultrastructure Studies
State University of New York, College of
Environmental Science and Forestry, Syracuse,
New York 13210*

INTRODUCTION

Wood structure can be studied at the gross level with the unaided eye, at the anatomical level with the light microscope, or at the ultrastructural level with the transmission electron microscope. In essence, resolving power of the optical system used defines the structural domain observed.

According to this guide, wood ultrastructure embraces such areas as cell wall organization and sculpturing, microfibrillar orientation, and the distribution of chemical constituents in the cell wall. The term "ultrastructure" bridges the gap in the realm of structure extending from anatomical features to molecular architecture in wood. In terms of dimensions it begins at the limit of resolving power of the light microscope (200 nm) and extends to approximately 0.5 nm, the present practical limit of resolving power of transmission electron microscopy, when specimen preparation methods and their limitations are considered.[8]

The guidelines of microscope resolving power and structural limits have been altered somewhat in recent years by increasing interest in and application of the scanning electron microscope (SEM) to wood structure studies. Now that the resolving power of the SEM is better than 0.1 nm, structures in wood that could be examined only through the tedious process of ultra-thin sectioning, or of replication, can be studied more easily and with more realistic perspective than with the transmission electron microscope (TEM).

It is appropriate, therefore, that advantage be taken of the three-dimensional appearance of scanning electron micrographs which have greater apparent depth of field than either photomicrographs or transmission electron micrographs. In effect, SEM complements the light and transmission electron microscopes and the results discussed in the review that follows include evidence from all three types of microscopy.

The objective of this paper is to bring together what is now known about the ultrastructure of both normal and reaction wood (compression and tension wood) and to relate to it the distribution of chemical constituents within the wood cells and the wood cell walls. Some of the chemical components of wood lend themselves to examination in place (following suitable treatment) better than others and it is not possible to provide graphic evidence for the distribution of all constituents. However, it is possible to determine with considerable accuracy where most of the various wood polymers are located in wood if other types of direct and indirect evidence are considered.

CHEMICAL COMPOSITION OF WOOD

The chemical composition of wood is well known and has been reviewed comprehensively by a number of authors.[5,44,52] These published results are based on standardized procedures for the analysis of whole wood which has been ground or milled to specified conditions. Unfortunately, this type of analysis does not reveal much about the distribution of the various constituents across the wood cell wall or whether the material was deposited in the cell lumen. Nevertheless, these data are useful as guides for estimating the probable make-up of certain types of cells especially if specific isolation procedures have been followed, for example the separation of the ray tissue.

In evaluating data on summative analyses of wood it is important that consideration be given to the question of statistical sampling. Wood is a natural polymeric material that is extremely variable in structure and therefore in chemical composition as well. Two trees of the same species can have somewhat different percentages of the principal constituents in single samples of each, depending on the source of the sampling in each tree. Such differences generally average out as the number of samples and sampling sites increases. The percentages of chemical constituents given in Tables 1 and 2 are based on adequate sampling.

Comparisons should be made between the values for the five angiosperms and those for the five gymnosperms. The cellulose content of both groups is relatively similar, but the wood of conifers contains considerably more lignin than that of broadleaved species. The hemicelluloses found in the two groups vary both as to type and quantity. As indicated, hardwoods are characterized by their high content of partly acetylated, acidic xylan. This amounts to approximately 20 to 35% of the wood. Glucomannan, another hemicellulose, is found only in small amounts.

In the softwoods, a partly acetylated galactoglucomannan amounts to as much as 20% while the xylan is in the range of 10%. From the species analyzed it can be seen that these are woods from temperate zone trees. Hardwoods from the tropics have lignin contents which can exceed that of many conifers.

A number of constituents present in much smaller amounts and removed with organic solvents or with water include the pectic materials and starch. Ash content in these woods rarely exceeds 0.5%. In other woods, particularly tropical species, the ash content can be much higher.

PHYSICAL NATURE OF CHEMICAL CONSTITUENTS

If the principal constituents of the wood cell wall are classified in terms of structure, it is convenient to follow the terminology introduced by Wardrop.[46] Cellulose, the material responsible for tensile strength in wood (and in other materials such as cotton), is classed as a *framework* substance. The hemicelluloses which are formed at the same time as the cellulose can be called *matrix* substances. These

Table 1. Chemical composition of wood from five angiosperms[*].
All values given in percentage of extractive-free wood.

Component	*Acer rubrum* L.	*Betula papyri- fera* Marsh.	*Fagus grandi- folia* Ehrh.	*Populus tremu- loides* Michx.	*Ulmus ameri- cana* L.
Cellulose	45	42	45	48	51
Lignin	24	19	22	21	24
O-Acetyl-4-*O*- methyl-glucu- rono-xylan	25	35	26	24	19
Glucomannan	4	3	3	3	4
Pectin, starch, ash, etc.	2	1	4	4	2

[*]From Kollmann and Côté.[29] Results provided by T. E. Timell.

Table 2. Chemical composition of wood from five conifers[*].
All values given in percentage of extractive-free wood.

Component	*Abies balsamea* (L.) Mill	*Picea glauca* (Moench) Voss	*Pinus strobus* L.	*Tsuga canaden- sis* (L.) Carr.	*Thuja occi- denta- lis* L.
Cellulose	42	41	41	41	41
Lignin	29	27	29	33	31
Arabino-4-*O*-methyl- glucurono-xylan	9	13	9	7	14
O-Acetyl-galacto- glucomannan	18	18	18	16	12
Pectin, starch, ash, etc.	2	1	3	3	2

[*]From Kollmann and Côté.[29] Results provided by T. E. Timell.

non-cellulosic polysaccharides may be responsible for the
rigidity of the cell until the walls have been lignified.
Lignin is then categorized as an *encrusting* substance the
precursors of which diffuse across the cell wall from the
intercellular region beginning near the cell corners.[46]
Although direct proof may be lacking, evidence from several
sources suggests that lignin is associated with or is combined with the matrix substances but not with the framework
substance.[47]

The form in which cellulose, hemicelluloses and lignin
exists in whole wood is obviously critical to a graphic
demonstration of the distribution of the polymeric materials
across the cell wall. Since cellulose is of greatest importance volumetrically if not also in terms of the ultimate
behavior of wood in use, it will be considered first.

Cellulose. Cellulose is organized into filamentous
microfibrils which are in turn packed in an orderly array
within lamellae which are assembled into cell wall layers.
The term *fibril* has been used for decades in connection with
the strand-like structure of the wood cell wall.[1,2,17,27,41]
The term *microfibril* evolved at some point early in the
development of electron microscopy[38] to differentiate between
the fibrils that were visible with the light microscope and
the smaller diameter strands resolved with the TEM.

Regarding the question of cellulose microfibrillar diameter in natural materials including wood, the past 20 years
have seen estimates shrink from 40 to 50 nm to become smaller
and smaller diameters. By 1964 there were reports of cellulose microfibril diameters of approximately 3.5 nm.[38] These
were termed *elementary fibrils* by Mühlethaler on the basis
that they were thermodynamically of optimal dimensions at
this size. Since that time, even smaller strands, some as
small as 1.7 nm, have been reported. Sullivan[43] applied the
term *protofibril* to his cellulose fibrils while Hanna[21,22,23]
used the name *sub-elementary fibril*.

Perhaps it is fortunate that we have now approached the
theoretical limit of size for a cellulose fibril which, in
the final analysis, must contain some minimum number of
cellulose chains. Otherwise, as technology increases the
resolving power of TEM, and as techniques for preparation
of specimens improve significantly, there could be a further
proliferation of terms describing the various levels of

cellulose fibrillar dimensions. Here, it is sufficient to
say that cellulose exists as strands of indefinite length
with diameters ranging widely from 1.7 nm to 50 nm or greater.
It may be safe to say that the larger strands are merely
aggregations of the smaller ones or that the small ones can
be created by various mechanical means from large strands.

The internal structure of the cellulose microfibrils is
not readily determined by electron microscopy. Crystallinity
is examined more effectively by x-ray diffraction and electron diffraction. Birefringence, as detected and measured
with the polarizing light microscope, remains a powerful tool
for anatomical levels of analysis. Speculation about the
existence of regions of less organization (reduced crystallinity) along the length of a cellulose microfibril is relevant to this discussion only from the point of view of
possible close association of lignin or of non-cellulosic
polysaccharide with cellulose in these less crystalline
regions. It has been proposed[18] in the past that a paracrystalline sheath could envelop the microfibril. However,
these questions border on molecular architecture and actually
fall outside of the domain of ultrastructure except with the
most advanced instrumentation.

Hemicelluloses. Meier[36] pointed out most astutely that
the term "hemicellulose" is a poor one indeed because these
non-cellulosic polysaccharides are not half-cellulose. However, he suggested that, in spite of its shortcomings, it is
preferable to older terms such as cellulosanes, polyuronides
and polyosanes, which have been wisely dropped in the modern
literature.

This group of polymers is not readily imaged in the
electron microscope as separate and identifiable material
such as cellulose. There is no evidence for fibrillar organization of the hemicelluloses in wood, and when cellulose is
removed chemically to leave the remainder of the cell wall
for observation, the hemicellulose is also removed. So far
there has not been a suitable method devised for the removal
of lignin and cellulose from the wood to leave a hemicellulose
"skeleton." Consequently, the distribution of hemicelluloses
in the cell wall must be determined by other means.

A method described by Meier[35] and also used by Kutscha[30]
offers an approach based on the separation of developing cells
into groups, using birefringence of the cellulosic layers as

a guide. In this technique, cells are separated into four
categories: cells with middle lamella (ML) and primary
wall (P) only (see Fig. 1); cells having an S1 layer in addition to ML and P; cells having the outer part of the S2 layer
as well as S1, ML and P; and cells having all cell wall layers
deposited, ML, P, S1, S2, and S3. The various fractions isolated from radial sections of developing tissue can then be
hydrolyzed and the sugars analyzed quantitatively. From these
percentages it is possible to deduce roughly what the proportion of each of the various polysaccharides is in each of the
cell wall layers since density or specific gravity of the
fractions, as well as their volumes, can be determined. While
the technique has shortcomings and is tedious, some conclusions can be drawn regarding the location of hemicelluloses
in the cell wall layers.

   Lignin. In contrast to cellulose, lignin does not exhibit birefringence when sections of wood are observed with a
polarizing light microscope. Cellulose behaves like a crystalline material due to the lattice-like orientation of the
cellulose chain molecules, but the chemical structure of lignin is known to be quite different.

   Ultra-thin sections of wood viewed in a TEM show regions
of greater electron density between wood cells; that is, the
intercellular region or middle lamella appears as a dark band
on an electron micrograph. This results in contrast between
cellulose and lignin in regions near cell boundaries, or
wherever lignin has high packing density. However, it must
be remembered that both the hemicelluloses and lignin are
distributed throughout the secondary cell wall. When electron
staining techniques are employed, lignified regions can be
emphasized as was demonstrated by Meyer.[37] He found potassium
permanganate particularly useful as an electron stain for
lignin and Mann's research[34] confirmed this specificity of
$KMnO_4$.

   Using ultraviolet and fluorescence optics, Frey-
Wyssling[19] demonstrated the uniformity of distribution of
lignin across the secondary cell walls as well as across the
compound middle lamella (primary wall-intercellular layer-
primary wall). The concentration of lignin in the compound
middle lamella is, of course, much higher than in the secondary cell wall. In fact, Frey-Wyssling concluded that in
the primary wall the concentration of lignin is more than

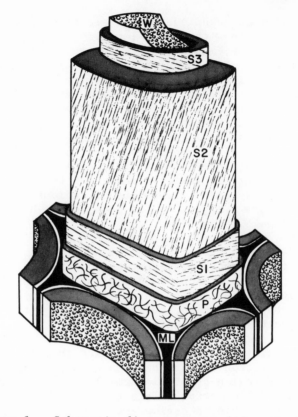

Figure 1. Schematic diagram summarizing the known
ultrastructure of a "model" wood cell. The wall layers are
primary wall (P), outer layer of secondary wall (S1), middle
layer of secondary wall (S2), inner layer of secondary wall
(S3), and the warty layer (W). The middle lamella (ML) is
the intercellular region. The generalized orientation of
the cellulose microfibrils in each of the layers is indicated on the drawing.

twice as high as in the secondary wall. He also concluded
"that the structural state of encrusted lignin is amorphous."

    For examining the distribution of lignin in wood there
is also the possibility of creating a lignin skeleton by the
chemical removal of the polysaccharides. Enzymatic removal
of cellulose and hemicellulose is done quite effectively by
selection of an appropriate wood decaying fungus which

metabolizes these components while leaving the lignin. The
major difficulty in relying on decay fungi for creation of
lignin skeletons is the time factor.[16] More on this subject
is presented in Chapter 9.

To shorten the specimen preparation time, the method
proposed by Sachs, Clark and Pew,[42] with some modifications,
produces lignin skeletons in small samples (8mm × 8mm × 100μm)
in a few days of treatment with increasing concentrations of
aqueous hydrofluoric acid with a final strength of 80%. The
detailed procedure may be found in Côté et al.[14] The results
of this approach appear to be relatively free of artifacts
in most instances, and have been confirmed by other workers.

Wood Extractives. Although present in relatively small
amounts, there is a large number of compounds, particularly
in heartwood, whose distribution is also of interest. The
important extractives such as terpenes and wood resins do not
have the structure required for imaging in electron microscopy. Some substances, such as polyphenols, could have some
electron density if present in sufficiently high concentrations or if a selective staining method could be used to
highlight their presence in an ultra-thin section. However,
in small amounts, this is difficult if not impossible. Similar comments can be made regarding tannins, tropolones, fatty
acids, and other extractives. Graphic demonstration of their
distribution at the ultrastructural level is not a simple
matter. At the light microscope level, fluorescence microscopy can be employed when concentrations of ultra-violet
fluorescent deposits are sufficiently high. In theory it
should also be possible to obtain cathodoluminescence scanning
electron micrographs of such substances *in situ*.

There has been no work reported in which the distribution
of extractives across the cell wall has been examined. It
would be a very difficult task, due to the chemical nature of
most extractives, and would require the analysis of separated
cell wall layers. It can only be said that heartwood contains
the largest proportion of such materials in most woods. For
general coverage of this topic, reference should be made to
"Wood Extractives," 1962, W. E. Hillis, Editor, Academic Press,
New York.

Inclusions. Inorganic constituents of wood which occur
at very low concentration in most species might be examined
by TEM following microashing techniques. However, to show

them within the wood cell wall is not feasible at the present time. Inclusions such as crystals, silica sand, druses, raphides, and other materials found within cell cavities may be recorded with the light microscope or the SEM. Energy-dispersive x-ray analysis (EDXA) of inorganic materials at adequately high concentrations is also possible.[20] Examples of this technique will be included in the section on distribution of chemical constituents in wood.

THE ULTRASTRUCTURE OF WOOD

Although this paper focusses on the ultrastructure of wood, in practice it is inadvisable to examine the sub-microscopic features without first establishing the anatomical areas within which the fine structure is located. In other words, bench marks provided by light microscopy are essential to effective interpretation of ultrastructure, whether in wood or some other material. A number of sources of information on wood anatomy are readily available in recent publications.[6,29,39] For a graphic review of wood structure as well as to offer examples of SEM applications to this area, two scanning electron micrographs (Figs. 2 and 3) are presented which contrast the anatomical patterns found in woods of conifers and broadleaved species.

From about 1950 to 1970, wood ultrastructure was a subject of great interest to anatomists and technologists. The TEM became more available and a number of techniques for specialized specimen preparation were developed for examination of wood. Chief among these methods were replication and ultra-thin sectioning. Papers were published in all parts of the world, but Australia, Japan, Germany, and the United States reflected great research activity.

Wood ultrastructure is still an active area of research.[7] However, since the introduction of the SEM, the orientation of studies has shifted somewhat from the basic structures which required high resolution to more practical and somewhat less demanding levels in terms of specimen preparation. Perhaps one reason for this is that the general ultrastructure of a "model" wood cell is considered to be quite well characterized already.[8] The structures being examined with SEM are certainly sub-light-microscopic, but the resolving power requirements are closer to 10 nm than to 1 or 1.5 nm, which was the case immediately prior to the commercialization of scanning

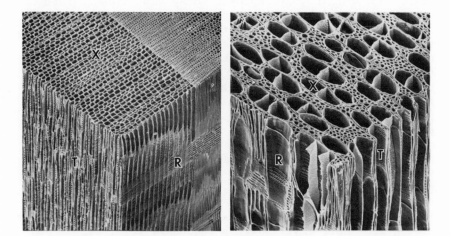

Figure 2 (Left).   Scanning electron micrograph of a small cube of the wood of a typical conifer, western larch (*Larix occidentalis* Nutt.) showing the transverse (X), radial (R) and tangential (T) surfaces.   Note the radial rows of longitudinal tracheids.

Figure 3 (Right).   Scanning electron micrograph of a small cube of the wood of Eastern cottonwood (*Populus deltoides* Bartr.) showing the transverse (X), radial (R) and tangential (T) surfaces.   In hardwoods the vessels and other elements are distributed in patterns characteristic for the species, but not in radial rows as in conifers.

electron microscopy. Because scanning micrographs are so
effective in yielding structural evidence with a "three-
dimensional-look," a number of them are included in this
review.

   Organization of the Wood Cell Wall. Our current knowledge of the organization of the wood cell wall is based on
the contributions of many investigators, not only since the
development of electron microscopy but during the 1800's as
well.   The terminology employed today was "standardized" by
Bailey and Kerr[1] but various aspects of it depended on the
proposals of earlier anatomists. Nevertheless, the Bailey
and Kerr descriptions of the secondary cell wall as having
a three-layered structure with the fibrillar orientation of
the inner and outer layers at approximately 90 degrees to
the cell axis and the middle layer more or less parallel with
the long axis of the cell remain valid today, some forty
years later. The light microscope was used to its maximum
potential by such anatomists. Techniques such as swelling
of the cell wall expanded the structure so that "fibrils"
could be resolved. Polarization microscopy was also employed
effectively and obviously the results were interpreted correctly.

   The commonly used designations in describing the outer,
middle and inner layers of the secondary wall, S1, S2 and
S3, depicted in Figure 1, are believed to have originated
with Wardrop and Dadswell. The outer envelope of the cell,
produced at the cambium during cell division, is the primary
wall and is designated as P. The intercellular region is
also called the middle lamella, ML. If the primary walls
and the true middle lamella are taken together, the term *compound middle lamella* applies. A layer discovered *via* electron
microscopy,[28,31] the warty layer, is labeled W or WL.

   Surface views of various cell wall layers may be obtained
by replication techniques following maceration, sectioning,
splitting, or any procedure that will help expose a particular
portion of the cell. This does not allow a view in depth
through the wall as is possible with the light microscope and
"optical sectioning." However, since the cellulose microfibrils are generally resolvable, the characteristics of each
layer can be determined through good planning or fortuitous
display of appropriate areas in a single micrograph. The change
in orientation between wall layers can be demonstrated and
occasionally the transition region between layers can be
observed (Fig. 4).

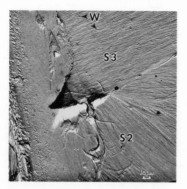

Figure 4.   The organization of the wood cell wall
revealed by a replica of a torn tracheid wall from spruce
pine (*Pinus glabra* Walt.).   The S1 is not visible in this
transmission electron micrograph, but the S2 and some transition lamellae between S2 and S3 can be seen.   A few scattered
warts (W) appear over the S3.

A mild chemical treatment will generally remove enough
of the matrix or encrusting material so that the microfibrils
can be profiled effectively by shadowcasting.   Figure 5 is an
example of the typical structure of the primary wall with its
characteristic randomly distributed microfibrils and comparatively loose texture.

Figure 6 was recorded following chlorite treatment which
removed some of the material masking the structure of the S3
layer in a Douglas-fir (*Pseudotsuga menziesii* (Mirb.) Franco)
tracheid.   Helical thickenings (HT) appear in the upper left
and lower right corners of the micrograph as well as diagonally
from lower left to upper right.   The slight criss-crossed
pattern of the microfibrils is observed quite frequently on
the surface of the S3 layer.

For electron microscopic studies of wood there is one
other specimen preparation technique that yields convincing
evidence of shifts in microfibrillar orientation within the
cell wall.   Ultra-thin sectioning is a method that required
many years of refinement for the preparation of general biological specimens.   The glass knife created a breakthrough and
with modern glass-breaking instruments, excellent cutting edges
can be produced consistently.   However, for cutting wood, glass

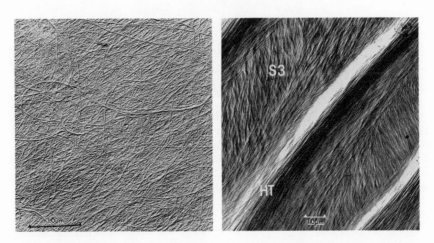

Figure 5 (Left). Primary wall from a tracheid of
tamarack compression wood (*Larix laricina* (Du Roi) K. Koch).
The primary walls have the same appearance in both normal
and reaction wood. This electron micrograph was made from
a replica. In order to emphasize the cellulose microfibrils
and their random pattern of distribution, matrix and encrusting substances were removed chemically.

Figure 6 (Right). Lumen lining of a tracheid of Douglas-
fir (*Pseudotsuga menziesii* (Mirb.) Franco) showing the S3
layer and portions of three helical thickenings (HT). Mild
chlorite treatment removed some of the masking substances so
that the loose texture of the S3 could be observed. Note
the criss-cross pattern of microfibrils at left side of
micrograph.

simply cannot hold an edge long enough for efficient and
effective sectioning of polymethacrylate or epoxy embedments.
The polished diamond knife made the ultra-thin sectioning of
wood feasible as a routine technique. Figure 7 is a cross-
section which shows portions of three Douglas-fir tracheids.
The three-layered structure of the secondary wall is clear
in such an electron micrograph. At greater enlargement it
is also possible to see the primary walls. The middle
lamella with its greater electron density, and the helical
thickenings, characteristic of Douglas-fir, are shown at the
outer and inner cell wall boundaries, respectively.

When lignin skeletons are prepared according to the
method described earlier, they can be embedded and sectioned
like untreated wood. Figure 8 illustrates this approach
with the lignin skeleton from portions of three balsam fir
(*Abies balsamea* (L.) Mill.) tracheids. The S2 layer is
always the predominating layer in the cell wall, but of
course in latewood tracheids the S2 is thicker, as shown
here, than in earlywood cells. The higher packing density
of the lignin in the middle lamella is obvious. Both the
S1 and the S3 are revealed as relatively thin layers and
careful scrutiny will also show the remnants of the warty
layer over the S3 and lining the cell cavity. This layer,
present in many coniferous woods, but not in all species,

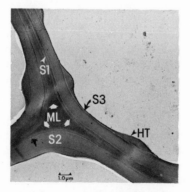

Figure 7. Ultra-thin section of Douglas-fir (*Pseudotsuga menziesii* (Mirb.) Franco) showing portions of three
tracheids in cross-section. The cell wall organization follows the pattern illustrated in Figure 3 with the addition
of helical thickenings (HT) which are typical for Douglas-
fir wood. TEM.

behaves remarkably like lignin during hydrofluoric acid
treatment.   The true chemical nature of these protuberances
has not been established definitively.

A higher magnification view of a lignin skeleton sectioned longitudinally (Fig. 9) reinforces the claims of some
workers that the fibrillar orientation of each lamella within
a layer is slightly different from its neighbor.   This residual parallel structure, in eastern larch tracheids in this
case, is interpreted as possibly reflecting such lamellar
orientations.   The porous regions adjacent to the compound
middle lamella are the S1 layers.   Before their removal, the
cellulose microfibrils were oriented approximately perpendi-
cular to those in the S2, in effect, emerging from the plane
of the micrograph.   Finally, this illustration shows a very
narrow denser band in the center of the compound middle
lamella.   This is the true middle lamella and the less electron dense portion on either side of it is primary wall.

Since wood is a natural polymeric substance, it is not
surprising that deviations from this "model cell" pattern of
structure can be found.   Even in quite normal wood there have
been variations from this three-layered organization noted.[12]
Figure 10 is an ultra-thin section of the wood of a southern
yellow pine which has several layers where one expects to
find only the S1 layer.   In this case it might be explained
as being several lamellae of the S1, a layer which has been
reported to have many alternating lamellae.[46]   However, it
might be argued just as strongly that such lamellae would not
have the thickness that is apparent in this micrograph.   Many
other variations from the normal have been found which are
not considered to be reaction wood, an abnormal form that
will be discussed later.

For the longitudinal elements of both softwoods and
hardwoods, there is general agreement that most cell walls
are organized in the three-layered pattern presented here.
The few studies reported on hardwood cell wall structure
indicate that vessel segments, fibers, vasicentric tracheids,
and other cell types follow the pattern found in softwood
tracheids.

It is more difficult to determine the organization of
ray parenchyma cell walls due to the complex sculpturing that
is often found in heavily pitted areas.   Yet published studies, which again are relatively few, report no major

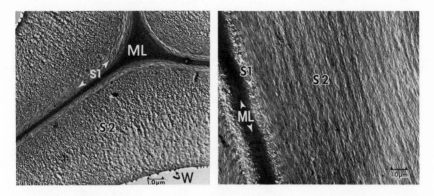

Figure 8 (Left). Transverse section of a lignin skeleton of normal balsam fir (*Abies balsamea* (L.) Mill.) showing
the uniform distribution of lignin in the secondary wall of
longitudinal tracheids, plus the higher packing density of
lignin in the middle lamella region. Warts (W) remain even
after this drastic HF-acid treatment. TEM.

Figure 9 (Right). Lignin skeleton of normal tamarack
(*Larix laricina* (Du Roi) K. Koch) again demonstrating the
apparent uniformity of lignin distribution in the secondary
cell wall. The true middle lamella (in the center of the
compound middle lamella) has a greater electron density than
the primary wall area, as viewed on the original electron
micrograph. Note the lamellar pattern in the S2 layer.

deviations from the structure of longitudinal elements.
Harada[24],[25] examined representatives of both angiosperms and
gymnosperms. He found that the microfibrillar orientation
of the S2 layer in these cells is parallel with the cell
axis. In longitudinal elements it is more usual to find
that the S2 forms some angle from the cell axis, perhaps as
great as 20 degrees in normal tracheids. In compression
wood it exceeds this.

Reaction Wood. Both hardwoods and softwoods may contain
zones of reaction wood, usually in the eccentric portion of
a tree cross-section. In conifers the reaction wood develops
on the lower side of the tree stem and in hardwoods on the
upper side of a leaning tree. Although reaction wood can
form in trees that are grown quite straight, it finds its

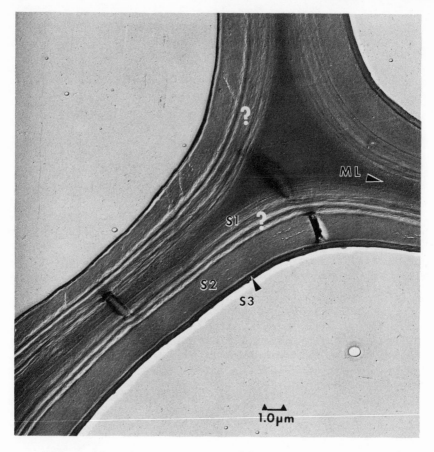

Figure 10.   Cross-section of portions of three tracheids in spruce pine (*Pinus glabra* Walt.) showing an anomalous cell wall structure between the S1 and S2 layers.   TEM of ultrathin section.

greatest development in strongly leaning trees or in branches.   Softwood reaction wood is called *compression wood* and in hardwoods this abnormal tissue is called *tension wood*.[11]

   The physiological, biochemical and structural aspects of reaction wood formation have been studied and reported in depth by a number of wood scientists.[48,49,50,51]   In 1964 and 1965 Wardrop[46,47] added new insights to all phases of

reaction wood studies. This subject continues to be of
great interest as evidenced by the current literature.
Recently Timell published a major work in this area.[45] In
the present review only the structural deviations from normal wood are considered, along with related variations in
chemical composition.

Compression wood can be recognized at the gross level
by its non-lustrous, "dead" appearance. It may be dark red
compared with the lighter tone of normal softwood. For its
density, it has lower tensile strength, modulus of elasticity
and impact strength than normal wood. Its longitudinal
shrinkage may range as high as 6 or 7 percent while normal
wood exhibits negligible shrinkage in that direction. Its
tendency to break suddenly under load is a major reason for
identifying and separating compression wood from structural
material although it can be used where strength is not
required.

The light microscope is adequate to confirm the presence
of compression wood. A cross-section reveals that it has
rounded tracheids and intercellular spaces. The S2 layer
usually contains helical checks or cavities that can be
detected without an electron microscope. When the ultrastructure is examined, however, it is obvious that the S3
layer is lacking in compression wood tracheids. The S1 layer
of the cell wall is often thicker than in normal cells and
the microfibrillar orientation of the S2 approaches 45 degrees,
a much flatter angle than in tracheids from normal wood.

The use of the polarizing microscope provides a clear
comparison between the cell wall organization of a normal
tracheid and one from the eccentric portion of a leaning
tree stem. Photomicrographs (Figs. 11 and 12) of cross-
sections of such woods also demonstrate that the S3 is lacking in compression wood tracheids. Electron micrographs
leave no doubt about the ultrastructure of the cell walls in
reaction wood tracheids. Figure 13 is an ultra-thin section
of red spruce (*Picea rubens* Sarg.) compression wood with a
tracheid that has a thick S1 and an S2 with deep helical
checks. The S3 is lacking. The SEM is particularly effective
in providing an overall view of a sample of compression wood
(Fig. 14). The intercellular spaces, rounded tracheids and
helical checks of this Douglas-fir reaction wood are readily
observed.

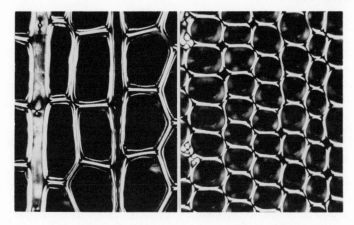

Figure 11 (Left). Polarized light photomicrograph of normal softwood cross-section revealing the three layered structure of the cell wall due to the birefringence of cellulose microfibrils. The S1 and S3 layers appear bright while the S2 is at total extinction, thus indicating the 90-degree shift in microfibrillar orientation from S1 to S2 and S2 to S3. 370X.

Figure 12 (Right). Compare this polarized light photomicrograph with Figure 11. This is a cross-section of compression wood. The bright S1 layer is present, but the S3 layer is lacking. The diffuse appearance of the S2 is caused by a flatter orientation of the microfibrils than in normal wood. 490X.

Tension wood is often more difficult to recognize than compression wood. In dressed lumber it has a silvery sheen. On green-sawn boards, zones of tension wood appear as woolly or fibrous areas on the surface. Tension wood does not shrink longitudinally as severely as compression wood, but nevertheless it shrinks more than normal hardwood.

The structural feature that characterizes tension wood at the anatomical level is the so-called "gelatinous fiber," now properly called a tension wood fiber. These cells contain a cell wall layer that was named the G-layer and it

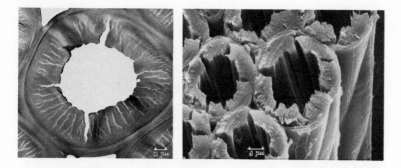

Figure 13 (Left). Ultra-thin section of red spruce
(*Picea rubens* Sarg.) compression wood. This micrograph con-
firms that the S3 layer is lacking and shows the helical
checks in the S2 layer of coniferous reaction wood tracheids.
Note also that an electron-dense zone appears at outer S2
where there is a higher concentration of lignin than in normal wood. TEM.

Figure 14 (Right). Scanning electron micrograph of
compression wood of Douglas-fir (*Pseudotsuga menziesii* (Mirb.)
Franco). This view emphasizes the rounded nature of the
tracheids, the intercellular spaces and the helical checks
in the S2 layer.

retains this designation. As will be emphasized in the following section, the special layer is unlignified and apparently pure cellulose which looks gel-like in light microscope sections. Its microfibrils are oriented parallel to
the long axis of the fiber and the layer appears to be only
loosely attached to the remainder of the cell wall.[9] Microtoming an unembedded sample seems to push all of the G-layers
in the same direction as can be seen in the photomicrograph
of poplar tension wood (Fig. 15).

The fascinating feature of tension wood fibers is that
the G-layer follows the deposition of any one of the usual
three wall layers, S1, S2 and S3. A tension wood fiber can
consist of P, S1 and G; or it may be P, S1, S2, and G; in
some cases all three layers of the secondary wall are deposited before the G-layer. The latter possibility is illustrated in Figure 16, a transmission electron micrograph of
an ultra-thin section of hackberry (*Celtis occidentalis* L.)

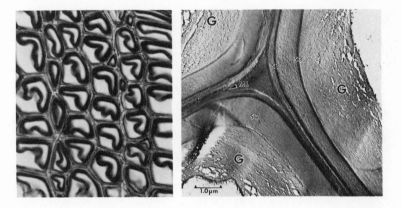

Figure 15 (Left). Photomicrograph of a cross-section
of tension wood in poplar (*Populus* sp.) demonstrating the
loose attachment of the G-layer (dark stain), and the tendency to be moved in the cutting direction during microtoming. The cellulosic, unlignified G-layer stains bright
green while the lignified cell wall layers stain red with
fast green/safranin staining combination. 400X.

Figure 16 (Right). Transmission electron micrograph
of ultra-thin section of hackberry (*Celtis occidentalis* L.)
through portions of three tension wood fibers. Two of the
fibers have G-layers deposited following the S2 while the
third fiber has an S3 layer before the G-layer.

tension wood. Figure 17 shows beech (*Fagus grandifolia*
Ehrh.) tension wood in which the G-layer was deposited following the S2. In hardwood reaction wood the S1 may be
thinner than normal while the primary wall appears normal.
A greater appreciation of the nature of tension wood fibers
can be developed by studying the scanning electron micrograph of poplar in Figure 18.

Sculpturing of the Wood Cell Wall. While cellulosic
fibers of plants such as cotton are relatively simple,
smooth-walled composites of lamellae, in wood the cell walls
are almost invariably interrupted by gaps (pits) and by
thickenings or other sculptured features. In fact, fibrillar
orientation is not easily measured because the microfibrils
often follow streamline paths around wall cavities or ana-
stamose into complex structures.

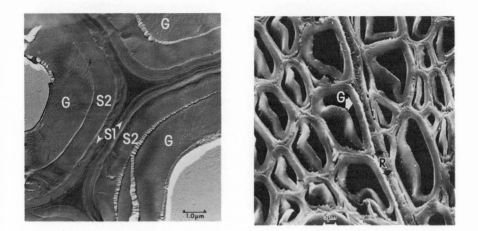

Figure 17 (Left). Tension wood of American beech (*Fagus grandifolia* Ehrh.) showing portions of three fibers all of which have G-layers deposited after the S2. Note the tenuous connection between G and S2 and the microfibrils stretching across the interface between them. TEM. Ultra-thin section.

Figure 18 (Right). Scanning electron micrograph of tension wood fibers in poplar (*Populus* sp.) with the thick G-layer attached loosely to the remainder of the secondary wall. A ray (R) stretches across the specimen surface.

Helical thickenings, illustrated earlier, are found in Douglas-fir tracheids as well as in several other species of conifers. In hardwoods similar thickenings may be seen in the vessels of basswood, maples and many other species. In Figure 19, a micrograph taken with the SEM, the wall of a basswood (*Tilia americana* L.) vessel element has been treated mechanically to expose the inner layers of the secondary wall. At the top of the micrograph both helical thickenings and intervessel pits interrupt the inner layer. In the lower half the bordered pit chambers of the adjoining cell, with some fragments of pit membranes, can be seen. At the center of the picture the microfibrils weave around the pit chambers. These form part of the cell nearer the viewer.

Another example of wall sculpturing is shown in Figure 20, a scanning micrograph of yellow birch (*Betula alleghaniensis* Britton) which has vessel walls covered with an intricate

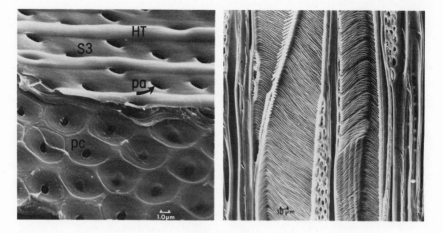

Figure 19 (Left). A torn vessel wall in basswood
(*Tilia americana* L.). The undamaged wall shows helical
thickenings (HT) and pit apertures (pa) sculpturing the S3
wall layer which lines the vessel. The area exposed by the
removal of part of the vessel wall is covered with inter-
vessel bordered pits of the neighboring vessel element.
Fragments of primary wall which forms part of the pit membrane may be seen in some of the pit chambers (pc). SEM.

Figure 20 (Right). SEM of sculptured vessel walls in
yellow birch (*Betula alleghaniensis* Britton). Ends of rays
are also visible on this tangential surface view.

design of thickenings. The streamlining effect is particularly striking around bordered pit apertures in softwood
tracheids such as in Figure 21, a micrograph taken with the
TEM from a replica of a southern yellow pine. The microfibrils are masked somewhat in this instance because of a
warty layer. The strands in the pit aperture form part of
the bordered pit torus which will be described below.

Pits are gaps in the secondary cell wall. Each pit has
a closing membrane at what was originally primary wall.
Generally pits occur in pairs, the gaps of two adjacent cells
coinciding and the pit membrane consisting of the primary
walls of each cell plus the intercellular substance.
Undoubtedly because of strength requirements, each cell produces a secondary wall whose cellulose microfibrils sweep

around pits at each layer and yet are still organized to
construct a pit chamber. It is a remarkable display of
architectural principles as may be appreciated from Figures
19-21.

The organization of the pit membrane in hardwoods is
simply that of the primary wall; that is, a random distribution of cellulose microfibrils and no visible openings in
the membrane. However, in coniferous bordered pit pairs
the membranes are evidently specialized to operate in a
valve-like manner. Many species have a thickening added to
the center of the membrane. This is disk-like and is called
a *torus*. The torus appears to be suspended from fine cellulosic strands which form an open membrane or *margo* around
the torus. Liquid flow is possible from tracheid to tracheid
through this membrane provided that the torus is not pressed
against one aperture or the other. In seasoned lumber the
torus is usually found in the aspirated position which probably restricts liquid movement to diffusion except where
the torus forms an imperfect seal. Figure 22 was selected
to show the array of pit apertures lining coniferous tracheids. The dome-like structures are bordered pits while
those in the raised area across the center of the scanning
micrograph are half-bordered pits leading to the ray parenchyma cells in contact with the tracheids. Membranes can be
seen in the latter, but Figure 23 is needed to show those of
the bordered pit pairs.

In a few species of hardwoods the pit chambers and pit
apertures are literally "decorated" by outgrowths which are
typical for vestured pits. These structures are more common
in tropical woods, but they occur normally in the wood of
black locust (*Robinia pseudoacacia* L.), Kentucky coffee-
tree (*Gymnocladus dioicus* (L.) K. Koch) and some other North
American species. These sculptured protuberances bear a
strong resemblance to warts, referred to earlier and below,
but they may be branched and much larger (Fig. 24). In one
paper it was suggested that they are of similar origin and
there has been some agreement to this proposal.[10]

The structure of warts has been an interesting topic of
study for more than 20 years. They were identified as normal ultrastructures by Harada[26] in Japan and Liese[32,33] in
Germany almost simultaneously. Warts are found consistently
in redwood (*Sequoia sempervirens* (D. Don) Endl.), balsam fir,
southern yellow pines and other hard pines, and a number of

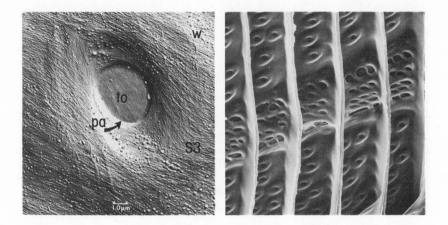

Figure 21 (Left). Bordered pit dome in a longitudinal tracheid of table-mountain pine (*Pinus pungens* Lamb.). The lumen walls of these southern yellow pines are usually covered with warts (W). Evidently this is an aspirated pit since the torus (to) is pressed tightly against the pit aperture (pa). Note how the microfibrils of the S3 layer form a streamline·pattern around the pit, disrupting their straight orientation. TEM, replica technique.

Figure 22 (Right). The longitudinal tracheid walls in western larch (*Larix occidentalis* Nutt.), as in all conifers, are sculptured by many pit apertures. In this SEM view, bordered pit domes appear above and below the bulging region in the center of the micrograph. The pits in that region lead to a ray located adjacent to the tracheids. The half-bordered pit pairs between longitudinal tracheids and ray parenchyma have characteristic shapes and sizes for many of the genera and species.

other woods, but not in all conifers. They occur in hardwoods, notably beech, but they appear to be less common than in softwoods, at least in North American species. As can be seen in Figure 24, tropical hardwoods having vestured pits will often have a warty structure on the wall of the cell lumen. The size, shape, distribution, and perhaps chemical composition of warts are variable. However, within a single species the nature of the warts is very similar from tree to tree. Figure 21 offers an example of wart structure in a hard pine.

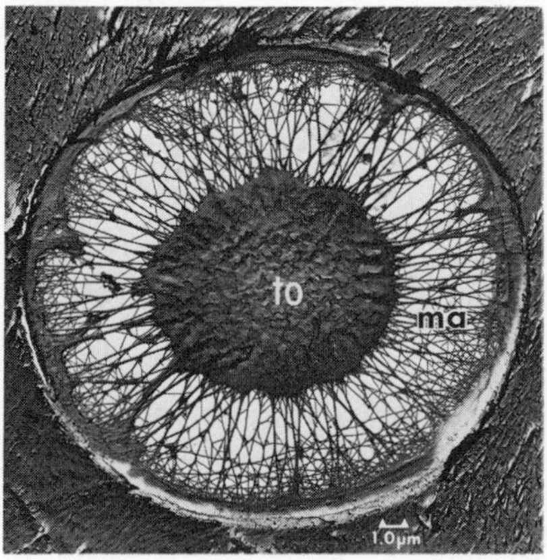

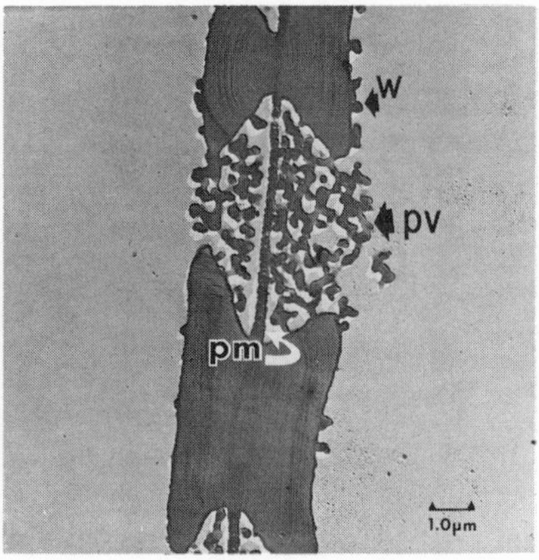

Figure 23 (Left). The bordered pit membrane in this
TEM view is in the unaspirated state because it was solvent
dried from the green condition. Note the openings in the
margo (ma) and the disk-like torus (to). Eastern hemlock
(*Tsuga canadensis* (L.) Carr.). Micrograph by G. L. Comstock.

Figure 24 (Right). Vestured pits are believed to be
related to the warty layer on the basis of transmission
electron micrographs such as this one from *Parashorea plicata* Brandis.   The pit vestures (pv) look like outgrowths
from the vessel walls and inter-vessel bordered pit chambers.
Warts (W) appear to be very similar in nature. Note the
fine openings in the pit membrane (pm).

DISTRIBUTION OF CHEMICAL CON-
STITUENTS IN WOOD

     Some indication of the distribution of the principal
chemical constituents of wood within its structure has already been given. It must be obvious from the discussion
of techniques employed to obtain information of this kind
that arriving at accurate figures is not easy at the ultrastructural level. In fact, part of the data available is
based on light microscopic control, with the SEM and TEM
evidence being primarily on lignin and polysaccharide skeletons.

Cellulose. The distribution of cellulose is probably the easiest to demonstrate graphically because cellulose molecules are aggregated into microfibrils which are easily resolved with the TEM, particularly after partial removal of encrusting and matrix substances from the surface of the specimen. From observations of electron micrographs it appears that cellulose is quite evenly distributed throughout the secondary cell wall. Microfibrils have not been found in the middle lamella, but they are found in a rather loose and random arrangement in the primary wall.

The technique employed by Meier[35,36] showed that in birch the polysaccharides of combined M (middle lamella) and P (primary) layers consisted of 41.4% cellulose. Sl was composed of 49.8%, the outer part of S2, 48%, and the inner part of S2 combined with S3, 60%. In spruce (*Picea abies* (L.) Karst.), Meier reported 33.4% for M + P, 55.2% for Sl, 64.3% for S2 outer part, and 63.6% for S2 inner part plus S3. It must be noted that this method requires judgment in separating the fractions by means of polarizing microscope observation which indicate the presence or absence of certain wall layers. Also, the volume ratios of the various layers estimated from relative thickness as measured on electron micrographs are subject to some error and cannot possibly account for extremes in variation. Specific gravity determinations of each of the fractions described above were carried out by Kutscha[30] to be certain that density variations between layers were not widely different. For normal wood the differences were not significant and, as in Meier's work, it was assumed that the weight of each wall layer was proportional to its thickness.

For an overview of the distribution of the total cellulose across the cell wall, Table 3 shows that the M + P layer contains but 1% and the S3 only 2%. The remainder of the cellulose is found in the S2 and Sl layers with the former containing the major proportion. These values are for normal wood of Scots pine.

The distribution of polysaccharides in reaction wood differs from normal wood, as would be expected because of the ultrastructural differences. For compression wood, a comparison of the values obtained by Kutscha[15,30] for balsam fir (Table 4) with those of Meier for normal wood (Table 3) is interesting. These tables do not list lignin content so it must be pointed out that compression wood has abnormally low cellulose content and abnormally high lignin content.

Table 3. Distribution of polysaccharides over the cell wall
layers of tracheids from normal wood of Scots pine (*Pinus
sylvestris* L.).*

Cell Wall Layer	Galac- tan	Cellu- lose	Galacto- gluco- mannan	Arabi- nan	Arabino- 4-*O*-methyl- glucurono- xylan
M + P	20	1	1	30	1
S1	21	11	7	5	12
S2 outer	Nil	47	39	Nil	21
S2 inner	59	39	38	21	34
S3	Nil	2	15	44	32

*All values in relative percentage of total amount of
each polysaccharide and recalculated from data of Meier.[35]
From Côté *et al.*[15]

Table 4. Distribution of polysaccharides over the cell wall
layers of tracheids from compression wood.*

Cell Wall Layer	Galac- tan	Cellu- lose	Galacto- gluco- mannan	Arabi- nan	Arabino- 4-*O*-methyl- glucurono- xylan
M + P	2	1	1	32	3
S1	32	21	23	7	29
S2 outer	49	35	39	16	33
S2 inner	16	43	37	45	35

*All values in relative percentage of total amount of
each polysaccharide.[15] Based on studies of balsam fir
(*Abies balsamea* (L.) Mill.) by N. P. Kutscha.[30]

Tension wood has a larger percentage of cellulose than normal wood and the percentage of lignin may be less than half of that for normal hardwood. This is due to the G-layer which is often quite thick and which is unlignified. From preliminary results yet unpublished, it appears that the cellulose concentrations of the S1, S2 and S3 layers, when present, are in the normal range. Evidently there have been no studies on hardwood of the type done by Meier and Kutscha on coniferous woods.

Hemicelluloses. Although hemicelluloses are essentially linear molecules with some short side chains, they are not resolved in the electron microscope as separate physical elements such as cellulose microfibrils. Suggestions that both hemicelluloses and lignin may enter the paracrystalline portions of cellulose microfibrils are difficult to confirm through optical instrumentation. In only one case has it been possible to see a hemicellulose in wood and identify it as such. Arabinogalactan is evidently quite different structurally from the other wood hemicelluloses and will be discussed separately.[13]

The distribution of hemicelluloses across the cell wall can best be summarized in table form. The results obtained by Meier for normal Scots pine (*Pinus sylvestris* L.) are listed in Table 3 and may be compared with the hemicellulose distribution for compression wood of balsam fir in the results obtained by Kutscha, Table 4. From these tables it can be seen that most of the galactan is found in the S1 and outer portion of S2 in compression wood. The small volume of M + P, although high in concentration of galactan in normal wood, does not affect the distribution in the other layers significantly, when comparing normal and compression wood. In the normal wood, M + P contained much of both galactan and arabinan. The S1 layer of compression wood tracheids contains larger proportions of the other polysaccharides than is found in normal tracheids in that region because the S1 is much thicker in reaction wood tracheids.

Evidently there have been no other studies published in this line of research. Meier's papers[35,36] include percentages of each cell layer fraction in terms of total polysaccharide content not only for pine, but also for spruce and birch (*Betula verrucosa* Ehrh.). These figures do not yield across-the-wall distribution for the cellulose and hemicelluloses without considerable interpretation. However, they provide a useful starting point for future studies.

The effect of cell wall thickness on polysaccharide content in various layers was investigated by examining the springwood and summerwood tracheids in Scots pine. Due to the thicker S2 layer in summerwood cells, Meier found a higher glucomannan content (24.8%) in these elements than in springwood tracheids which contained 20.3% of this hemicellulose. It follows that there was a lower glucurono-arabinoxylan content in the summerwood since the S3 makes up a smaller proportion of the total cell wall in this instance.

The special case of arabinogalactan mentioned earlier represents an interesting departure from the pattern of other non-cellulosic polysaccharides. This hemicellulose is found in small quantities in many conifers and in a few hardwoods. However, in the heartwood of eastern larch (or tamarack, *Larix laricina* (Du Roi) K. Koch) and western larch (*Larix occidentalis* Nutt.), and in other members of the genus *Larix*, arabinogalactan is present in large quantities as an extra-cellular polysaccharide. Its commercial recovery is feasible because it makes up from 5% to 40% by weight of the wood.[13]

Arabinogalactan is a heavily branched molecule which occurs as a fine, amorphous, water-soluble powder. It is found in the cell lumens of the tracheids, ray cells and epithelial cells, but because it can be dissolved through the usual water-softening procedures used before microtoming wood, special care is required in specimen preparation for microscopic observation. Figures 25 and 26 show transverse and longitudinal views of tracheids filled with arabino-galactan.

In the case of tension wood, apparently no data exist to indicate the distribution of hemicelluloses.

Lignin. For a period of about fifty years, beginning with the early work of Ritter,[41] it was thought that in the wood of conifers 75% of the lignin is located in the compound middle lamella. The results of most of the early workers in the field were reviewed by Berlyn and Mark[4] who re-evaluated these accepted figures in the light of more recent evidence. They claim that less than 40% of the total lignin in softwood is in the middle lamella, most of it being found in the

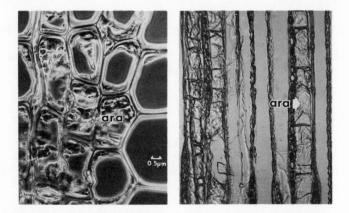

Figure 25 (Left). Deposits of arabinogalactan (ara) in the longitudinal tracheids of heartwood of western larch (*Larix occidentalis* Nutt.). In this photomicrograph of a cross-section many of the cells are filled.

Figure 26 (Right). Longitudinal section of same wood as Figure 25 showing the extent of distribution of arabino-galactan (ara) in tracheids of woods of the genus *Larix*. 100X.

secondary wall of coniferous tracheids. In the past ten years a number of papers have appeared which support this new position and several of them were based on lignin skele-tonizing techniques and TEM.[3,14,16,40]

In normal wood of tamarack, for example, the three layers of the secondary wall, S1, S2 and S3, are lignified to approxi-mately the same extent.[16] For this species, previous evidence that the S3 layer should contain a higher proportion of lignin than other layers was not confirmed. The compound middle lamella, although exhibiting a high packing-density of lignin, appears to account for no more than the figure proposed by Berlyn and Mark.[4] An electron micrograph of a cross-section of a lignin skeleton from tamarack is included as Figure 27. The uniform lignin network in the secondary wall region is clear, as is the indication of cell wall organization for each of the three layers.

For normal wood of loblolly pine (*Pinus taeda* L.),[40] using the same hydrofluoric acid treatment, the S3 region of

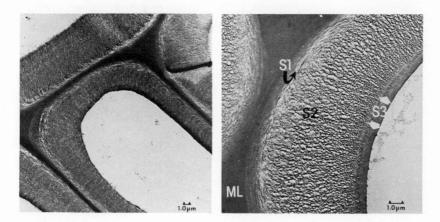

Figure 27 (Left). The uniform web of lignin in the
secondary wall can be appreciated in this TEM view of a
cross-section of tamarack (*Larix laricina* (Du Roi) K. Koch)
which was skeletonized using the HF-acid technique.

Figure 28 (Right). In loblolly pine (*Pinus taeda* L.),
the S3 layer appears to be more compact than in other conifers in lignin skeletons. TEM. Ultra-thin section.

both earlywood and latewood tracheids retained a dense network of unhydrolyzable material which may be interpreted as
lignin (Fig. 28).

Compression wood of tamarack studied in the same manner
had only a small proportion of the total lignin in the compound middle lamella. Intercellular spaces account for part
of this, but the high concentration of lignin in the outer
portion of S2 is apparently the main reason. The S1 is lignified to a much lower extent than outer S2. Inner S2, the
region containing the helical cavities, is uniformly encrusted
with lignin, and of course there is no S3 in compression wood
tracheids. Loblolly pine compression wood averaged about 7%
more in lignin content than did normal wood.[40] Lignin distribution was similar to that found in tamarack reaction wood.
Electron micrographs of compression wood lignin skeletons
offer convincing evidence not only for lignin distribution
but also for cell wall structure (Figs. 29 and 30). A complementary micrograph of a delignified compression wood specimen
adds another degree of confidence in the techniques employed

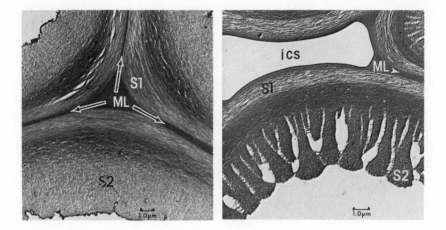

Figure 29 (Left). The nature of the secondary wall of compression wood tracheids is revealed more clearly with lignin skeletons such as this one of balsam fir (*Abies balsamea* (L.) Mill.). The dense lignin deposit in outer S2 even retains its fibrillar appearance after the removal of the cellulose microfibrils.

Figure 30 (Right). Transmission electron micrograph of lignin skeleton of compression wood of tamarack (*Larix laricina* (Du Roi) K. Koch). Large intercellular spaces (ics) and helical checks in the S2 typify this reaction wood. Note the wide S1 layer and the dense lignin zone in outer S2.

(Fig. 31). Finally, the SEM provides proof of the continuity of the lignin skeleton when small blocks of the type used for embedding, ultra-thin sectioning and TEM observation are studied in bulk form. Figure 32 shows the integrity of the lignin skeleton in earlywood of a southern yellow pine, while Figure 33 gives similar evidence with latewood of Douglas-fir. The more compact nature of the lignin in the compound middle lamella is particularly striking in these views.

Ray parenchyma cell walls of both normal and reaction wood of conifers are more heavily lignified than the longitudinal tracheids. The dense lignin network is distributed uniformly throughout the secondary wall (Fig. 34). At higher magnification the lignin skeletons reflect the lamellar nature of the ray parenchyma cell walls.

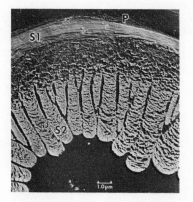

Figure 31. A carbohydrate skeleton of compression wood from the same species as Figure 30 shows how the evidence is complementary. There are relatively few fragments of cellulose microfibrils in the outer S2 which was shown as a heavily lignified area in the lignin skeleton. The orientation of microfibrils in the thick S1 layer is clear. Note the sparse microfibrillar remnant in the primary wall area. TEM.

Normal and tension wood of red maple (*Acer rubrum* L.), American beech, quaking aspen (*Populus tremuloides* Michx.), and American elm (*Ulmus americana* L.) were selected for lignin studies using the same lignin skeletonizing technique as for the softwoods.[3] It was found that with hardwoods swelling can occur in the course of acid hydrolysis which may lead to distorted lignin skeletons. However, in micrographs that appear relatively artifact-free, vessel walls in both normal and tension woods are highly lignified, much like conifer tracheid walls. Hardwood fibers have a looser, more open lignin network which agrees with the findings of Sachs, Clark and Pew[42] (Fig. 35). In normal wood the S1 layer has a relatively high lignin concentration which appears to be higher than in the S2. It could not be determined whether S3 is more highly lignified than S1 or S2.

In tension wood, the vessels and rays appear to be lignified as in normal hardwood. In the tension wood fibers, the non-gelatinous layers of the cell wall are highly lignified, with S1 probably slightly more lignified than that in normal wood. The S2 layer can be either more or less lignified than S1, but almost invariably it is more lignified than

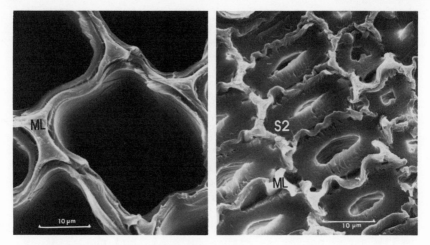

Figure 32 (Left). The integrity of the cell walls is retained in spite of the removal of the polysaccharide fraction by HF-acid treatment. This lignin skeleton was made from earlywood of southern yellow pine. SEM.

Figure 33 (Right). In Douglas-fir (*Pseudotsuga menziesii* (Mirb.) Franco), a latewood sample that was lignin skeletonized retains its basic structure. The difference in the lignin of the middle lamella region from that of the secondary wall is apparent in this SEM view.

S2 of normal wood fibers. The thickness of the G-layer seems to affect the degree of lignification of the remainder of the cell wall which would be expected in view of the similar amounts of lignin in both types of wood. Therefore, a thin G-layer is related to a heavily lignified S2. Figure 36 is a cross-section of maple tension wood that has been skeletonized to remove the polysaccharide fractions. A longitudinal section of a different species, American elm, also demonstrates how little lignin is found in the G-layer region of tension wood fibers (Fig. 37).

Inclusions. In a strict sense, starch, resin, gums, and similar substances can be classed as wood inclusions. However, these materials are found in the cell cavities of many wood species, and hence are outside the coverage of the present review.

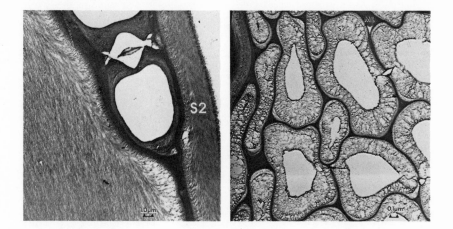

Figure 34 (Left). Both normal and reaction wood ray cell walls are heavily lignified. This lignin skeleton of normal tamarack (*Larix laricina* (Du Roi) K. Koch) allows a comparison with the secondary wall layers of neighboring longitudinal tracheids. TEM.

Figure 35 (Right). In red maple (*Acer rubrum* L.) and other hardwoods, lignin skeletons of the secondary wall are looser or more open textured. The S1 may be more lignified than the S2. It is difficult to judge the nature of the S3 because it is so narrow in these TEM micrographs.

Crystals of acicular and styloid shapes, druses, raphides, crystal sand, and silica in other forms also fall into the category of inclusions, but because they can now be analyzed using SEM and EDXA (energy-dispersive x-ray analysis), they are briefly described. Many domestic woods occasionally have crystals in ray or longitudinal parenchyma cells and in some instances their presence is of diagnostic value. In tropical woods, crystals are more common and sometimes affect the utilization of a particular species, especially if silica is present in high concentration because it dulls tools rapidly. The SEM and EDXA combination can be used to detect, analyze and map the distribution of elements on the atomic chart beginning with sodium. That lighter elements cannot be detected is advantageous for wood research because they would otherwise create interference.[20] Examples of SEM images and their corresponding element maps for crystal

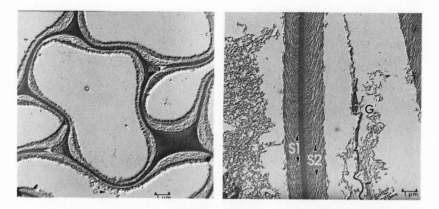

Figure 36 (Left). The remains of the G-layer in these tension wood fibers from red maple (*Acer rubrum* L.) are limited to very small amounts. The G-layer is unlignified and HF-treatment leaves little trace of it although in Figure 37 some fragments can be seen. TEM.

Figure 37 (Right). Lignin skeleton of tension wood fibers from American elm (*Ulmus americana* L.). S1, S2 and the middle lamella are well preserved. Evidently there is no S3 in these cells and only remnants of the G-layer remain with perhaps a trace of terminal lamella. TEM.

inclusions are shown in Figures 38 and 39. Published proceedings of the annual Illinois Institute of Technology Research Institute SEM Symposium bring together many varied uses of these tools in wood research.

SUMMARY

This review has attempted to outline highlights of wood ultrastructure and to relate chemical composition of wood with its submicroscopic structure. Although progress has been made in this broad area, it is obvious that much more is required in both breadth and depth. Some topics remain unexplored because of the lack of appropriate instrumentation or technique while for others there is significant new information being made available. In terms of breadth of research coverage, the proliferation of ultrastructure laboratories in wood research institutions offers promise for broad

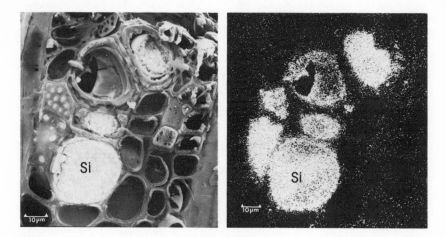

Figure 38 (Left). Tangential surface of a sample of teak (*Tectonia grandis* L.) showing ray cells filled with un- identified deposits. Following SEM/EDXA analysis, silicon was found in many of the cells. See Figure 39 for an ele- ment map of Si which can be superimposed over this micro- graph.

Figure 39 (Right). Si element map of the specimen area shown in the scanning electron micrograph of Figure 38. Note that in the cell at extreme left Si was detected even though it was covered by the end wall.

coverage of the many species of wood that have not been studied. Depth of study requires that more imagination and new technology be developed and applied to wood as a natural polymeric material. The tools available in other areas of materials science might have applicability to this material.

ACKNOWLEDGMENTS

Much of the work discussed in this paper, apart from the review of the general literature, was done over a period of about twenty years with various collaborators. Chief among them were Arnold C. Day and Dr. Tore E. Timell. Gilbert L. Comstock, Richard L. Gray, Robert B. Hanna, Norman P. Kutscha, Paul T. Mann, Robert W. Meyer, and Russell A. Parham, each mentioned in this review, made significant contributions

to the field of ultrastructure of wood during their doctoral
studies. The able technical support of Arlene Bramhall,
Lorraine Mann and John J. McKeon made much of this research
possible. I owe each of them a debt of gratitude for the
opportunity to work in an atmosphere of freedom and friend-
ship. To the countless students, staff, and colleagues who
contributed ideas and support during the same period go my
heartfelt thanks.

REFERENCES

1. Bailey, I. W. and T. Kerr. 1935. The visible struc-
    ture of the secondary wall and its significance in
    physical and chemical investigations of tracheary
    cells and fibers. *J. Arnold Arboretum XVI:*273-300.
2. Bailey, I. W. and M. R. Vestal. 1937. The orientation
    of cellulose in the secondary wall of tracheary cells.
    *J. Arnold Arboretum XVIII:*185-195.
3. Bentum, A. L. K., W. A. Côté, Jr., A. C. Day, and T. E.
    Timell. 1969. Distribution of lignin in normal and
    tension wood. *Wood Science and Technology 3:*218-231.
4. Berlyn, G. P. and Mark, R. E. 1965. Lignin distribu-
    tion in wood cell walls. *Forest Prod. J. 15:*140-141.
5. Browning, B. L., ed. 1963. The Chemistry of Wood.
    Interscience Publishers, Div. of John Wiley & Sons,
    New York.
6. Core, H. A., W. A. Côté, and A. C. Day. 1976. Wood
    Structure and Identification. Syracuse Univ. Press,
    Syracuse, N.Y.
7. Côté, Wilfred A., Jr., ed. 1965. Cellular Ultra-
    structure of Woody Plants. Proceedings of National
    Science Foundation sponsored Advanced Science Seminar,
    Pinebrook, Sept. 1964. Syracuse Univ. Press. 603 pp.
8. Côté, Wilfred A., Jr. 1967. Wood Ultrastructure--An
    Atlas of Electron Micrographs. University of Washing-
    ton Press, Seattle. 64 pp.
9. Côté, Wilfred A., Jr., and A. C. Day. 1962. The G-
    layer in gelatinous fibers--Electron microscopic stu-
    dies. *Forest Prod. J. 12:*333-339.
10. Côté, W. A., Jr., and A. C. Day. 1962. Vestured pits--
    Fine structure and apparent relationship with warts.
    *Tappi 45:*906-910.
11. Côté, W. A. and A. C. Day. 1965. Anatomy and ultra-
    structure of reaction wood. *In:* Cellular Ultrastruc-
    ture of Woody Plants (W. A. Côté, ed.), pp. 391-418.
    Syracuse Univ. Press, Syracuse, N.Y.

12. Côté, W. A. and A. C. Day. 1969. Wood Ultrastructure
    of the Southern Yellow pines. Tech. Publ. No. 95,
    State Univ. of New York, College of Forestry at
    Syracuse University, Syracuse, New York. 70 pp.
13. Côté, W. A., Jr., A. C. Day, B. W. Simson, and T. E.
    Timell. 1966. Studies on larch arabinogalactan I.
    The distribution of arabinogalactan in larch wood.
    *Holzforschung 20:*178-192.
14. Côté, W. A., A. C. Day, and T. E. Timell. 1968. Dis-
    tribution of lignin in normal and compression wood
    of tamarack (*Larix laricina* (Du Roi) K. Koch). *Wood
    Science & Technology 2:*13-37.
15. Côté, W. A., Jr., N. P. Kutscha, B. W. Simson, and
    T. E. Timell. 1968. Studies on compression wood VI.
    Distribution of polysaccharides in the cell wall of
    tracheids from compression wood of balsam fir (*Abies
    balsamea* (L.) Mill.). *Tappi 51:*33-40.
16. Côté, W. A., T. E. Timell, and R. A. Zabel. 1966.
    Studies on compression wood. Part 1. Distribution
    of lignin in compression wood of red spruce, *Picea
    rubens* Sarg. *Holz als Roh-Werkstoff 24:*432-438.
17. Frey, A. 1926. Die submikroskopische Struktur der
    Zellmambranen. *Jahrb. Wiss. Bot. 65:*195-223.
18. Frey-Wyssling, A. 1959. Die Pflanzliche Zellwand.
    Springer-Verlag. Berlin/Gottingen/Heidelberg.
    367 pp.
19. Frey-Wyssling, A. 1964. Ultraviolet and fluorescence
    optics of lignified cell walls. *In:* The Formation of
    Wood in Forest Trees (M. H. Zimmermann, ed.), pp. 153-
    167. Academic Press, New York.
20. Gray, Richard L. and W. A. Côté. 1974. SEM/EDXA as a
    diagnostic tool for wood and its inclusions. *IAWA
    Bulletin 3:*6-11.
21. Hanna, Robert B. 1971. The interpretation of high
    resolution electron micrographs of the cellulose ele-
    mentary fibril. *J. Polymer Sci.,* Part C, *36:*409-413.
22. Hanna, Robert B. 1973. Cellulose sub-elementary
    fibrils. Ph.D. Dissertation, SUNY College of Environ-
    mental Science and Forestry, Syracuse, New York.
23. Hanna, R. B. and W. A. Côté, Jr. 1974. The sub-
    elementary fibril of plant cell wall cellulose.
    *Cytobiologie 10:*102-116.
24. Harada, H. 1965. Ultrastructure and organization of
    gymnosperm cell walls. *In:* Cellular Ultrastructure
    of Woody Plants (W. A. Côté, Jr., ed.), pp. 215-233.
    Syracuse Univ. Press, Syracuse, N.Y.

25. Harada, H. 1965. Ultrastructure of angiosperm vessels
    and ray parenchyma. *In:* Cellular Ultrastructure of
    Woody Plants (W. A. Côté, Jr., ed.), pp. 235-249.
    Syracuse Univ. Press, Syracuse, N.Y.
26. Harada, H. and Y. Miyazaki. 1952. The electron-
    microscopic observation of the cell wall of conifer
    tracheids. *J. Japan Forestry Soc. 34:*350-352.
27. Harlow, W. M. 1939. Contributions to the chemistry
    of the plant cell wall. *Paper Trade J. 109:*38-42.
28. Kobayashi, K. and N. Utsumi. 1951. Electron micro-
    scopy of conifer tracheids (in Japanese). *Committee
    Note on Electron Microscopy, No. 56,* p. 93.
29. Kollmann, F. F. P. and W. A. Côté, Jr. 1968. Princi-
    ples of wood science and technology, Vol. I. Solid
    Wood. Springer-Verlag, Berlin.
30. Kutscha, Norman P. 1968. Cell wall development in
    normal and compression wood of balsam fir, *Abies
    balsamea* (L.) Mill. Ph.D. Dissertation, SUNY College
    of Forestry at Syracuse University, Syracuse, N.Y.
31. Liese, W. 1951. Demonstration elektronenmikroskopischer
    Aufnahmen von Nadelholztüpfeln. *Ber. Deut. Botan.
    Ges. 64:*31-32.
32. Liese, W. 1965. The warty layer. *In:* Cellular Ultra-
    structure of Woody Plants (W. A. Côté, ed.), pp. 251-
    269. Syracuse Univ. Press, Syracuse, N.Y.
33. Liese, W. and M. Hartmann-Fahnenbrock. 1953. Elek-
    tronenmikroskopische Untersuchungen über die Hoftüpfel
    der Nadelhölzer. *Biochim. Biophys. Acta 11:*190-198.
34. Mann, Paul T. 1972. Ray parenchyma cell wall ultra-
    structure and formation in *Pinus banksiana* and *Pinus
    strobus*. Ph.D. Dissertation, State University of New
    York College of Forestry at Syracuse University,
    Syracuse, New York.
35. Meier, Hans. 1961. The distribution of polysaccharides
    in wood fibers. *J. Polymer Sci. 51:*11-18.
36. Meier, H. 1964. General chemistry of cell walls and
    distribution of the chemical constituents across the
    walls. *In:* The Formation of Wood in Forest Trees
    (M. H. Zimmermann, ed.), pp. 137-151. Academic Press,
    New York.
37. Meyer, Robert W. 1967. Ultrastructural ontogeny of
    tyloses in *Quercus alba* L. Ph.D. Dissertation, State
    University of New York College of Forestry at Syracuse
    University, Syracuse, New York.

38. Mühlethaler, Kurt. 1965. The fine structure of the cellulose microfibril. *In:* Cellular Ultrastructure of Woody Plants (W. A. Côté, ed.), pp. 191–198. Syracuse Univ. Press, Syracuse, N.Y.

39. Panshin, A. J. and Carl de Zeeuw. 1970. Textbook of Wood Technology, Vol. I, Third Edition. McGraw–Hill Book Co., New York.

40. Parham, R. A. and W. A. Côté, Jr. 1971. Distribution of lignin in normal and compression wood of *Pinus taeda* L. *Wood Science and Technology* 5:49–62.

41. Ritter, G. J. 1934. Structure of the cell wall of wood fibers. *Paper Industry* 16:178–183.

42. Sachs, I. B., I. T. Clark, and J. C. Pew. 1963. Investigation of lignin distribution in the cell wall of certain woods. *J. Polymer Sci.* Part C. 2:203–212.

43. Sullivan, J. 1968. Wood cellulose protofibrils. *Tappi.* 51:501–507.

44. Timell, T. E. 1965. Wood and bark polysaccharides. *In:* Cellular Ultrastructure of Woody Plants (W. A. Côté, Jr., ed.), pp. 127–156. Syracuse Univ. Press, Syracuse, N.Y.

45. Timell, T. E. 1973. Ultrastructure of the dormant and active cambial zones and the dormant phloem associated with formation of normal and compression woods in *Picea abies* (L.) Karst. Tech. Publ. No. 96, SUNY College of Environmental Science and Forestry, Syracuse, N.Y.

46. Wardrop, A. B. 1964. The structure and formation of the cell wall in xylem. *In:* The Formation of Wood in Forest Trees (M. H. Zimmermann, ed.), pp. 87–137. Academic Press, New York.

47. Wardrop, A. B. 1965. Cellular differentiation in xylem. *In:* Cellular Ultrastructure of Woody Plants (W. A. Côté, Jr., ed.), pp. 61–124. Syracuse Univ. Press, Syracuse, N.Y.

48. Wardrop, A. B. and H. E. Dadswell. 1948. The nature of reaction wood. I. The structure and properties of tension wood fibres. *Australian J. Sci. Res.,* Ser. B 1:3–16.

49. Wardrop, A. B. and H. E. Dadswell. 1950. The nature of reaction wood. II. The cell wall organization of compression wood tracheids. *Australian J. Sci. Res.,* Ser. B 3:1–13.

50. Wardrop, A. B. and H. E. Dadswell. 1955. The nature of reaction wood. IV. Variations in cell wall organization of tension wood fibres. *Australian J. Botany 3:* 177–189.

51. Wardrop, A. B. and G. W. Davies. 1964. The nature of reaction wood. VIII. The structure and differentiation of compression wood. *Australian J. Botany 12:* 24-38.
52. Wise, L. E. and E. C. Jahn. 1952. Wood Chemistry. Reinhold Publishing Corp., New York.

Chapter Two

THE BIOSYNTHESIS OF CELLULOSE AND OTHER PLANT CELL WALL

POLYSACCHARIDES

DEBORAH P. DELMER

MSU/ERDA Plant Research Laboratory
Michigan State University
East Lansing, Michigan 48824

INTRODUCTION

In recent years we have observed very rapid progress
toward achieving an understanding of the structure of plant
cell walls. In large part, this seems to be due to wise
application by researchers of many newly-developed techniques
for analysis of complex carbohydrates. Unfortunately, the
area of cell wall biosynthesis cannot yet claim comparable
progress. It is difficult to study the biosynthesis of
polymers whose structures and interconnections have been
poorly understood until quite recently. However, new
knowledge concerning cell wall structure, availability of
many new radioactive precursors, and recent advances in the
isolation of plant organelles, all indicate that the subject
of plant cell wall biosynthesis will be an exciting field
of research in the years ahead.

This presentation briefly reviews the types of polysaccharide structures which exist in plant cell walls and
then discusses the current status of our knowledge about
the biosynthesis of these polymers. It is in no way
intended as a complete, comprehensive literature survey of

the entire field. Rather, it presents a general survey of the state of the art, points out current limitations on our knowledge, and emphasizes where attention should be focussed in the future. Considerably more emphasis is given to the biosynthesis of cellulose than to the biosynthesis of other cell wall constituents, some of which are covered in subsequent chapters.

STRUCTURAL CONSIDERATIONS RELATIVE TO BIOSYNTHESIS

It is really not possible to describe a "typical plant cell wall." Plants are composed of a variety of tissues and, within these tissues, of various cell types. The type of cell wall a plant cell possesses is very often one of the major criteria used to distinguish it from other cell types. Nevertheless, a few generalizations can be made. Nearly all higher plant cell walls consist of a network of cellulose microfibrils embedded in a matrix of non-cellulosic polymers, as described in Chapter 1. The thin wall of young, undiffer- entiated cells, the so-called primary cell wall, may well be quite similar in a wide variety of plants. From studies on the composition of the cell walls derived from cells of a number of dicotyledonous and monocotyledonous plants grown in tissue culture,[20,105] Albersheim's group has concluded that the primary cell walls of several unrelated dicotyle- donous plants are quite similar; the primary cell walls of the cultured monocotyledonous plant cells (all grasses) they examined were also all quite similar to each other, but distinctly different in at least one respect from dicot cells. However, these studies need to be extended to primary cell walls of cells that have developed on the plant before any final conclusions are reached.

Table 1 lists the major types of polysaccharides which have been identified in plant cell walls. This list is incomplete in detail and the reader is referred to several recent reviews on cell wall structure for more detailed information.[3,5,6,59,70] The glycoproteins of cell walls are considered in Chapter 3. Table 1 identifies components commonly associated with primary cell walls and those from thicker secondary cell walls which are characteristic of mature, differentiated plant cells. The major points from this table to be considered relative to the problem of bio- synthesis are as follows:

Table 1.   Carbohydrate polymers of plant cell walls

Cell wall type	Cell wall fraction	Specific polymer type(s)
Primary walls	Cellulosic	Low and heterogeneous DP* (2,000–6,000) of β-1,4-glucan
	Hemicellulosic	Xyloglucan (dicots) Arabinoxylan (grasses)
	Pectic	Galacturonans (e.g. rhamnogalacturonan, apiogalacturonan) Galactan Arabinan Arabinogalactan
Secondary walls	Cellulosic	High and homogeneous DP* (13,000–16,000) of β-1,4-glucan
	Hemicellulosic	Xylan Glucomannan

*DP, degree of polymerization.

(1) Extracts prepared from whole seedlings, even organs or tissues, reflect a variety of cell types, of which each was engaged at the time of harvest in synthesis of its own pattern of cell wall polymers. This problem is minimized by use of relatively homogeneous cell types derived from tissue culture, from specialized single cells obtainable in large quantities such as cotton fibers or pollen tubes, or from certain tissues such as elongating coleoptiles rich in a single cell type.

(2) The problem of studying the biosynthesis of secondary wall components is complicated by the difficulty of adequately breaking thick-walled cells such that the integrity of organelles which may be needed for normal biosynthesis is preserved. Whether such care is crucial or not for biosynthesis is not altogether clear. In any case, most studies to date have been performed using extracts

prepared from young tissues, where the majority of products
would be expected to resemble the types of polysaccharides
found in primary cell walls.

(3) Many of the polymers listed contain more than one
type of monosaccharide unit. Listings in Table 1 are over-
simplified since we know that many of these polymers also
contain small amounts of other types of sugars. In fact, in
Albersheim's model of the primary cell wall of cultured
sycamore maple cells,[3,4,8,50,93] all polymers except cellulose are proposed to be covalently linked to each other,
making the cell wall matrix one gigantic macromolecule.
Most biosynthetic studies to date have concerned the
incorporation of a single type of labeled precursor into
polysaccharide, a condition which almost certainly cannot
mimic the *in vivo* situation. Structural considerations
strongly suggest that in the future more emphasis should
be placed on how combinations of various substrates interact
in biosynthesis. Current knowledge of wall structure provides
clues as to which combinations are most effective. The
listings of Table 1 do not specify exact linkages or ratios
of sugars in these polymers, although many of these are now
known with some certainty. Such details are quite important
to the enzymologist since, for example, the same monosaccharide linked in two different ways in a polymer almost
certainly is polymerized by different enzymatic reactions.

(4) Little is known about the regulation of cell wall
biosynthesis. Are the types of polymers synthesized by
various cell types controlled at the level of transcription
or translation of the enzymes, by the levels of substrates,
by small molecule activators or inhibitors, or by hormones?
These questions are largely unanswerable, and will remain
so until the biosynthetic processes are understood, at
which time regulation of cell wall biosynthesis will become
an exciting area for studies on developmental regulation in
plants.

SYNTHESIS OF THE MATRIX COMPONENTS

The In vitro *Approach*. It is now generally agreed
that nucleoside diphosphate sugars are the general class of
sugar donors for polysaccharide synthesis.[45] The usual
enzymological approach to cell wall biosynthesis in plants
has been to grind tissues, to isolate (usually) membrane

preparations, to add a simple type of radioactive precursor, and then to assay for the production of some insoluble product. Usually, but not always, an attempt is made to characterize the reaction product at least partially. Table 2 lists some examples of the types of *in vitro* syntheses which have been reported using plant extracts. These reports indicate that enzymes utilize various nucleotide sugars for the production of polysaccharide-like products. In nearly all cases reported in the literature, products were not purified, molecular size was not determined, and the degree of polymerization of the radioactive product was not established. In a few cases linkage was determined. Although there is no doubt that such studies provide a foundation from which to work, this approach may have exhausted its usefulness. The early work showing the interaction of GDP-glucose and GDP-mannose in glucomannan synthesis[29,98] and the recent preliminary results using UDP-glucose and UDP-xylose as precursors to a xyloglucan type of polymer[79,102] are examples of good integration of structural knowledge and enzymology, which deserve emulation.

Table 2.   Examples of *in vitro* synthesis of plant polysaccharides

Substrate	General type of product	Reference
UDP-galacturonic acid	Polygalacturonic acid	63, 103
TDP-galacturonic acid	Polygalacturonic acid	63
S-adenosylmethionine	Methyl esters of pectin	49
UDP-arabinose	Araban	73
UDP-xylose	Xylan	12, 73
UDP-xylose + UDP-glucose	Xyloglucan	79, 102
UDP-apiose	Apiogalacturonan	54
UDP-galactose	Galactan	75
GDP-mannose	Mannan	38
GDP-mannose + GDP-glucose	Glucomannan	29

There is also serious need for better characterization
of reaction products. Our laboratory has put considerable
effort into development of techniques of methylation analysis
and into ways of adapting these procedures to the study of
in vitro-synthesized radioactive products. Such techniques
applied to products that are carefully purified before
analysis should prove fruitful. Of course analysis of small
quantities of polysaccharide, particularly those which are
not water soluble, is not an easy task. Use of special
solvents such as urea, cuprammonium salts or dimethylsulfoxide, which do not degrade polysaccharides, should be
explored further for solubilizing difficult polymers. Once
products are in solution, use of sizing columns of porous
glass beads, described by Villemez,[99] the properties of
which are unaffected by many solvents, may prove useful for
purification and for determining the size of the products
synthesized.

Except for the intriguing reports suggesting a role for
glucosylinositol as primer for glucan synthesis,[51,52]
essentially nothing is known about the type of endogenous
acceptors present in our preparations. Information on this
rather crucial point may only come when products of *in vitro*
reactions are successfully isolated and characterized.
Improved techniques for organelle isolation and characterization should help in using *in vitro* studies to identify the
intracellular sites of synthesis of cell wall polymers; but
such studies cannot be properly interpreted until it is
established that the products synthesized *in vitro* truly
are wall precursors in size, composition, and structure.
Finally, awareness of the possibilities of interconversions
of substrates[73] is needed, both because of the possibility
of competing reactions, and because of the possibility of
internal degradation of the reaction products by the
presence of hydrolases.

Another question as yet unresolved by *in vitro* studies
is the question of whether lipid intermediates may play a
role in the synthesis of matrix polysaccharides. There is
now clear evidence in bacterial and mammalian systems that
phosphorylated polyprenols (see structure below) can mediate
as carriers between the level of nucleoside diphosphate
sugars and that of the final polysaccharides or glycoproteins.[61]

$$H-\left[CH_2-\overset{\overset{\textstyle CH_3}{|}}{C}=CH-CH_3\right]_x-O-\overset{\overset{\textstyle O}{\|}}{\underset{\underset{\textstyle O-}{|}}{P}}-OH$$

The number of isoprene units (x) in the lipid varys from 11 (in bacterial systems) to 22 in mammalian systems. Polyprenols (x = 6-13) have been isolated from higher plants,[104] but other than some preliminary, as yet inconclusive, reports of the synthesis of such compounds[2],[15],[48],[49],[62] there is still little evidence for a role for such compounds in the biosynthesis of cell wall polysaccharides in plants. Using membrane preparations from cotton fibers, Forsee and Elbein[35] demonstrated the synthesis of glucosyl- and mannosyl-lipids with properties closely resembling glycosyl-phosphoryl-polyprenols, and recently reported that mannosyl-lipid serves as a donor to an oligosaccharide-lipid and to glycoprotein.[36] Recent data of Ericson and Delmer[32] also strongly indicate a role for such intermediates in glycoprotein synthesis in developing bean cotyledons. Conceivably, such lipids serve as intermediates for cell wall biosynthesis as well, and inability to demonstrate their existence conclusively lies in the fact that they turn over very rapidly and have low steady-state levels. If mutants or specific inhibitors of wall synthesis existed such as those which exist for bacteria, such questions could be explored more easily.

The In vivo *Approach.* A quite different approach to the study of the biosynthesis of the matrix polysaccharides has been to feed radioactive precursors such as ^{14}C-glucose to living tissues and to follow patterns of incorporation into cellular organelles and cell wall fractions. Pulse-chase experiments permit one to monitor, either by electron microscopy or by biochemical fractionations, the flow of carbon from membranous sites of polymer synthesis to final incorporation into the wall. Such studies are therefore best suited for determining the intracellular sites of synthesis and for determining the size and structures of precursor polymers synthesized *in vivo*.

From such studies has evolved the concept that matrix polymers (or polysaccharides destined for secretion such as root cap slime[76]) are synthesized in internal membrane systems, either in the Golgi apparatus, or the endoplasmic reticulum, or both. From these sites the polymers are moved to the plasmalemma and deposited in the wall or secreted as slime. A good review of our current knowledge concerning these processes is found in a recent article by Northcote.[71] From the studies cited therein, it is clear that polysaccharides with compositions resembling those of cell wall polymers are found both in membranes of the endoplasmic reticulum and in the Golgi. Whether these two organelles represent separate sites of synthesis for distinct polymers or, in accord with our current concepts of membrane flow, whether synthesis of all mature polymers is initiated in the endoplasmic reticulum, then further modified or completed in the Golgi and transported to the wall is unresolved.

It seems that there is great potential now for using such *in vivo* studies for the isolation and structural characterization of wall precursors. It is clear that high molecular weight polysaccharides can be isolated from membranes of growing plant cells,[13,14] and by employing new, sensitive techniques for structural characterization, the future should see exciting advances in our knowledge of how the cell wall is assembled.

THE SPECIAL PROBLEM OF
CELLULOSE SYNTHESIS

Precisely what mechanisms are involved in the biosynthesis of cellulose, the world's most abundant organic compound, remains one of the major unresolved questions in plant biochemistry. At present, the literature is so cluttered with incomplete, conflicting data on the topic that it is difficult to present a coherent status of the field. To summarize briefly from all of what is to follow in this section, it is clear that no one has yet succeeded in achieving *in vitro* synthesis of true microfibrillar cellulose. This is in contrast to a recent, exciting report of the *in vitro* synthesis of chitin microfibrils in a fungal system.[84,85] It is true that syntheses of materials which contain β-1,4-glucose residues have been reported but the physiological significance of these results with respect to cellulose synthesis is questionable. In the sections below, the nature of the problem is outlined, and the nature of the confusions is clarified, in presenting suggestions as to future research.

Considerations of the Structure of Cellulose Relevant to Biosynthetic Studies. Cellulose, as it exists in its native form in the cell walls of plants, is an association of chains of β-1,4-glucan. These chains are associated into fibrils which have a high degree of order and crystallinity, and are highly insoluble. The reader is referred to the works of Blackwell[41],[42],[57],[58] for detailed descriptions and other references on structure as deduced from X-ray crystallography.

Until recently, there was much debate over whether the chains of cellulose were parallel or antiparallel. If the chains were antiparallel, and if growth of fibrils occurred from one end only, it would seem necessary to postulate two types of enzymes, one which polymerized glucose residues from the reducing end and one which polymerized from the non-reducing end. Two recent papers[42],[86] which provide strong evidence that chains of native cellulose (cellulose I) are parallel may have resolved the dilemma. It is clear, however, that regenerated cellulose (cellulose II) has quite a different X-ray pattern which is most consistent with antiparallel chains,[58],[86] but this form of cellulose does not seem to exist in cell walls.

As mentioned previously, the chains are invariably tightly associated into fibrils by hydrogen bonding. The so-called elementary fibrils[40] of approximately 3.5 nm diameter, containing about 32 chains, further associate into microfibrils of varying size. In cotton fiber, the secondary wall microfibrils have average diameters of about 20 nm.[55] An intact microfibril may contain up to 600 individual cellulose chains. Furthermore, there is some evidence that microfibrils are further aggregated in cotton into larger so-called macrofibrils.[28],[57] The existence of multichain fibrils suggests that cellulose may be synthesized by a large multi-subunit complex, with each subunit responsible for synthesis of a single chain.

Every second glucose residue of a chain is rotated 180° compared to its nearest neighbor.[81] Could this type

of stereochemistry present problems for synthesis? Might
some activated form of cellobiose (the basic repeating unit
of the chain) be the precursor rather than an activated form
of glucose?

Marx-Figini[66,67] has shown that, in the cotton fiber
at least, the degree of polymerization (DP) of primary
wall cellulose is low and relatively heterogeneous (2,000
to 6,000 residues), whereas the DP of secondary wall cellulose is high (approximately 14,000) and remarkably homogeneous, and she has suggested that separate enzyme systems may
be involved in the synthesis of primary and secondary wall
cellulose. At present there is no information as to how
the degree of polymerization is controlled so precisely
during secondary wall synthesis. Furthermore, orientation
of microfibrils, particularly in secondary walls, is very
precise. Some evidence suggests that microtubules are
involved in controlling orientation[46,47,69] but precisely
how this might occur is not known.

The presence of sugars other than glucose associated
with "purified" cellulose has been demonstrated.[1,68] It is
not yet known whether these small amounts of sugars are
covalently linked to the cellulose (and thus really a part
of the molecule to be synthesized) or whether they are
simply residual portions of other polysaccharides strongly
bound to cellulose by non-covalent interactions.

The Site of Cellulose Synthesis. With one exception,
significant levels of β-1,4-glucans have never been
detected during analyses of intracellular membrane-associated
polysaccharides in plant cells. The exception is the rather
unusual alga *Pleurochrysis* which synthesizes cellulosic
scales within a single large Golgi apparatus.[16] This lack,
coupled with electron microscopic observations (see ref. 71
and references therein) suggest that, in probable contrast
to the matrix components, cellulose is synthesized on the
plasmalemma at the cell surface. Recently Brown and
Montezinos[17] have obtained beautiful electron micrographs
of granules associated as a linear complex on the plasmalemma of *Oocystis*, an alga which deposits large quantities
of cellulosic microfibrils in its cell wall. The biosynthesis
of the microfibril is visualized as being in association
with these granular complexes. Other reports of similar
granules in association with microfibrils on the plasmalemma
of plants have appeared, further supporting the concept

that large, multisubunit complexes on the plasmalemma are the sites of cellulose synthesis.[72,82,83] To account for the existence of the reported β-1,4-glucan synthetase activities found in internal membrane systems,[80] Kiermayer and Dobberstein[53] have proposed that these are cellulose synthetases in the process of being synthesized and transported to the plasmalemma.

Possible Substrates for Cellulose Synthesis. Some controversy still exists regarding the substrate for cellulose synthesis. The major candidates remain UDP-glucose and GDP-glucose. Hassid's group[7,30] was the first to achieve *in vitro* synthesis of β-1,4-glucan with extracts from a higher plant. In this case GDP-glucose, and no other nucleoside diphosphate glucose, was effective as substrate. Since that time a number of other reports of *in vitro* synthesis of β-1,4-glucan from GDP-glucose, using plant extracts have been reported,[21,26,31,34,87] and for a number of years, one could see GDP-glucose documented in review articles and textbooks as the substrate for cellulose synthesis in higher plants. Unfortunately, the situation is not quite so simple. A number of years ago, UDP-glucose was reported to serve as substrate for cellulose synthesis in the bacterium *Acetobacter xylinum*.[43] Furthermore, a number of reports[18,22,39,60,64,74,80,87,88,92,94,97,99,100] appeared which claimed that UDP-glucose could also serve as a substrate for cellulose synthesis by enzyme preparations from higher plant tissues. Many of these reports have since been disputed and others lacked sufficient characterization of the reaction product to allow a positive judgement that cellulose was indeed the reaction product. However, a few of these reports seem to be reliable, and so we are left with the conclusion from the enzyme work that one, or both, of these substrates can be a likely precursor to cellulose in higher plants. Some reports, however, indicate that, depending on the concentration of UDP-glucose used, mixed glucans containing β-1,3 and β-1,4 linkages can result and that the higher the UDP-glucose concentration, the higher the proportion of β-1,3 linkages synthesized.[77,89] It is also possible that the low levels of β-1,4-glucan synthesized in some of these cases may be a backbone for other polymers such as xyloglucan (see section on Matrix Polysaccharides).

Much of the confusion in the literature resulted from inadequate proof of structure of the reaction product, and

especially from the by-now-disproven notion that insolubility of the product of the reaction in hot alkali is
sufficient evidence to claim cellulose synthesis. This has
become important because a variety of plant extracts contain
a highly active UDP-glucose:glucan synthetase[21,27,31,33,90,
94,95] which catalyzes the synthesis not of cellulose, but of
β-1,3-glucan, commonly called callose. It has been a
common misconception that β-1,3-glucans are soluble in hot
alkali, but it is now quite clear that, at least under
certain conditions, this is not so.

When I initiated work on cellulose synthesis several
years ago, having read Marx-Figini's suggestion of two
different enzyme systems for cellulose synthesis, it occurred
to me that one of these two nucleotide sugars might serve
as substrate for primary wall synthesis (some preliminary
data suggested GDP-glucose might be the most likely candidate[66]) while the other nucleotide sugar (UDP-glucose?)
could serve as substrate for secondary wall cellulose.
The developing cotton fiber offered an excellent model
system for testing this hypothesis, and indeed, for
studying cellulose synthesis in general. Cotton fibers are
single cells which elongate synchronously on the boll and
the events of primary and secondary wall synthesis are
separate in time. Furthermore, the secondary wall of the
cotton fiber is essentially pure cellulose. Sufficient
material can be obtained for biochemical analyses and,
thanks to the ingenious and careful work of C. A. Beasley
at the University of California at Riverside, a technique
had been developed to culture cotton ovules with their
associated fibers in a simple synthetic medium.[9,10,11]
Figure 1 outlines the general developmental sequence of
fiber elongation and cell wall synthesis as it occurs under
our experimental system. Concomitant with our enzyme work,
we are studying the composition and structure of the cotton
fiber cell wall during development, and we find that the
composition and changes in the cell wall associated with
development are remarkably similar in cultured fibers to
those grown on the plant.[68] In general, we use plant-grown
fibers for our enzyme work and culture-grown fibers for *in
vivo* labeling studies. The proven similarities in the two
systems permit us to compare these results with confidence.

Our first result[26] was highly promising (Fig. 2). By
incubating detached fibers of various ages with GDP-^{14}C-glucose

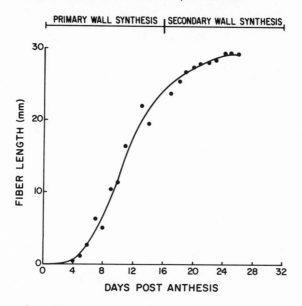

Figure 1.   Time course of development of the cotton fiber.   The developmental pattern shown is for *Gossypium hirsutum* Acala SJ-1 as determined under conditions of growth in environmentally-controlled growth chambers (13.5 hr day length, 33 C daytime temperature, 22 C at night).

or UDP-^{14}C-glucose, we observed incorporation of radio-activity into a hot alkali-insoluble product which was highly dependent upon fiber age.   GDP-glucose was active as substrate only during the period of elongation and primary wall synthesis, but declined to undetectable levels follow-ing the onset of secondary wall synthesis.   UDP-glucose incorporation followed a reverse pattern, lowest during primary wall synthesis and steadily increasing during secondary wall synthesis.   Further analyses of the reaction product confirmed that the product synthesized from GDP-glucose was indeed β-1,4-glucan,[26] but, to our surprise, over 90% of the radioactivity incorporated from UDP-glucose was soluble in chloroform/methanol.   At that time we speculated that this glucolipid could be a precursor for cellulose synthesis,[26] but we have since shown that over 95% of this glucolipid can be accounted for as acetylated and nonacetylated steryl glucosides.   Occasionally we see a trace of a product resembling a glucosyl-phosphoryl-

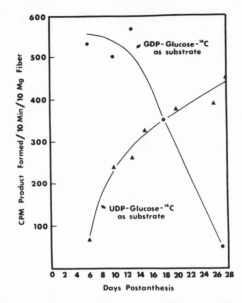

Figure 2. Capacity of cotton fibers of various ages for the biosynthesis of hot alkali-insoluble product from GDP-^{14}C-glucose or UDP-^{14}C-glucose.[26] (Reproduced with permission of Plant Physiology.)

polyprenol of the type reported by Forsee and Elbein[35] to be synthesized with membrane preparations from cotton fibers. The levels are so low under all conditions, however, that we have been unable to draw any conclusions regarding its function or significance.

As for matrix polysaccharides, the whole matter of possible lipid intermediates in cellulose synthesis remains an open question. The observations of cellobiosyl-lipid produced from UDP-glucose by extracts from *Acetobacter xylinum* are intriguing[23,56] particularly since, as mentioned previously, for steric reasons cellobiose is an attractive unit to serve as a precursor. However, these reports await further confirmation before any definite conclusions can be drawn.

From the results described above, it was possible to speculate that GDP-glucose could be the precursor for primary, but not secondary wall cellulose. However, futher results both from our own and other laboratories, shed doubts on even this conclusion, once again pointing out that, in the field of cellulose biosynthesis "the closer you look, the more confused you get." The first thing that made us suspicious were reports that suggested that GDP-glucose incorporation in plant extracts was stimulated by GDP-mannose,[29,98,101] and the further suggestion that this enzyme system functions primarily for the synthesis of glucomannan. Although the stimulation is variable, we have been able to observe this phenomenon in cotton fibers as well. The second suspicious observation (Fig. 3a) was the unusual kinetics of the reaction using membrane preparations from cotton fibers; the reaction was linear for only a brief time and both the initial rate and the final plateau were proportional to the amount of enzyme present. It seemed as if the enzyme might be terminating endogenous chains and that the concentration of acceptor present determined the final plateau. The fact that addition of boiled extract raised the plateau supported this possibility (Fig. 3b). Our suspicion was reenforced by our observation that capacity for synthesis of GDP-glucose in plants seemed to be very low. Table 3 lists possible routes of synthesis of these nucleotide sugars and Table 4 compares levels of UDP-glucose pyrophosphorylase and GDP-glucose pyrophosphorylase in extracts of cotton fibers and mung bean seedlings. These data show that the apparent capacity for synthesis of UDP-glucose via its specific pyrophosphorylase as judged by *in vitro* activity is at least 10,000 times higher than that for the synthesis of GDP-glucose. In fact, we have never, under any conditions, detected any GDP-glucose pyrophosphorylase activity in cotton fiber extracts. The other possible route of synthesis of GDP-glucose, via sucrose synthetase, also seems highly unlikely since this enzyme, from mung bean seedlings, when purified to homogeneity shows a much greater preference for UDP than for GDP as substrate.[24]

Results of an entirely different type of experiment also fail to support (but do not totally exclude) the concept of two separate enzyme systems for primary and secondary wall cellulose synthesis. Coumarin has been reported to inhibit cellulose synthesis,[44] and we have found that, in short term experiments, it specifically inhibits incorporation of radioactivity into the cellulosic fraction of cell walls of cultured fibers which were pulsed for short periods of time (2 hrs) with ^{14}C-glucose, (Table 5). Using this knowledge, that 1.0 mM coumarin seems equally

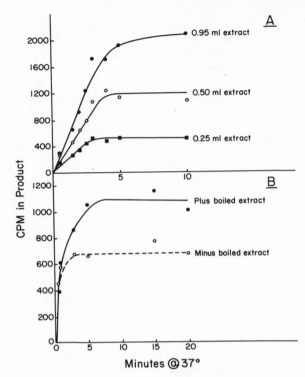

Figure 3. Kinetics of incorporation of radioactivity from GDP-^{14}C-glucose into β-1,4-glucan by membrane preparations of cotton fibers. (a) Kinetics as a function of the amount of membrane preparation assayed. (b) Effect of addition of boiled membrane preparation on the kinetics of the reaction.

Table 3. Possible routes of synthesis of UDP-glucose and GDP-glucose

---

1. Pyrophosphorylases: UTP + Glucose-1-P → UDP-glucose + PP$_i$
                        GTP + Glucose-1-P → GDP-glucose + PP$_i$
   (separate enzymes for each nucleotide sugar)

2. Sucrose synthetase: Sucrose + nDP → nDP-glucose + fructose
   (single enzyme where n may be uridine, adenosine, thymidine, guanosine, or cytosine)

Table 4.   Comparison of levels of activity of UDP-glucose
and GDP-glucose pyrophosphorylase

Tissue source	Activity UDP-glucose pyrophosphorylase	GDP-glucose pyrophosphorylase
	(nmoles/hr/mg protein)	
Etiolated mung beans[25]	68,000	4.5
Cotton fibers*		
7 days post-anthesis	39,800	Not detectable
14 days post-anthesis	103,000	Not detectable
21 days post-anthesis	61,800	Not detectable
28 days post-anthesis	62,000	Not detectable

*
Enzymes were prepared and assayed as described by Delmer
and Albersheim.[25]

Table 5.   Effect of coumarin on incorporation of ^{14}C-glucose
into cotton fiber cell walls.

Unfertilized cotton ovules (*Gossypium hirsutum* Acala S.J.-1)
with their associated fibers were cultured as described by
Beasley and Ting.[10]   At the age indicated, fibers were
pulsed with ^{14}C-glucose in the absence or presence of
coumarin (0.5 mM or 1.0 mM).   Glucose uptake was greater
than radioactivity recovered in the fibers because uptake
occurred also into the ovule cells.   Cell walls were fractionated as described by Hara *et al.*[44]

| | 16 day-old fibers | | |
	No coumarin added, cpm	0.5 mM coumarin cpm	Inhibition %
Glucose uptake	1,290,000	1,230,000	5
Soluble pools	544,770	523,980	4
Cell walls			
Pectic fraction	33,270	33,095	0
Hemicellulosic fraction	50,944	49,308	3
Cellulosic fraction	69,524	44,200	36

Table 5. Continued

| | 7 day-old fibers | | |
| | | 0.1 mM | |
	No coumarin added, cpm	coumarin cpm	Inhibition %
Glucose uptake	744,562	953,437	–
Soluble pools	396,332	391,888	2
Cell walls			
Pectic fraction	14,538	13,690	6
Hemicellulosic			
fraction	20,120	20,080	0
Cellulosic fraction	10,302	3,786	63

effective in inhibiting cellulose synthesis in cultured
ovules synthesizing primary wall cellulose (6–12 days of
age) and those actively depositing secondary wall (greater
than 12 days of age). These results (Table 6) indicate
that both primary and secondary wall cellulose synthesis
share a common coumarin-sensitive step. Since we do not
yet know the mode of action of this inhibitor, it is still
possible that this step is not the final polymerization
reaction.

From the above and a report by Franz[37] which indicates
that UDP-glucose is the major soluble nucleoside diphosphate
glucose in the cotton fiber, we feel that UDP-glucose is
the most logical candidate for precursor to cellulose. Yet
in our early studies, we observed essentially no incorporation from UDP-glucose into a polymer even resembling cellulose. Further work by Heiniger *et al.*[27] resulted in the
demonstration of a highly active UDP-glucose:glucan synthetase in these fibers. Particularly at low concentrations
of substrate, activity is dependent upon addition of β-
linked disaccharides to the reaction. Cellobiose or laminaribiose (3-*O*-β-glucosyl glucose) are equally effective as
activators. Activity can be measured either by assaying
detached fibers directly[26] or by using an isolated particulate fraction from the fibers. Under our conditions of
isolation, the enzyme, although particulate in nature, is
not associated with any single, definable organelle fraction. The enzyme shows substrate activation by UDP-glucose
(Fig. 4) and the effect of cellobiose is to shift the

Table 6.   The effect of coumarin on synthesis of primary
and secondary wall cellulose in culture-grown cotton fibers
(D. Delmer and T. Skokut, unpublished results)

Ovules plus fibers were grown and pulsed as described in
Table 5.   Coumarin (Cou) was present as indicated at 1.0 mM.
Cellulose was isolated by treatment with acetic/nitric
reagent as described by Updegraff.[96]   Secondary wall synthesis
begins between 12 and 14 days post-anthesis.[68]   Results are
the average of triplicate incubations.

Fiber age (days post-anthesis)	Treatment	Cpm in cellulose of fiber wall (per ovule)	% Inhibition by coumarin
6	−Cou	1079	83
	+Cou	186	
7	−Cou	450	
	+Cou	109	76
9	−Cou	820	
	+Cou	304	62
12	−Cou	877	
	+Cou	315	64
14	−Cou	2776	
	+Cou	878	68
16	−Cou	2604	
	+Cou	699	73
19	−Cou	1466	
	+Cou	260	82

range in which this activation occurs to lower concentrations of UDP-glucose.   Figure 5 shows a comparison of the
measurable levels of activity of this glucan synthetase
with the *in vivo* rate of cellulose deposition in the fiber.
(The latter rate was calculated from data obtained on the
content of cellulose per unit length of fiber during
development.[68])   The *in vitro* rates of glucan synthesis in
these detached fibers during development is within the
range of the *in vivo* rate of cellulose synthesis.   Is this
product cellulose?   Although the glucan product synthesized

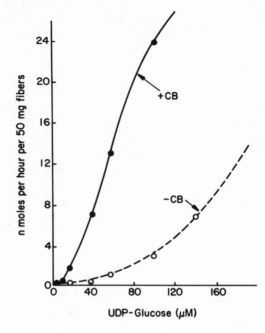

Figure 4.   Velocity of cotton fiber β-1,3-glucan
synthetase reaction as a function of UDP-glucose concentration.   Detached fibers (50 mg, harvested 19 days post-
anthesis) were incubated in 250 μl of TES buffer pH 7.5
containing 6 mM $MgCl_2$ and UDP-^{14}C-glucose as indicated in
the absence or presence of 10 mM cellobiose.   Reactions
were terminated by the addition of chloroform:methanol (1:1).
Radioactivity insoluble in water and chloroform:methanol
represents incorporation into β-1,3-glucan.   Reproduced
with the permission of Plant Physiology (in press).

by detached fibers is insoluble in hot alkali, at least
95% of the product is definitely *not* cellulose, but rather
a linear β-1,3-glucan.   Table 7 summarizes the information
we have obtained to support this conclusion.

The Non-cellulosic Glucan of Cotton Fibers.   At the
same time we were obtaining information on this glucan
synthetase, the composition and structure of the cell wall
of the developing cotton fiber was also being studied.[68]

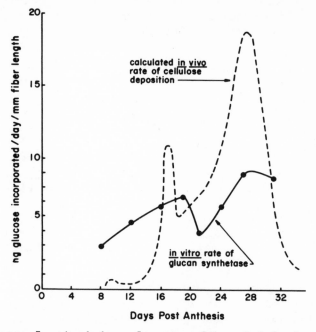

Figure 5. Activity of cotton fiber β-1,3-glucan synthetase as a function of fiber age. Assays were performed as described in the legend to Figure 4 except that fiber age was varied. Cellobiose was present at 10 mM; UDP-^{14}C-glucose at 1.0 mM. The *in vivo* rate of cellulose deposition was calculated from our data of cellulose content in the fiber cell wall as a function of fiber age.[68]

Some of the data obtained offered a probable explanation for the physiological role of the β-1,3-glucan synthetase of cotton fibers. The relevant observation was that, just prior to the onset of secondary wall cellulose, a large rise in non-cellulosic glucan is observed in cell wall fractions of the cotton fiber (Fig. 6). Preliminary methylation analyses of unfractionated cell walls show a corresponding rise in 3-linked glucose, suggesting that this polymer is a 3-linked glucan. Although it is still premature to draw any conclusions about the function of this glucan, it is interesting that its rise is nearly coincident with the onset of secondary wall cellulose synthesis. Three-linked glucans are not often found as structural components of higher plant cell walls. There are, however, data in the

Table 7. Structural analyses of the product of the cotton
fiber glucan synthetase reaction

Treatment	Results
1N NaOH, 100 C, 15 min	>75% of product remains insoluble.
Acetic/nitric reagent,[96] 100 C, 30 min	95% solubilized (cellulose remains insoluble).
Total acid hydrolyses	95% hydrolyzed; all radioactivity released as glucose
α-Amylase or pronase	No effect.
β-Glucanases or partial acid hydrolysis	Hydrolysis with release of glucose, laminaribiose and laminaritriose.
Periodate oxidation	Resistant.
Methylation and separation of derivatives by GLC	Majority of radioactivity found in 3-linked glucose. Small amount of terminal glucose; no substantial amount of branched residues indicated.

literature which suggest that non-cellulosic glucans may
serve a function as reserve polymers. Callose (β-1,3-
glucan) can be mobilized as a reserve in pollen,[91] and
other data can be interpreted as indicating a possible
reserve function for the mixed β-1,3-β-1,4-glucans found
in cell wall preparations of various plant tissues.[19,55]
Also of interest is the observation of a specific decrease
in non-cellulosic glucan (which is accompanied by an almost
compensating increase in cellulose) in the cell walls of
Avena coleoptiles undergoing IAA-induced elongation in the
absence of external carbon sources.[65,78] We are currently
testing to see whether the non-cellulosic glucan of cotton
fibers exhibits turnover, and, if so, whether it may serve
as a specific precursor to cellulose. Two recent papers
report the existence of a "cell surface" cellulose synthetase
in pea stem sections.[87,88] The enzyme exhibits many of the
properties expected of a cellulose synthetase, including
the observation that hormone treatment alters the intracellular

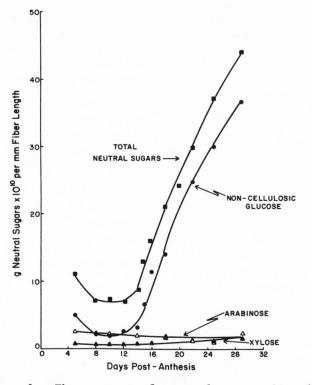

Figure 6. The content of neutral sugars in polysaccharides of the cotton fiber cell wall as a function of fiber age. (From Meinert.[68])

location of this enzyme. We find the product of this enzyme in peas is a 3-linked glucan rather than cellulose. This is still of great interest if this enzyme is functioning to produce a precursor to cellulose.

We have no evidence yet which specifically proves any precursor function for non-cellulosic glucan. Another, perhaps bizarre, possibility is that these widespread and highly active β-1,3-glucan synthetases are really cellulose synthetases gone awry due to rough treatment during isolation and assay. The great difference in the stereochemistry of 1,3 and 1,4 linkages makes such a possibility rather unlikely but it is an intriguing thought and might provide

an explanation for the often-observed wound response of
plants--that of callose production. If we assume that
cellulose synthetase is altered in its stereospecificity by
the assault of wind, hail, insects, or enzymologists, the
immediate result could be the production of callose.

In any case, it is abundantly clear that we lack some
fundamental key needed to unlock the pathway of cellulose
synthesis in higher plants. For some reason which we do
not understand--perhaps extreme lability of enzyme, inappro-
priate substrates, or missing essential cofactors--we have
not been convincingly successful in achieving the *in vitro*
synthesis of cellulose. We have preliminary data which
indicate that, by increasing the external osmotic pressure
of the medium surrounding cultured fibers a few bars above
normal, *in vivo* cellulose synthesis is rapidly and significantly depressed. If the process is so sensitive, we may
be in for a long struggle, but the journey so far has been
fun. and promises to be so in the future.

ACKNOWLEDGMENTS

The research data presented herein with respect to the
biosynthesis of cellulose in the cotton fiber was the result
of my fruitful collaboration with a number of individuals
who deserve my special thanks and recognition. Some of the
early work was a result of my own post-doctoral research
(supported by an NIH Postdoctoral Fellowship) with Peter
Albersheim (University of Colorado), and later of a joint
project with C. A. Beasley, University of California at
Riverside (supported in part by Cotton Incorporated). The
more recent work has been done at the MSU/ERDA Plant Research
Laboratory, Michigan State University and was supported by
ERDA contract number E(11-1)-1338. My colleagues at MSU
in this work have been Ursula Heiniger (glucan synthetase),
Maureen Meinert (cell wall structure), Tom Skokut (coumarin
studies), and Carl Kulow, our able technician, who contri-
butes in one way or another to just about everything we do.

REFERENCES

1. Adams, G. A., and C. T. Bishop. 1955. Polysaccharides
    associated with alpha-cellulose. *Tappi* *38*:672.

2. Alam, S. S. and F. W. Hemming. 1973. Polyprenol phosphates and mannosyl transferases in *Phaseolus aureus*. *Phytochemistry* *12*:1641.

3. Albersheim, P. 1975. The walls of growing plant cells. *Sci. Amer.* *232*:81.

4. Albersheim, P., W. D. Bauer, K. Keegstra, and K. W. Talmadge. 1973. The structure of the wall of suspension-cultured sycamore cells. *In* "Biogenesis of Plant Cell Wall Polysaccharides" (F. Loewus, ed.), pp. 117–147. Academic Press, New York.

5. Aspinall, G. O. 1970. Pectins, plant gums and other plant polysaccharides. *In* "The Carbohydrates", Vol. IIB (W. Pigman and D. Horton, eds.), pp. 515–536. Academic Press, New York.

6. Aspinall, G. O. 1973. Carbohydrate polymers of plant cell walls. *In* "The Biogenesis of Plant Cell Wall Polysaccharides" (F. Loewus, ed.), pp. 95–116. Academic Press, New York.

7. Barber, G. A., A. D. Elbein, and W. Z. Hassid. 1964. The synthesis of cellulose by enzyme systems from higher plants. *J. Biol. Chem.* *239*:4056.

8. Bauer, W. D., K. W. Talmadge, K. Keegstra, and P. Albersheim. 1973. The structure of plant cell walls. II. The hemicellulose of the walls of suspension-cultured sycamore cells. *Plant Physiol.* *51*:174.

9. Beasley, C. A. and I. P. Ting. 1973. The effects of plant growth substances on *in vitro* fiber development from fertilized cotton ovules. *Am. J. Bot.* *60(2)*:130.

10. Beasley, C. A. and I. P. Ting. 1974. Effects of plant growth substances on *in vitro* fiber development from unfertilized cotton ovules. *Am. J. Bot.* *61*:188.

11. Beasley, C. A., I. P. Ting, A. E. Linkens, E. H. Birnbaum, and D. P. Delmer. 1974. Cotton ovule culture: a review of progress and a preview of potential. *In* "Tissue Culture and Plant Science 1974" (H. E. Street, ed.), pp. 169–192. Academic Press, New York and London.

12. Ben-Arie, R., L. Ordin, and J. L. Kindinger. 1973. A cell-free xylan synthesizing enzyme from *Avena sativa*. *Plant and Cell Physiol.* *14*:427.

13. Bowles, D. J. and D. H. Northcote. 1974. The amounts and rates of export of polysaccharides found within the membrane system of maize root cells. *Biochem. J.* *142*:139.

14. Bowles, D. J. and D. H. Northcote. 1976. The size
     and distribution of polysaccharides during their
     synthesis within the membrane system of maize root
     cells. *Planta 128*:101.
15. Brett, C. T. and D. H. Northcote. 1975. The formation
     of oligoglucans linked to lipid during synthesis of
     β-glucan by characterized membrane fractions isolated
     from pea. *Biochem. J. 148*:107.
16. Brown, R. M., Jr., W. W. Franke, H. Kleinig, H. Falk,
     and P. Sitte. Cellulosic wall component produced by
     the Golgi apparatus of *Pleurochrysis scherffelii*.
     Science 166:894.
17. Brown, R. M., Jr. and D. Montezinos. 1976. Cellulose
     microfibrils: visualization of biosynthetic and
     orienting complexes in association with the plasma
     membrane. *Proc. Nat. Acad. Sci. USA 73*:143.
18. Brummond, D. O. and A. P. Gibbons. 1964. The enzymatic
     synthesis of cellulose by the higher plant *Lupinus
     albus*. *Biochem. Biophys. Res. Commun. 17*:156.
19. Buchala, A. J. and K. C. B. Wilkie. 1971. The ratio
     of β-(1→3) to β-(1→4) glucosidic linkages in non-
     endospermic hemicellulosic β-glucans from oat plant
     (*Avena sativa*) tissues at different stages of
     maturity. *Chemistry 10*:2287.
20. Burke, D., P. Kaufman, M. McNeil, and P. Albersheim.
     1974. The structure of plant cell walls. VI. A
     survey of the walls of suspension-cultured monocots.
     Plant Physiol. 54:109.
21. Chambers, J. and A. D. Elbein. 1970. Biosynthesis of
     glucans in mung bean seedlings. Formation of β-(1→4)-
     glucans from GDP-glucose and β-(1→3)-glucans from
     UDP-glucose. *Arch. Biochem. Biophys. 138*:620.
22. Clark, A. F. and C. L. Villemez. 1972. The formation
     of β-(1→4)-glucan from UDP-α-D-glucose catalyzed by
     a *Phaseolus aureus* enzyme. *Plant Physiol. 50*:371.
23. Dankert, M., R. García, and E. Recondo. 1972. Lipid-
     linked intermediates in the biosynthesis of polysac-
     charides in *Acetobacter xylinum*. *In* "Biochemistry
     of the Glycosidic Linkage" (R. Piras and H. G.
     Pontis, eds.), pp. 199-206. Academic Press, New York.
24. Delmer, D. P. 1972. The purification and properties
     of sucrose synthetase from etiolated *Phaseolus
     aureus* seedlings. *J. Biol. Chem. 247*:3822.
25. Delmer, D. P. and P. A. Albersheim. 1970. The bio-
     synthesis of sucrose and nucleoside diphosphate
     glucose in *Phaseolus aureus*. *Plant Physiol. 45*:782.

26. Delmer, D. P., C. A. Beasley, and L. Ordin. 1974. Utilization of nucleoside diphosphate glucoses in developing cotton fibers. *Plant Physiol.* *53*:149.

27. Delmer, D. P., U. Heiniger, and C. Kulow. 1976. Glucan synthesis from UDP-glucose in cotton fibers. *Plant Physiol.* *57*, Supplement #162.

28. Dolmetsch, H. H. and H. Dolmetsch. 1969. The fibrillar bundling of cellulose molecules in cotton. *Textile Res. J.* *39*:568.

29. Elbein, A. D. 1969. Biosynthesis of a cell wall glucomannan in mung bean seedlings. *J. Biol. Chem.* *244*:1608.

30. Elbein, A. D., G. A. Barber, and W. Z. Hassid. 1964. The synthesis of cellulose by an enzyme system from a higher plant. *J. Amer. Chem. Soc.* *86*:309.

31. Elbein, A. D. and W. T. Forsee. 1973. Studies on the biosynthesis of cellulose. *In* "Biogenesis of Plant Cell Wall Polysaccharides" (F. Loewus, ed.), pp. 259-295. Academic Press, New York and London.

32. Ericson, M. C. and D. P. Delmer. 1976. Glycoprotein synthesis in plants: lipid intermediates. *Plant Physiol.* *57*, Supplement #237.

33. Feingold, D. S., E. F. Neufeld, and W. Z. Hassid. 1958. Synthesis of a β1,3-linked glucan by extracts of *Phaseolus aureus* seedlings. *J. Biol. Chem.* *233*:783.

34. Flowers, H. M., K. K. Batra, J. Kemp, and W. Z. Hassid. 1969. Biosynthesis of cellulose *in vitro* from guanosine diphosphate D-glucose with enzymic preparations from *Phaseolus aureus* and *Lupinus albus*. *J. Biol. Chem.* *244*:4969.

35. Forsee, W. T. and A. D. Elbein. 1973. Biosynthesis of mannosyl- and glucosyl-phosphoryl polyprenols in cotton fibers. *J. Biol. Chem.* *248*:2858.

36. Forsee, W. T. and A. D. Elbein. 1975. Glycoprotein biosynthesis in plants. Demonstration of lipid-linked oligosaccharides of mannose and N-acetylgluco-samine. *J. Biol. Chem.* *250*:9283.

37. Franz, G. 1969. Soluble nucleotides in developing cotton hair. *Phytochemistry* *8*:737.

38. Franz, G. 1973. Biosynthesis of salep mannan. *Phytochemistry* *12*:2369.

39. Franz, G. and H. Meier. 1969. Biosynthesis of cellulose in growing cotton hairs. *Phytochemistry* *8*:579.

40. Frey-Wyssling, A. and K. Mühlethaler. 1963. Die Elementarfibrillen der Cellulose. *Makromol. Chemie* *62*:25.

41. Gardner, K. H. and J. Blackwell. 1974. The hydrogen
     bonding in native cellulose. *Biochem. Biophys.
     Acta 343*:232.
42. Gardner, K. H. and J. Blackwell. 1974. The structure
     of native cellulose. *Biopolymers 13*:1975.
43. Glaser, L. 1958. The synthesis of cellulose in cell-
     free extracts of *Acetobacter xylinum*. *J. Biol. Chem.
     232*:627.
44. Hara, M., N. Umetsu, C. Miyamoto, and K. Tamari. 1973.
     Inhibition of the biosynthesis of plant cell wall
     materials, especially cellulose biosynthesis by
     coumarin. *Plant Cell Physiol. 14*:11.
45. Hassid, W. Z. 1969. Biosynthesis of oligosaccharides
     and polysaccharides in plants. *Science 165*:137.
46. Heath, I. B. 1974. A unified hypothesis for the role
     of membrane bound enzyme complexes and microtubules
     in plant cell wall synthesis. *J. Theor. Biol. 48*:
     445.
47. Hepler, P. K. and B. A. Palevitz. 1974. Microtubules
     and microfilaments. *Ann. Rev. Plant Physiol. 25*:309.
48. Kauss, H. 1969. A plant mannosyl-lipid acting in
     reversible transfer of mannose. *FEBS Letters 5*:81-84.
49. Kauss, H. 1974. Biosynthesis of pectin and hemicellu-
     loses. *In* "Plant Carbohydrate Chemistry" (J. B.
     Pridham, ed.), pp. 191-205. Academic Press, New
     York.
50. Keegstra, K., K. W. Talmadge, W. D. Bauer, and P.
     Albersheim. 1973. The structure of plant cell
     walls. III. A model of the walls of suspension-
     cultured sycamore cells based on the interconnections
     of the macromolecular components. *Plant Physiol. 51*:
     188.
51. Kemp, J. and B. C. Loughman. 1973. Chain initiation
     in glucan synthesis in mung beans. *Biochem. Soc.
     Trans. 1*:446.
52. Kemp, J. and B. C. Loughman. 1974. Cyclitol gluco-
     sides and their role in the synthesis of a glucan
     from uridine diphosphate glucose in *Phaseolus aureus*.
     Biochem. J. 142:153.
53. Kiermayer, O. and B. Dobberstein. 1973. Membrankom-
     plexe dictyosomaler Herkunft als "Matrizen" für die
     extraplasmatische Synthese und Orientierung von
     Mikrofibrillen. *Protoplasma 77*:437.
54. Kindel P. 1973. Occurrence and metabolism of D-apiose
     in *Lemna minor*. *In* "The Biogenesis of Plant Cell
     Wall Polysaccharides" (F. Loewus, ed.), pp. 85-94.
     Academic Press, New York.

55. Kivilaan, A., R. S. Bandurski, and A. Schulze. 1971. A partial characterization of an autolytically solubilized cell wall glucan. *Plant Physiol. 48*: 389.

56. Kjosbakken, J. and J. R. Colvin. 1973. Biosynthesis of cellulose by a particulate enzyme system from *Acetobacter xylinum*. *In* "Biogenesis of Plant Cell Wall Polysaccharides" (F. Loewus, ed.), pp. 361–371. Academic Press, New York.

57. Kolpak, F. J. and J. Blackwell. 1975. Deformation of cotton and bacterial cellulose microfibrils. *Textile Res. J. 45*:568.

58. Kolpak, F. J. and J. Blackwell. 1976. Determination of the structure of cellulose II. *Macromolecules 9*:273.

59. Lamport, D. T. A. 1970. Cell wall metabolism. *Ann. Rev. Plant Physiol. 21*:235.

60. Larsen, G. L. and D. O. Brummond. 1974. β-(1→4)-D-glucan synthesis from UDP-[^{14}C]-D-glucose by a solubilized enzyme from *Lupinus albus*. *Phytochemistry 13*:361.

61. Lennarz, W. J. and M. G. Scher. 1973. The role of lipid-linked activated sugars in glycosylation reactions. *In* "Membrane Structure and Mechanisms of Biological Energy Transduction" (J. Avery, ed.), pp. 441–453. Plenum Press, New York.

62. Lezica, R. P., C. T. Brett, P. R. Martinez, and M. A. Dankert. 1975. A glucose acceptor in plants with the properties of an α-saturated polyprenyl-mono-phosphate. *Biochem. Biophys. Res. Commun. 66*:980.

63. Liu, T. Y., A. D. Elbein, and J. C. Su. 1966. Substrate specificity in pectin synthesis. *Biochem. Biophys. Res. Commun. 22*:650.

64. Liu, T. Y. and W. Z. Hassid. 1970. Solubilization and partial purification of cellulose synthetase from *Phaseolus aureus*. *J. Biol. Chem. 245*:1922.

65. Loescher, W. and D. Nevins. 1972. Auxin-induced changes in *Avena* coleoptile cell wall composition. *Plant Physiol. 50*:556.

66. Marx-Figini, M. 1966. Comparison of the biosynthesis of cellulose *in vitro* and *in vivo* in cotton bolls. *Nature 210*:754.

67. Marx-Figini, M. and G. V. Schulz. 1966. Über die Kinetik und den Mechanismus der Biosynthese der Cellulose in den höheren Pflanzen (nach Versuchen an den Samenhaaren der Baumwolle). *Biochim. Biophys. Acta 112*:74.

68. Meinert, M.   1975.   Biochemical changes in cell wall
    composition associated with *in vivo* and *in vitro*
    fiber development in *Gossypium hirsutum*.   M.S. Thesis,
    Michigan State University.
69. Newcomb, E. H.   1969.   Plant microtubules.   *Ann. Rev.
    Plant Physiol. 20*:253.
70. Northcote, D. H.   1972.   Chemistry of the plant cell
    wall.   *Ann. Rev. Plant Physiol. 23*:113.
71. Northcote, D. H.   1974.   Sites of synthesis of the
    polysaccharides of the cell wall.   *In* "Plant Carbo-
    hydrate Chemistry" (J. B. Pridham, ed.), pp. 165-
    181.   Academic Press, New York.
72. Northcote, D. H. and D. R. Lewis.   1968.   Freeze-
    etched surfaces of membranes and organelles in the
    cells of pea root tips.   *J. Cell Sci. 3*:199.
73. Odzuck, W. and H. Kauss.   1972.   Biosynthesis of pure
    araban and xylan.   *Phytochemistry 11*:2489.
74. Ordin, L. and M. A. Hall.   1968.   Cellulose synthesis
    in higher plants from UDP-glucose.   *Plant Physiol.
    43*:473.
75. Panayotatos, N. and C. L. Villemez.   1973.   The forma-
    tion of a β-(1→4)-D-galactan chain catalyzed by a
    Phaseolus aureus enzyme.   *Biochem. J. 133*:263.
76. Paull, R. E. and R. L. Jones.   1975.   Studies on the
    secretion of maize root cap slime.   II. Localization
    of slime production.   *Plant Physiol. 56*:307.
77. Péaud-Lenoël, C. and M. Axelos.   1970.   Structural
    features of the β-glucans enzymatically synthesized
    from uridine diphosphate glucose by wheat seedlings.
    FEBS Letters 8:224.
78. Ray, P. M.   1963.   Sugar composition of oat coleoptile
    cell walls.   *Biochem. J. 89*:144.
79. Ray, P. M.   1975.   Golgi membranes form xyloglucan
    from UDPG and UDP-xylose.   *Plant Physiol. 56*,
    Supplement #84.
80. Ray, P. M., T. L. Shininger, and M. M. Ray.   1969.
    Isolation of β-glucan synthetase particles from
    plant cells and identification with golgi membranes.
    Proc. Nat. Acad. Sci. USA. 64:605.
81. Rees, D. A. and W. E. Scott.   1969.   Conformational
    analysis of polysaccharides: stereochemical signifi-
    cance of different linkage positions in β-linked
    polysaccharides.   *Chemical Communications Comm. No.
    1003.*   Pub. by the Chemical Society, Burlington
    House, London.   p. 1037.

82. Robinson, D. G. and R. D. Preston. 1972. Plasmalemma
     structure in relation to microfibril biosynthesis in
     Oocystis. *Planta 104*:234.
83. Roland, J. A. and P. E. Pilet. 1974. Implications du
     plasmalemme et de la paroi dans la croissance des
     cellules végétales. *Experentia 30*:441.
84. Ruiz-Herrera, J. and S. Bartnicki-Garcia. 1974.
     Synthesis of cell wall microfibrils *in vitro* by a
     "soluble" chitin synthetase from *Mucor rouxii*.
     Science 186:357.
85. Ruiz-Herrera, J., V. O. Sing, W. J. Van der Woude, and
     S. Bartnicki-Garcia. 1975. Microfibril assembly
     of granules of chitin synthetase. *Proc. Nat. Acad.
     Sci. USA 72*:2706.
86. Sarko, A. and R. Muggli. 1974. Packing analysis of
     carbohydrates and polysaccharides. III. *Valonia*
     cellulose and cellulose II. *Macromolecules 7*:486.
87. Shore, G. and G. A. MacLachlan. 1975. The site of
     cellulose synthesis. Hormone treatment alters the
     intracellular location of alkali-insoluble β-1,4-
     glucan (cellulose) synthetase activities. *J. Cell
     Biol. 64*:557.
88. Shore, G., Y. Raymond, and G. A. MacLachlan. 1975.
     The site of cellulose synthesis. Cell surface and
     intracellular β-1,4-glucan (cellulose) synthetase
     activities in relation to the stage and direction
     of cell growth. *Plant Physiol. 56*:34.
89. Smith, M. M. and B. A. Stone. 1973. β-Glucan synthesis
     by cell-free extracts from *Lolium multiflorum* endo-
     sperm. *Biochem. Biophys. Acta 313*:72.
90. Southworth, D. and D. B. Dickinson. 1975. β-1,3-
     Glucan synthetase from *Lilium longiflorum* pollen.
     Plant Physiol. 56:83.
91. Stanley, R. G. and H. F. Linskens. 1974. Pollen:
     biology, biochemistry, management. Springer-Verlag,
     p. 135.
92. Stepanenko, B. N. and A. V. Morozova. 1970. O vozmozh-
     nosti biosinteza tsellulozy iz UDF-glucozi u
     khlopchatnika. *Fiziologia Rastenii 17*:302.
93. Talmadge, K., K. Keegstra, W. D. Bauer, and P. Alber-
     sheim. 1973. The structure of plant cell walls.
     I. The macromolecular components of the walls of
     suspension-cultured sycamore cells with a detailed
     analysis of the pectic polysaccharides. *Plant
     Physiol. 51*:158.

94. Tsai, C. M. and W. Z. Hassid. 1971. Solubilization
     and separation of uridine diphospho-D-glucose: β-(1→4)
     glucan and uridine diphospho-D-glucose: β-(1→3)
     glucan glucosyltransferase from coleoptiles of *Avena
     sativa. Plant Physiol.* 47:740.
95. Tsai, C. M. and W. Z. Hassid. 1973. Substrate activa-
     tion of β-(1→3)-glucan synthetase and its effect on
     the structure β-glucan obtained from UDP-glucose and
     particulate enzymes of oat coleoptiles. *Plant
     Physiol.* 51:998.
96. Updegraff, D. M. 1969. Semi-micro determination of
     cellulose in biological materials. *Anal. Biochem.*
     32:420.
97. Van der Woude, W. J., C. A. Lembi, D. J. Morré, J. I.
     Kindinger, and L. Ordin. 1974. β-Glucan synthetases
     of plasma membrane and golgi apparatus from onion
     stem. *Plant Physiol.* 54:333.
98. Villemez, C. L. 1971. Rate studies of polysaccharide
     biosynthesis from guanosine diphosphate α-D-glucose
     and guanosine diphosphate α-D-mannose. *Biochem. J.*
     121:151.
99. Villemez, C. L. 1973. The relation of plant enzyme-
     catalyzed β-1,4-glucan synthesis to cellulose bio-
     synthesis *in vivo*. *In* "Plant Carbohydrate Chemistry"
     (J. B. Pridham, ed.), pp. 183-189. Academic Press,
     New York.
100. Villemez, C. L., G. Franz, and W. Z. Hassid. 1967.
      Biosynthesis of alkali insoluble polysaccharide from
      UDP-glucose with particulate enzyme preparation from
      Phaseolus aureus. Plant Physiol. 42:1219.
101. Villemez, C. L. and J. S. Heller. 1970. Is guanosine
      diphosphate-D-glucose a precursor of cellulose?
      Nature 227:80.
102. Villemez, C. L. and M. Hinman. 1975. UDP-glucose
      stimulated formation of xylose containing polysac-
      charides. *Plant Physiol.* 56, Supplement #79.
103. Villemez, C. L., T. Y. Lin, and W. Z. Hassid. 1965.
      Biosynthesis of the polygalacturonic acid chain of
      pectin by a particulate enzyme preparation from
      Phaseolus aureus seedlings. *Proc. Nat. Acad. Sci. USA*
      54:1626.
104. Wellburn, A. R. and F. W. Hemming. 1966. Gas-liquid
      chromatography of derivatives of naturally-occurring
      mixture of long-chain polyisoprenoid alcohols.
      J. Chromatog. 23:51.

105. Wilder, B. M. and P. Albersheim. 1973. The structure
      of plant cell walls. IV. A structural comparison of
      the wall hemicellulose of cell suspension cultures
      of sycamore (*Acer pseudoplatanus*) and of red kidney
      bean (*Phaseolus vulgaris*). *Plant Physiol. 51*:889.

Chapter Three

STRUCTURE, BIOSYNTHESIS AND SIGNIFICANCE OF CELL WALL
GLYCOPROTEINS

DEREK T. A. LAMPORT

MSU/ERDA Plant Research Laboratory
Michigan State University
East Lansing, Michigan 48824

INTRODUCTION

    Glycoproteins are interesting molecules because they
play important roles in morphogenesis. But glycoproteins
are not always as sweet as their name sounds; they are often
very slippery both conceptually and in experimenters' hands.[33]
We can if we wish wrestle with the problem of deciding between
Eylar's suggestion[13] that sugar residues are the passport for
secretion, and the proposal of Winterburn and Phelps[50] that
sugars are a code not for secretion *per se* but for topo-
graphical location of the "secreted" glycoproteins. Such
unifying concepts are always good for a few years of controversy! In this chapter are presented several additional suggestions which may help in understanding the origin of the
extracellular matrix in plants and animals.

    First is described the structure of extensin, the
hydroxyproline-rich glycoprotein in the cell walls of higher
plants. These follow a description of a new deglycosylation
technique, solvolysis in anhydrous hydrogen fluoride, enor-
mously helpful in the purification and structural analysis of

the polypeptide backbone from heavily glycosylated glycoproteins. Then extensin first is compared with its possible
soluble relatives in higher plants and secondly with its
possible counterpart in some primitive algae, finally push-
ing this comparison to its extreme by comparing plant
hydroxyproline-rich glycoproteins with analogous animal glycoprotein, collagen. Is it really possible that Aaronson
is correct in suggesting[1] that extensin and collagen share
a common ancestry? New evidence is presented to support
Aaronson's suggestion and to support the idea that if all
proteins are evolutionary relics, then each protein or enzymic process should be traceable to its original organismic
type. Plants are, according to the endosymbiont capture
hypothesis,[30] quadrigenomic and hence possible origins are
suggested for extensin and collagen, proline hydroxylase,
and the enzymes involved in biosynthesis of the asparagine-
N-acetyl glucosamine glycopeptide linkage.

STRUCTURE OF EXTENSIN

      We have mainly used primary cell walls isolated after
sonic disruption of tomato suspension cultures which were
grown for four weeks at 27 C. These walls contain about
1.7% (dry wt.) bound hydroxyproline which can be released in
macromolecular form only through degradative procedures.
Early work[24] showed that a crude mixture of enzymes possessing carbohydrase and protease activities releases hydroxyproline-rich glycopeptides from the cell walls, while pure
proteases are relatively ineffective. These glycopeptides
(Table 1) contain arabinose in a linkage stable to alkali,
so that alkaline hydrolysis releases a family of hydroxyproline arabinosides (Fig. 1). Most of the hydroxyproline
residues in the tomato wall are O-substituted by a tri- or
tetrasaccharide of arabinose (Fig. 2). Treatment of cell
walls at pH 1 for 1 hr at 100 C completely cleaves these
arabinosyl linkages and renders the remaining hydroxyproline-
rich polypeptide backbone very much more labile to proteases.
Thus trypsin releases high and low molecular weight peptides.[2]
The low molecular weight fraction accounts for about 1/3 of
the wall-bound hydroxyproline and accounts for 5 tryptic
peptides (Table 2) whose sequence totals 48 residues. These
peptides probably represent either the complete extensin
sequence or about one-third of the sequence (Fig. 3). The
peptides represent the complete sequence if all the high
molecular weight non-sequenced material has the same sequence

as the low molecular weight but has become crosslinked during "maturation," similar for example, to collagen or elastin crosslinks.

The complete sequence of extensin is not easily determined. Here are the problems: first, the high degree of glycosylation; second, the repetitive sequence; third, the fact that residues between the repeat units are most often serine, which gives a low yield on Edman degradation; and fourth, the insolubility of extensin as it exists in the wall.

Table 1. Tomato glycopeptides obtained by digestion of tomato cell walls with "cellulase"

		NH_2-Terminus
1.	ara_{25} gal_6 hyp_{10} ser_3 tyr	SER
2.	ara_{14} gal_3 hyp_{10} ser_3 lys_2 thre, val	LYS
3.	ara_{20} gal_4 hyp_9 ser_3 lys, tyr	LYS
4.	ara_{16} gal_4 hyp_9 ser_3 tyr	SER
5.	ara_{16} gal_2 hyp_9 ser_3 lys_3 val, tyr	LYS

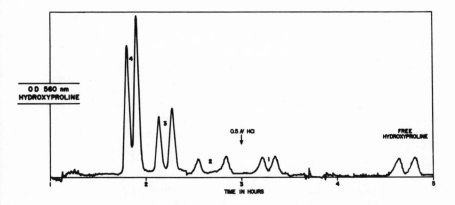

Figure 1. Separation of hydroxyproline arabinosides by chromatography on chromobeads B.[24]

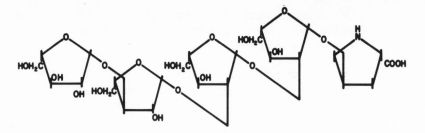

Figure 2.   Hydroxyproline tetraarabinoside.

Table 2.   Extensin tryptides

A.   SER-HYP-HYP-HYP-HYP-SER-HYP-SER-HYP-
     HYP-HYP-HYP-("TYR"-TYR)-LYS
     Exists with 3, 2, or 1 galactose residues.

B.   SER-HYP-HYP-HYP-HYP-SER-HYP-LYS
     Exists with 2, 1, or 0 galactose residues.

C.   SER-HYP-HYP-HYP-HYP-THR-HYP-VAL-TYR-LYS
     Exists with 1 or 0 galactose residues.

D.   SER-HYP-HYP-HYP-HYP-LYS
     Exists with 1 or 0 galactose residues.

E.   SER-HYP-HYP-HYP-HYP-VAL-"TYR"-LYS-LYS
     Exists with 1 or 0 galactose residues.

     Two important facts about extensin appear when we
sequence the tryptic peptides. The pentapeptide Ser-Hyp-
Hyp-Hyp-Hyp occurs at least once in all extensin peptides
examined (including sycamore-maple), and galactosylated
serine residues occur frequently.[26] These results are important because they may help in our quest both for the complete
sequence of extensin and the hypothetical region of polysaccharide attachment.

     Galactosyl serine may be especially significant in this
quest. Theoretically the presence of galactosyl serine should
assist determination of the amino acid sequence. For example,

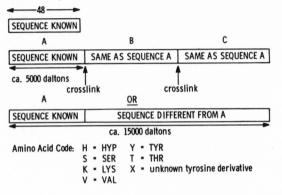

48 residues sequenced accounting for ca. 33% cell wall hydroxyproline

[SHHHHSHSHHHHXYK] [SHHHHSHK] [SHHHHTHVYK] [SHHHHK] [SHHHHVXKK]

Empirical Formula: HYP_{27} SER_8 LYS_6 VAL_2 TYR_2 X_2 THR_1

Possible interpretation -- tryptic peptides represent 1/3 of the complete sequence or the entire sequence:

Figure 3. Possible amino acid sequences in extensin.

alkaline borohydride treatment of extensin should lead to the β-elimination of galactose residues from serine followed by reduction of dehydroalanine to alanine. Alanine, unlike serine, gives excellent yields on Edman degradation. The presence of galactosyl serine also raises the possibility of its involvement in polysaccharide attachment to extensin, rather than *via* the hydroxyproline arabinosides as suggested earlier.

We will consider current work in the following two sections.

β-Elimination of Serine Residues in Isolated Tomato Cell Walls and Evidence for Polysaccharide Attachment to Serine. Although the advantages to sequence analysis are self-evident, base catalyzed β-elimination of galactosyl residues from peptide-bound serine (Fig. 4) does not occur nearly as easily in extensin as the elimination of xylose from serine in chondroitin sulphate. The serine residues in *isolated tryptic peptides* of extensin only underwent significant elimination when we increased the severity of the reaction conditions[26] (5 hr at 50 C in 0.2 M NaOH/0.2 M $NaBH_4$). However, when we applied these same reaction

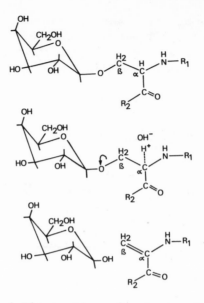

Figure 4.   β-Elimination of galactosyl serine.

conditions to the isolated cell walls*, the extent of the
reaction, judged by loss of serine (Fig. 5), was disappoint-
ingly small. Unfortunately any further increase in the
severity of the reaction conditions, judging from control
experiments with glucagon, leads both to non-specific elimination of the serine hydroxyl group and, perhaps more
seriously from our point of view, peptide bond cleavage.
Fortunately this impasse was not total. Downs *et al.*[12] showed
that semiaqueous conditions considerably enhanced the elimination of sugar residues from serine in submaxillary mucin,
although a few peptide bonds broke. We applied these same
reaction conditions to cell walls and then made one important
modification, namely the inclusion of molar $NaBH_4$ in the
reaction mixture. The semiaqueous conditions alone greatly
enhanced the loss of serine (Fig. 5) and gave a concomitant

---

*We wanted to eliminate the extensin serine residues.
where they were maximally glycosylated (>80%), namely in
the wall, rather than in the tryptic peptides where some
loss of galactose seems to have occurred.

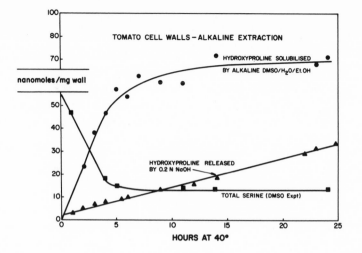

Figure 5. Alkaline extraction of tomato cell walls.

solubilization of hydroxyproline-rich material, which could be taken as support for the idea that at least one of the links binding extensin in the wall involves serine. The inclusion of the molar NaBH$_4$ in the reaction mixture had the unforeseen advantage of preventing the solubilization of wall components until *after* removal of the DMSO reaction mixture by filtration. Washing the walls with 96% ethanol removed most of the remaining DMSO reaction mixture. Then the addition of water solubilized 70 to 90% of the wall-bound hydroxyproline. Fractionation of this water-soluble material, referred to hereafter as *wall elimination product* or WEP, on Ultrogel AcA54 yielded two high molecular weight fractions and a major, relatively low molecular weight fraction which voids a Sephadex G-25 column but is almost completely retarded by Sephadex G-50. The low molecular weight fraction (Low MW WEP) was surprisingly soluble in 80% ethanol.* This provided a simple, quick method for initial fractionation of the

---

*Others have also noted the enhanced solubility of arabinose-rich polymers.[14]

crude water soluble WEP (Fig. 6). Further fractionation by
chromatography on the cation exchanger sulfopropyl Sephadex
C-25 (Fig. 7), isoelectric focusing (Fig. 8) and chromatography on anion exchanger DEAE cellulose (Fig. 9) yielded
three reproducible[*] major fractions which, on further amino
acid and sugar analysis, gave some significant data (Table 3).

Table 3. Low molecular weight WEP analysis of glycopeptides
after fractionation on Sephadex C-25

		nMoles Amino Acid Residues/30 HYP				
		(Fractions described in Figure 6)				Tryptic Peptides of Table 2
		I[*]	II[**]	III	IV	
	HYP	30	30	30	30	27
	ASP	.6	.7	.6	1.2	0
	THR	.5	.4	.5	1.2	1
	SER	2.3	2.1	2.2	1.8	8
	GLU	0	.3	.4	.7	0
	GLY	1.2	.6	1.1	1.7	0
	ALA	.8	.9	1.1	1.8	0
	VAL	.9	1.3	2.2	5.0	2
	ILU	0	.3	.3	.6	0
	LEU	0	.2	.3	.6	0
	TYR	.6	2.1	2.4	3	2
	LYS	.8	3.2	4.8	8.5	6
TYR Deriv.		.5	1.2	1.8	1.9	2
N_2H_4 Labile	SER	1.0	0.8	1.2	0.3	
ACN[***] Labile	SER	1.2	n.d.	0.4	1.6	
	ARA/HYP	2.8	5.7	3.3	4.6	
	GAL/SER	4.9	2.0	1.9	0.9	
Galacturonic Acid		0	0	0	0	

[*]After further fractionation on isoelectric focusing
and DEAE cellulose. [**]Peak II is a minor fraction.
[***]Amino terminus determined by a subtractive method involving reaction with acrylonitrile.

---

[*]The reproducibility of low MW WEP yields depends on
rigorous temperature control during the DMSO treatment of
the walls.

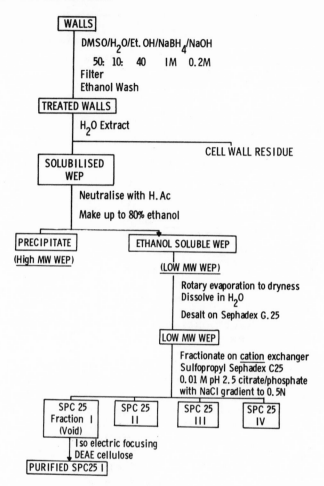

Figure 6.   Preparation of WEP (wall elimination product) from tomato cell walls.

First, there are far fewer serine residues (only about 25% of
the serine residues are left) compared with the tryptic peptides of extensin; second, the *average* number of galactose
residues per serine residue varies from one to five.   Because

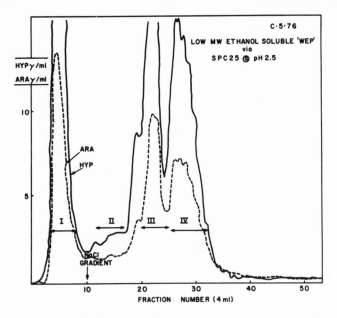

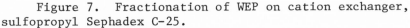

Figure 7.   Fractionation of WEP on cation exchanger,
sulfopropyl Sephadex C-25.

galactose residues are *not* attached to hydroxyproline in
wall bound extensin, we conclude that we have at last obtained
direct evidence for galacto-oligosaccharide attachment to the
serine residues of extensin. We estimated the actual number
of galactose residues attached to each serine residue by mea-
suring the number of hydrazine-sensitive serine residues.
This is a measure of the number of glycosylated serine residues because, as we have shown elsewhere,[26] only the unsubstituted serine residues yield a hydrazide which can regene-
rate free serine on acid hydrolysis. Substituted serine
residues are degraded during hydrazinolysis probably by β-
elimination induced by N_2H_2 at 100 C. On this basis it was
concluded earlier that hydrazinolysis does indeed cleave the
O-glycosidic linkages of serine and threonine, invalidating
the conclusion by others[18] that hydrazinolysis released glycopeptides from sycamore-maple cell walls, without cleavage
of glycosidic linkages.*

---

*We also observed that crude pectin is especially sensi-
tive to hydrazinolysis and rapidly depolymerizes in N_2H_4 at
100 C, again presumably *via* β-elimination.

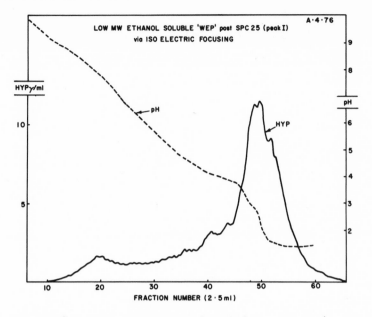

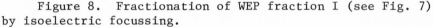

Figure 8.   Fractionation of WEP fraction I (see Fig. 7)
by isoelectric focussing.

      Because some of the serine residues of the WEP glyco-
peptides were stable to hydrazinolysis we conclude that there
is more galactose attached to the hydrazine-sensitive serine
residues than suggested by the average figure; for example,
the acidic WEP glycopeptide (Sephadex C-25 I) has an *average*
of ca. 5 moles galactose/serine (Table 3), but half of the
serine residues are hydrazine-sensitive.   Therefore, we estimate 10 galactose residues/serine residue, as a reasonable
figure based on a rough approximation of the molecular weight
of the glycopeptide through gel filtration and subtractive
N-terminal determination.   The *absence* of uronic acid from
this glycopeptide fraction is especially noteworthy, because
uronic acid would help verify covalent attachment of polyuronides to extensin as Albersheim's group suggests.[22]   Final
proof of oligosaccharide attachment to serine requires the
demonstration of galacto-oligosaccharide release, recovery
and characterization after further β-elimination of these
purified "WEP" glycopeptide fractions.   This work is in progress.   Of course, one must ask how any serine residues remain
glycosylated after elimination?   What happens to eliminated

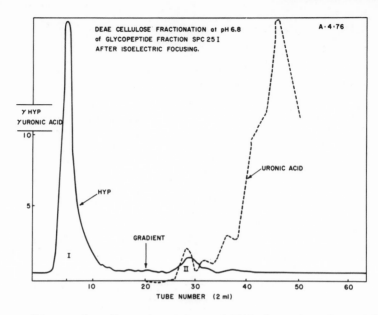

Figure 9.   Fractionation of WEP fraction I on DEAE
cellulose after isoelectric focussing.

serine residues?   And how much peptide bond cleavage occurs?
Presumably elimination is incomplete because N-terminal
serine residues are present (Table 3); these would undergo
elimination only very slowly.[26]   The eliminated serine residues are classically supposed to yield dehydroalanine which
the borohydride reduces to alanine.   Our glycopeptides, however, contain glycine besides alanine.   As there is no glycine
in previously isolated extensin peptides, it looks as though
glycine in the WEP peptides arises from cleavage between the
α- and β-carbons of serine.   On further sequencing it will be
interesting to see whether the positions formerly occupied by
serine, really are heterogeneous or whether the conversion to
glycine or alanine is a strict function of position in the
peptide.   Because we consistently obtain three major low
molecular weight WEP glycopeptide fractions with reproducibly
similar amino acid and sugar compositions, purified to almost
constant composition, we consider that they are reasonably
homogenous glycopeptides.   However, the small number of these
fractions shows that non-specific peptide bond cleavage did
not occur during the elimination reaction.

Evidence for Polysaccharide Attachment to Hydroxyproline.
The current emphasis on the possible involvement of galactosyl serine rather than arabinosyl hydroxyproline as a site of
polysaccharide attachment to extensin needs some explanation.
The impossibility of removing significant amounts of extensin
from most* cell walls is explicable if extensin occurs in the
form of a crosslinked network. This cross-linking could
involve amino acids exclusively (as seen in collagen and
elastin) or glycosidic linkages or perhaps other compounds
such as the phenolics. The discovery of arabinosyl hydroxyproline[23] fitted our preconceptions nicely. Unfortunately
we were, until relatively recently, unable to isolate anything
larger than a tetrasaccharide attached to hydroxyproline.
Originally an alkaline labile linkage was postulated to con-
nect the tetrasaccharide with other wall polysaccharide. Although this sort of hypothesis is difficult to prove or dis-
prove, Talmadge *et al.*[46] methylated intact sycamore-maple
cell walls and showed that the fourth arabinose residue (1-3
linked) of hydroxyproline tetraarabinoside (Fig. 2) was indeed
terminal and therefore unlikely to be involved in further glycosidic linkages. (Because of the alkaline conditions, their
data do not rule out ester linkages.) More recent data from
a variety of sources reinforce the conclusion that polysaccharide is not attached to the hydroxyproline residues of
extensin *as it exists in the cell wall*. One can summarize
this new evidence by stating simply that macromolecules do
indeed occur in which *polysaccharide is directly attached to
the hydroxyl group of hydroxyproline*. Because we can demonstrate polysaccharide attachment directly to hydroxyproline
in material isolated from the cytosol or xylem sap or growth
medium (Fig. 10), but *not* in the cell wall, we conclude that
a similar hydroxyproline-polysaccharide linkage does *not* occur
in the wall. But these hydroxyproline-polysaccharide
(extensin-like?) macromolecules are interesting and significant because of their possible relationships to the cell wall,
disease resistance and development.

Fincher *et al.*[14] isolated a hydroxyproline-rich glycopeptide from wheat endosperm. This glycopeptide has a low
molecular weight (ca. 22,000 daltons) which is highly unusual
for a hydroxyproline-rich macromolecule. In fact they calculated that the peptide portion probably consists of only about

---

*There are exceptions,[4] but they do not help us to under-
stand how extensin is built into the cell wall.

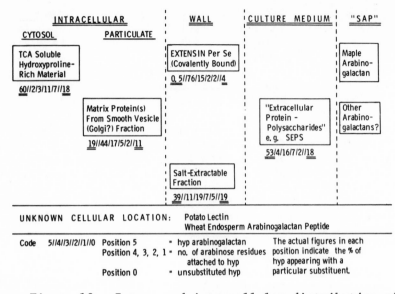

Figure 10. Inter- and intracellular distribution of various hydroxyproline-rich proteins.

20 residues raising the possibility that this material is not a protein *per se*, but derived perhaps by degradation of a glycoprotein. They suggested direct attachment of arabinogalactan to hydroxyproline on the basis of chromatographic coincidence between hydroxyproline and arabinogalactan after alkaline degradation of the intact glycopeptide. Unfortunately they did not indicate whether they assayed these alkaline hydrolysates directly for hydroxyproline or whether they assayed after further acid hydrolysis. This is important because non-peptidyl hydroxyproline reacts in the usual colorimetric assay[27] even when the hydroxyl group is substituted, whereas peptidyl hydroxyproline is unreactive.

The work of Dr. David Pope in our laboratory rekindled my interest in the possibility of polysaccharide attachment to hydroxyproline when he examined the extracellular hydroxyproline-rich polysaccharide complexes which cell suspension cultures secrete into their growth medium. We had earlier accepted the entirely reasonable suggestion of Keegstra *et al.*[22] that one could regard *soluble* glycoproteins found in the growth medium as a model for those present in the cell wall. This suggestion fitted in nicely with the work of Brysk

and Chrispeels[6] who claimed to have isolated an extensin precursor loosely bound to carrot cell walls. Pope continued
this work[41] hoping to obtain enough extensin precursor for
purification and chemical characterization. However, two
quite unexpected results forced us to reappraise the significance of glycoproteins secreted into the culture medium.
First, Pope incubated sycamore-maple cells with ^{14}C-proline
and chased with unlabeled proline but did *not* observe the
expected transfer of label from the so-called loosely wall-
bound extensin (equivalent to Brysk and Chrispeels' cell wall
salt-soluble fraction) to firmly bound extensin of the wall.
We conclude that extensin does *not* exist as a precursor
loosely bound to the wall. Second, Pope began the characterization of this loosely bound material as well as the hydroxyproline-rich macromolecules secreted into the growth medium
of sycamore-maple cells. When he attempted a routine determination of the hydroxyproline arabinoside profile by alkaline
hydrolysis of the crude soluble extracellular "polysaccharide"
an unknown peak appeared, which on direct assay for hydroxyproline, behaved as free non-peptidyl hydroxyproline, yet
voided a Sephadex G-25 column.[40] Further purification of this
fraction by gel filtration (Fig. 11) and isoelectric focusing
followed by chemical analysis showed it to be a single residue of hydroxyproline with an arabinogalactan substituent on
the hydroxyl group (Table 4). Unlike the hydroxyproline arabinosides, the arabinogalactan was rather resistant to acid
hydrolysis--partial acid hydrolysis with 0.1 N trifluoro-
acetic acid for 1 hr at 100 C yielded *galactosyl hydroxyproline*. But we have never observed galactosyl hydroxyproline
in cell walls of higher plants (although we have in *Chlamydomonas*).[32] It seems reasonable to conclude that the hydroxyproline-rich macromolecules in the growth medium are *not* precursors to extensin, are *not* excess extensin and are certainly
not models for cell wall polysaccharides! But what are they?
This question has no simple answer, but there are some clues.
One might dismiss the secretory products of tissue cultures
as irrelevant products--"mad molecules" from a deranged
genome--were it not that "all" suspension cultures seem to
produce these macromolecules, despite the fact that cultures
are notoriously poor producers of secondary products (i.e.,
irrelevant to the process of cell division and growth). If
we conclude that these secreted glycoproteins *are* relevant to
growth, then they should occur in plants grown under normal
conditions. They do. For example, the so-called arabinogalactan isolated by Adams and Bishop[2] from sugar maple sap,
in our hands, turns into at least two hydroxyproline-rich

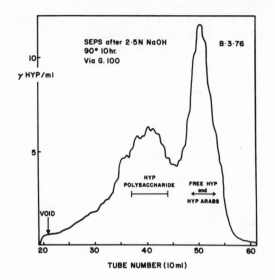

Figure 11. Fractionation of sycamore-maple extra-cellular polysaccharide (SEPS) on Sephadex G-100.

Table 4. Arabinogalactans compared

		Sugar Molar Ratios		
	Rham-nose	Arab-inose	Galac-tose	(Sugar moles/ HYP)
1. Sugar maple sap arabino-galactan[2]	1	8.8	10.2	(n.d.)
2. Sugar maple sap glycoprotein after Sephadex G-100	1	5.8	7.1	(50)
3. Hydroxyproline-polysaccharide obtained by alkaline hydro-lysis of HYP-rich glyco-protein, from culture medium of sycamore-maple	1	9.2	10.1	(49)
4. As (3) but from culture medium of tomato	1	9.9	10.0	(ca. 80)

glycoproteins (Figs. 12 and 13), one of which seems to have the arabinogalactan attached directly to hydroxyproline (Table 5). Perhaps arabinogalactans from other sources such as *Lillium* stigma exudate[29] will also turn out to be hydroxy-proline-rich glycoproteins. Allowing for the fact that we have not rigorously purified these glycoproteins, plus the fact that we are comparing *Acer pseudoplatanus* with *Acer saccharum*, there is good correspondence between the analytical data. Thus the amino acid analyses compare well (Table 6) (note especially that they are alanine-rich macromolecules) and sugar compositions are also in good agreement (Table 4). We were only able to obtain meaningful amino acid analyses of the high molecular weight SEPS (sycamore-maple extra-cellular polysaccharides) through purification *via* HF-deglyco-sylation followed by gel filtration. In our hands anhydrous hydrogen fluoride cleaves virtually all the O-glycosidic linkages of neutral sugars[34] within 1 hr at 0 C while peptide bonds remain stable.[28] Thus we initially concentrated SEPS by ultrafiltration (Amicon XM100A membrane) and followed this by desalting (Sephadex G-25) and freeze-drying. We placed the rigorously-dried crude mixture in a Kel-F reaction vessel which was then evacuated and cooled to -70 C so that HF distilled over from a reservoir. We allowed the temperature to rise to 0 C (364 mm Hg) and timed the reaction from that

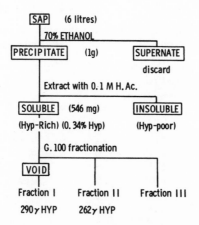

Figure 12. Preparation of hydroxyproline-rich glyco-protein (arabinogalactan) from sugar maple sap.

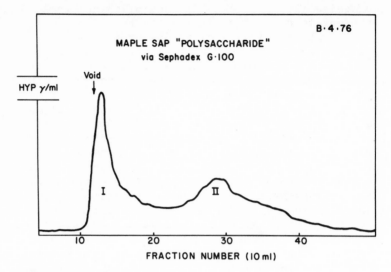

Figure 13. Separation of maple sap "polysaccharide"
on Sephadex G-100.

Table 5. Hydroxyproline-rich glycoprotein from maple sap
after ethanol precipitation and Sephadex G-100 filtration

Fraction 1 (= Void)   expt.:   B.4.76

nmoles of amino acids and sugars
normalized to 30 residues of hydroxyproline

Analysis 13176/1		Analysis 13176/1				Sugar moles/HYP
HYP	30	CYS	--	Rhamnose	102	3.4
ASP	10.3	MET	--	Arabinose	579	19.3
THR	12.4	ILU	2.1	Galactose	705	23.5
SER	21.2	LEU	9.8	Glucose	131	4.4
GLU	12.5	TYR	--			
PRO	n.d.	PHE	1.9	Sugar		50.6
GLY	10.3	LYS	9.5	Residues/HYP		
ALA	25.8	HIS	2.9			
VAL	6.9	ARG	1.5			

Table 6. Comparison of amino acid analyses of soluble extracellular hydroxyproline-rich glycoproteins after partial purification

Species	Sycamore-maple	Sugar maple	Tomato	Brassica napus[20]
Source of glycoprotein	Growth medium	Xylem sap	Growth medium	Seed lectin
Purification steps	1. Ultrafiltration via XM100A 2. HF 1 hr at 0 C 3. G-100	1. EtOH ppt. 2. G-100 (void)	1. Ultrafiltration via XM100A	
Analysis	23.6.75/2	13176/1	13176/3	
		nmoles of each amino acid normalized to 30 HYP		
HYP	30	30	30	30
ASP	18	10.3	24.4	23.5
THR	16	12.4	17.3	21.5
SER	27	21.2	26.6	32.1
GLU	13	12.5	19.7	17.8
PRO	--	--	--	13.4
GLY	15	10.3	21.2	23.0
ALA	29	25.8	27.7	34.1
VAL	13	6.9	21.7	15.0
CYST	--	--	--	6.7
MET	--	--	1.3	5.7
ILU	7	2.1	7.3	8.5
LEU	12	9.8	11.4	18.9
TYR	tr	--	2.7	6.7
PHE	7	1.9	5.9	8.8
LYS	13	9.5	8.5	11.9
HIS	6	2.9	2.2	2.8
ARG	5	1.5	4.3	6.5

point. We removed the HF *in vacuo* via a calcium oxide trap.
We then dissolved the HF-treated material in 0.1 M acetic
acid and fractionated the soluble material on Sephadex G-100
(Fig. 14). The major hydroxyproline-rich fraction was now
retarded (ca. 2x void) and contained less than 5% of the original sugars, and showed about a two-fold enrichment in
hydroxyproline based on amino acid analyses before and after
the HF treatment (Table 7).

Jermyn and Yeow[20] have also obtained evidence for
hydroxyproline-rich glycoproteins in a wide range of seeds.
They also find a strict correlation between the presence of
hydroxyproline, high alanine levels and the presence of arabinogalactan (Table 6), and suggest that these glycoproteins
are highly conserved from an evolutionary point of view.
They do not suggest a biological role for the glycoproteins
but point out that their wide occurrence implies a fundamental importance especially as they also show lectin-like
activity. Based on the location and structure of these

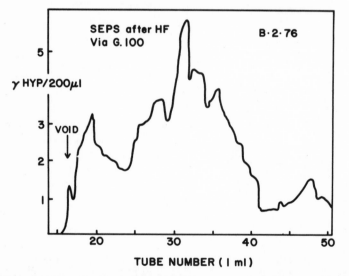

Figure 14.   Separation of HF-treated SEPS on Sephadex
G-100.

Table 7. Crude extracellular polysaccharides from
sycamore-maple (*Acer pseudoplatanus*) before and after
treatment with anhydrous HF

	(Amino acid residues)	
	Before[*]	After[**]
HYP	30	30
ASP	22	24
THR	37	19
SER	43	26
GLU	44	16
PRO	n.d.	n.d.
GLY	47	15
ALA	42	26
VAL	34	11
CYST	--	--
MET	5	--
ILU	22	7
LEU	35	11
TYR	5	1
PHE	23	7
LYS	32	10
HIS	15	5
ARG	14	4
Analysis Date	3376/3	4976/1

[*]The crude extracellular polysaccharide fraction
was obtained by ultrafiltration of the growth medium
through an Amicon XM100A membrane. The concentrated
retentate was gel filtered, freeze dried and analyzed.
[**]After HF-treatment the hydroxyproline-rich fraction was retarded on a Sephadex G-100 column.

glycoproteins, it should be possible to construct a working
hypothesis. They tend to be extracellular and are associated with active growth as in tissue cultures or periods
immediately preceeding active growth as in sugar maple sap
and seeds. They are present in low concentrations and are
therefore probably not related to energy metabolism. Their
presence in xylem sap presupposes that they are moving from

root[*] to shoot. Perhaps that is the message! We can rationa-
lize this by another hypothesis. These glycoproteins are
quite rich in hydroxyproline. Consider the possibility that
their polypeptide backbone is structurally related to extensin. On this basis there would be a hydroxyproline-
polysaccharide (hydrophilic) region and a hydrophobic alanine-
rich region (Fig. 15). Amphipathic molecules typify membrane
proteins,[49] suggesting novel possibilities for this new class
of glycoproteins in the regulation of growth and development.

If hydroxyproline-rich glycoproteins are so essential
to higher plants, the presence of similar glycoproteins in
lower plants would perhaps provide some useful clues to their
function--which leads to the question considered in the following section.

EXTENSIN IN LOWER PLANTS?

Comparative biochemistry helps to separate trivia from
essentials. The primary cell wall of higher plants has a
remarkably similar composition within major groups, consisting

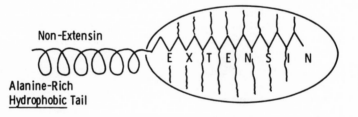

Figure 15. Postulated structure for hydroxyproline/
alanine-rich arabinogalactan glycoproteins.

---

[*]But we have not yet demonstrated their presence in
xylem sap obtained directly from roots.

of only six major polymers:[3] cellulose, polyrhamnogalactu-
ronans, 1-4 β-linked galactan, xyloglucan (or arabinoxylan
in monocotyledons), arabinan, and extensin. This probably
holds true from higher plants to mosses and liverworts. However, the green algae (Chlorophyceae) show more diversity
(especially the groups lacking cellulose) which are, from
our point of view, the most interesting because their cell
wall is a crystalline lattice[42] built virtually exclusively
from a hydroxyproline-rich glycoprotein. The large amounts
of hydroxyproline and serine almost certainly indicate homology between this glycoprotein and extensin,[33] while the
presence of hydroxyproline glycosides in the glycoprotein of
algal cell walls[27,32,33] should dispel most remaining doubts,
except perhaps whether or not one should call it extensin.

The presence of these glycoproteins in the algae raises
some difficult questions, but also supports some exciting
speculations. The cell wall of *Chlamydomonas* is an ana-
chronism. The glycoprotein subunits are not linked together
covalently, but electrostatically. Thus the wall can be
disassembled and reassembled *in vitro*.[19] The non-cellulosic
wall of *Chlamydomonas* is therefore much weaker than a covalently linked wall of comparable thickness such as that of
Chlorella, easily demonstrated when one attempts to isolate
such walls by sonic disruption--*Chlamydomonas* walls tending
to disintegrate completely into fragments while the wall of
Chlorella ruptures with difficulty. Indeed the presence of
a contractile vacuole in *Chlamydomonas* is probably a reflec-
tion of a weak wall unable to cope with sudden increases in
tugor pressure. Perhaps we have overlooked some advantage
of having a weak cell wall. For example, these algae are
motile under some ("favorable") conditions, but not others.
Optimally formulated media rarely occur in nature. The textbook description of so many Protists as unicellular, rele-
gates to a footnote the fact that many of these Protists
often "seek safety in numbers," and therefore have the advantages both of the motile* *and* the sedentary way of life.
Chlamydomonas grows perfectly well colonially on agar plates,
and sometimes even in liquid media giving the so-called palmella stage, which can also be artificially induced by

_____

*Cellulosic-type cell walls probably preclude the pre-
sence of flagella.

platinum[*] compounds.[36] The palmella stage is significant
because it arises by hypertrophy of the cell wall, thereby
creating an extracellular matrix around individual cells.
Is such the origin of multicellularity? The extracellular
matrices of prokaryotes lead nowhere, but the hydroxyproline-
rich matrix has ramifications, which, if Aaronson is correct,[1]
extend throughout the entire plant and animal kingdoms.
Restricting ourselves for the moment to *Chlamydomonas* and its
close relatives, one realizes intuitively that the classical
textbook series of increasing complexity running from *Chlamydomonas* via *Gonium*, *Pandorina*, *Eudorina*, and *Pleodorina* to
Volvox, results directly from the arrangement of individual
cells in an extracellular matrix which developed from the
original *Chlamydomonas* cell wall. If there is any proof for
this it resides in *Volvox*, the ultimate product of this evolutionary line. The extracellular matrix of *Volvox* consists
largely of a hydroxyproline-rich glycoprotein (Table 8) with
most of the hydroxyproline residues O-glycosylated by hetero-
saccharides containing arabinose, galactose, and mannose.
Using our new technique for chemical deglycosylation involving anhydrous hydrogen fluoride (worked out largely by Andrew
Mort in our laboratory), we partially purified the peptide
portion; it contains about 38 mole percent hydroxyproline
(Table 8) and is in this respect similar to extensin. Although we classically regard *Volvox* as an evolutionary cul-
de-sac, the presence of a matrix based on a hydroxyproline-
rich glycoprotein suggests that we should reexamine Ernst
Haeckel's comparison between *Volvox* and a blastula.[17] All
metazoan development passes through the single-cell-layered
hollow ball, or blastula, stage of development. All metazoa
also have an extracellular matrix based on one major
hydroxyproline-rich glycoprotein, collagen. The evolutionary
origins of the Metazoa, the blastula and collagen are still
obscure, but their occurrence together shows 100% correlation,
implying that if we know the origin of any one character we
know the origin of the other two. One could argue that *Volvox* comes close to satisfying two out of three of these criteri
(Table 9), while the third is perhaps on its way to collagen.

---

[*]All modern workers with platinum compounds are obviously looking for Ehrlich's "magic bullet" being inspired no
doubt by those immortal lines:

    I shoot the Hippopotamus with bullets made of platinum
    because if I used leaden ones, his hide would sure to
    flatten 'em!

Table 8. Soluble extracellular matrix of *Volvox carteri*

| | Amino acid analysis (residues) | | | Sugar analysis | |
	Before HF	After HF and G-100		Before HF (residues/HYP)	
HYP	30	30	MAN	8	
ASP	11	5	GAL	4	
THR	6	4	ARA	4	
SER	13	10	RHA	2	
GLU	7	4	GLC	1	
PRO	n.d.	2	XYL	0.5	
GLY	29	4			
ALA	9	4			
VAL	5	4			
CYST	--	0			
MET	--	0			
ILU	2	1			
LEU	6	3			
TYR	1	1			
PHE	2	2			
LYS	5	2			
HIS	1	0			
ARG	4	3			
Analysis Date	3376/1	20376/3		4276/6	

Table 9. Some metazoan characteristics of *Volvox*

1. An *individual*, not a colony--shows polarity and special- ized cells.

2. Cell division via *cleavage furrow*.

3. Embryology involves *cell movement* and *eversion* (similar to sponge embryology).

4. Extracellular matrix involves a hydroxyproline-rich glycoprotein.

5. Requires vitamin B_{12}.

DO EXTENSIN AND COLLAGEN SHARE
A COMMON ANCESTRY?

    Two branches of a phylogenetic tree may be so distant
from each other that no obvious homology remains. Thus
collagen is essentially a repeating tripeptide of Gly-Pro-X
or Gly-X-Pro, while extensin contains the frequently repeating pentapeptide Ser-Hyp-Hyp-Hyp-Hyp. Were it not for the
fact that both proteins possess hydroxyproline we would probably not consider the possibility of homology (Table 10).
However, hydroxyproline occurs in few proteins. Hydroxyproline is also unusual in that its biosynthesis requires
post-translational modification of proline and involves the
fixation of oxygen by a remarkable enzyme--proline hydroxylase. These facts suggest that hydroxyproline appeared relatively late on the evolutionary scene, well after codon
assignments, but probably not until the blue-green bacteria
became well established. There is no *a priori* reason for
assuming that only one protein would contain hydroxyproline
except for the fact that hydroxyproline is intimately involved
in stabilizing molecular shape. In collagen, this involves
hydrogen bond formation from the hydroxyproline hydroxyl
group of one peptide chain, to the glycine carbonyl of an
adjacent chain.[5] In extensin, hydroxyproline provides attachment sites for arabinose residues which in turn may stabilize
the peptide backbone by folding back and hydrogen bonding
with it. Shape or molecular conformation may be the clue to
origins here. It is axiomatic that the tertiary structure
of proteins is more highly conserved than their primary structure (sequence). How then does the tertiary structure of
collagen compare with that of extensin? Collagen consists
of three peptide chains coiled together, each chain existing
as a left handed helix with a pitch of 94 nm, similar to the
conformation of polyproline II. Isolated peptides of extensin
are so rich in hydroxyproline that one also expects extensin
to adopt the polyproline II conformation. Our preliminary
experiments support that conclusion. We have compared the
circular dichroic spectra of various extensin peptides, poly-
L-proline, and poly-L-hydroxyproline with bovine serum albumin (Fig. 16). The spectra of poly-L-hydroxyproline, poly-
L-proline, crude extensin, and two extensin tryptic peptides
share two common features, a 205 nm negative absorption maximum and the lack of the strong 195 nm absorption maximum
present in α and β helical structures.

Table 10. Do collagen and extensin share a common origin,
i.e., are they in the same protein "super family"?

1.	Position:	Extracellular matrix
2.	Function:	Structural
3.	Chemically:	Hydroxyproline-rich
4.	Structure:	Glycoproteins Helix type: Polyproline II (9.4 A pitch, 3 residues/turn)
5.	Biosynthesis:	*No* hydroxyproline codon (implies "late" evolution) Post-translational modification Involves very similar proline hydroxyl-ases utilizing molecular O_2 and co-factors Fe^{++}, ascorbate and α-keto-glutarate
6.	Occurrence:	Extensin-like proteins occur in most photosynthetic protists Collagen does *not* occur in protozoans, i.e., HYP occurrence correlates with photosynthetic ability
7.	Evolutionary Branch Point:	*Volvox*-like organism? (Origin of the blastula?)

The highly conserved tertiary structure in both extensin
and collagen strongly implies that they are indeed members of
the same protein "super family." There are also other more
subtle suggestions of affinity between the two proteins. For
example, collagen from lower animals[16,38] has much more associated carbohydrate than does collagen from higher animals,
although there are recent reports[9,39] that procollagen from
higher animals contains carbohydrate-rich regions cleaved
during maturation. Hydroxylysine occurs in collagen (often
as glucosyl galactosyl-Hyl)[7] but not as far as we know in
extensin of higher plants. There are, however, reports of

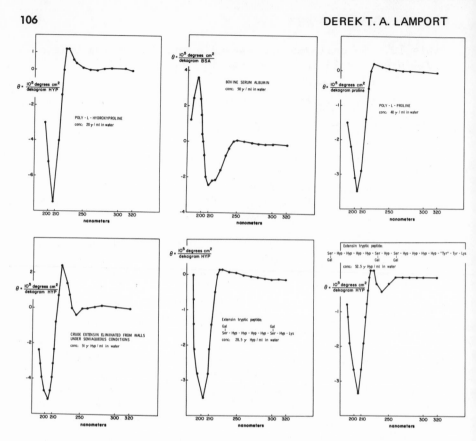

Figure 16. Circular dichroic spectra of poly-L-hydroxy-proline, poly-L-proline, bovine serum albumin, crude exten-sin, and two extensin tryptic peptides.

hydroxylysine[*] in the cell walls of some algae.[47] Galactosyl serine, a characteristic feature of extensin at least in higher plants,[26] also occurs in the cuticle collagen[35] of the common earthworm. Hydroxyproline glycosides occur uniquely in plants, but if collagen and extensin are members of the same super family, one would expect to find hydroxyproline glycosides somewhere among primitive animals such as the sponges or coelenterates.

A comparison of the biosynthesis of collagen and exten-sin also indicates affinity. Thus proline hydroxylase from

---

[*]Whether glycosylated or not is unknown for the moment.

plants[44] hydroxylates the polypeptide substrate prepared
from either plants or animals (Fig. 17). Apparently, the
biosynthesis and secretion of collagen does not involve the
Golgi apparatus.[50] There is no agreement about the precursor
pathway of extensin. Dashek[10] fractionated sycamore-maple
cells during pulse-chase experiments and concluded that the
biosynthesis and transport of extensin occurred in a smooth
membraneous component rather than in the classical Golgi
apparatus. Roberts and Northcote,[43] also using sycamore-
maple cells, arrived at a similar conclusion based on auto-
radiography. However, in the most recent work Gardiner and
Chrispeels[15] used a combination of pulse-chase labeling with
cell fractionation by isopycnic centrifugation and identification of marker enzymes. The Golgi marker enzyme, inosine
diphosphatase, coincided with a UDP arabinosyl transferase
which may or may not be identical with Karr's transferase
system[21] for stepwise addition of arabinose residues to the
hydroxyproline residues of extensin. Thus Golgi involvement
in the biosynthesis and transport of extensin is still in
doubt.

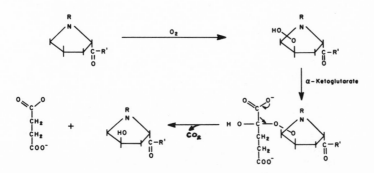

Figure 17. A hypothetical scheme for proline hydroxy-
lation. (Taken from G. J. Cardinale and S. Udenfriend. 1973.
Collagen proline hydroxylase and other α-ketoglutarate
requiring enzymes. *In* Oxidases and Related Redox Systems
(T. E. King, H. S. Mason, and M. Morrison, eds.), Univ. Park
Press, Baltimore, with permission of the publisher.)

GLYCOPROTEINS, A PRIMARY STEP IN
THE ORIGIN OF THE EUKARYOTES?

This chapter concludes with some speculations about the
overall significance of glycoproteins. How did they originate and what did this lead to? As a working hypothesis, we
may suggest that the origin of glycoproteins was a key factor in the transition from prokaryote to eukaryote organization. This is based on two observations: first, the extreme
rarity of glycoproteins in prokaryotes compared with their
ubiquity in eukaryotes; and second, the morphogenetic role of
glycoproteins as building blocks in the extracellular matrix.

If we ask what was the initial *chemical* difference which
initiated the eukaryotic line of evolution--preceeding the
capture of endosymbionts--the answer devolves upon the cell
surface, because the rigid peptidoglycan prevents phagocyto-
sis. A transition organism would have to lose its peptidoglycan layer and acquire a more flexible "protective" cell
surface but this could only occur in the absence of osmotic
stress. *Halobacterium salinarium* may well represent such a
transition organism. It grows in high salt levels, its cell
wall lacks a peptidoglycan layer, but consists instead of a
glycoprotein[31] which contains two glycopeptide linkages, one
of which can be considered the primitive[45] or archetypal
glycopeptide linkage of eukaryote glycoproteins, *viz* aspara-
gine-N-acetyl glucosamine, while the other (O-galactosyl threonine), occurs much less frequently.[35] The change-over from
peptidoglycan to glycoprotein may not be quite so drastic as
it seems at first sight; there are basic similarities perhaps
only obscured by time. Peptidoglycan and cell wall glycoprotein are both cell surface glycoproteins. A comparision
of their biosynthesis is even more instructive. Both pepti-
doglycans, and glycoproteins possessing the asparagine-linked
sugars, involve *block* rather than stepwise transfer of sugars
or glycopeptide to the main polymer backbone.[48] In both
instances the lipid cofactor involved in the block transfer
is an isoprenoid: undecaprenyl phosphate (11 isoprenoid units)
in bacteria and dolichol phosphate (16-21 isoprenoid units) in
plants and animals:[48]

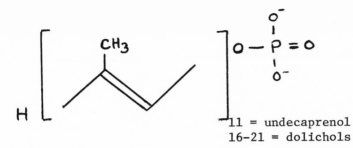

11 = undecaprenol
16-21 = dolichols

One could argue that because sterols are ubiquitous in eukaryotes but absent from prokaryotes,[37] sterols initiated eukaryotic evolution. However, molecular oxygen is directly involved in the biosynthesis of sterols, but not of iso-prenoids or asparagine-linked glycoproteins. Thus asparagine-linked glycoproteins probably arose before sterols, and, according to the most recent work,[31] in prokaryotes.

The enzymes and cofactors involved in biosynthesis of animal glycoproteins may therefore be relics of cell wall biosynthesis in bacteria. Because of increasing evidence for evolution by hybrid formation between grossly disparate organisms--the endosymbiont capture hypothesis[30]--it seems reasonable to suggest that multigenomic organisms still pos-sess relics of their previous ancestry as suggested in Table 11. Indeed it may even be possible to explain many examples of non-Mendelian inheritance on this basis, i.e., non-nuclear genes as remnants of a captured genome. (Some cell wall mutants of *Chlamydomonas* may be pertinent examples.[11])

Proline hydroxylase is another possible relic enzyme, probably arising in the blue-green bacteria where molecular oxygen first became available. This view implies that pro-line hydroxylase of animals was acquired *via* an intermediate photosynthetic eukaryote (Fig. 18). We do not know the iden-tity of this intermediate organism, but it was probably multi-cellular--there are no hydroxyproline-rich proteins reported among the protozoa! What sort of multicellular photosynthetic organism is unknown but the Volvocine line of evolution pro-duced an organism which looks remarkably like a metazoan blastula (Table 9). If the similarity is more than super-ficial, one can predict that *Volvox* will show other metazoan characteristics such as pathways involved in sterol biosyn-thesis,[37] the presence of enzymes such as DNA-polymerase-β

Table 11. Possible relic enzymes and pathways

Activity	Present distribution	Original source
Proline hydroxylase	All Metazoa and photo-synthetic Eukaryotes	Cyanophyta?
Sterol biosyn-thesis	All Metazoa and nearly all Eukaryotes	Cyanophyta
Glycoprotein biosynthesis	All Eukaryotes	Peptidoglycan pathway in bacteria
Transfer of N-acetyl gluco-samine oligo-saccharide *from* DOLICHOL-P-P-oligosaccharide *to* asparagine amide group		Transfer of N-acetyl muramyl pentapeptide *from* UNDECA-PRENYL-P-P etc., *to* pepti-doglycan
Tubulin → Micro-tubules	All Eukaryotes	Spirochaetes?

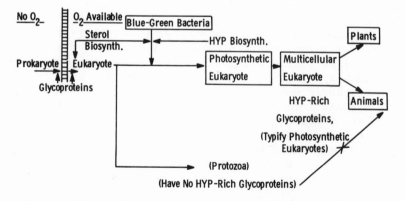

Figure 18. A hypothetical scheme for the molecular evolution of eukaryotes.

specific to the metazoa[8] and metazoan-like amino acid
sequences in key proteins such as cytochrome c.

Our study of cell wall glycoproteins has led us a long
way from our original starting point, the primary cell wall
of higher plants. It has, however, helped us to appreciate
the old aphorism that "All Life is One."

ACKNOWLEDGMENTS

I thank Michael Caughey for his highly skilled technical assistance, notably in preparing and analyzing the "wall
elimination product." I am grateful to Maria Makri-Shimamoto
who helped during the early stages of this work, and to
Joseph Hsung for help with the CD spectroscopy.

I am greatly indebted to other members of my research
team: to Joann Lamport for help with *Volvox* and general
forbearance, to Andrew Mort who developed the HF deglycosylation technique, to Dr. Joyce Clarke for SDS gel electrophoresis and recent work on the soluble extracellular glycoproteins, to Dr. David Pope who made them come alive, and
finally to High School Honors Science Program student, Yiet
Wong who showed that the extracellular matrix of *Volvox carteri* is probably *not* glycine-rich.

This research program is supported by the U. S. Energy
Research and Development Administration through contract
E(11-1)-1338 and the National Science Foundation, grant num-
ber PCM76-02549.

REFERENCES

1. Aaronson, S. 1970. Molecular evidence for evolution in
      the algae: A possible affinity between plant cell walls
      and animal skeletons. *Ann. N.Y. Acad. Sci. 175*:531.
2. Adams, G. A. and C. T. Bishop. 1960. Constitution of an
      arabinogalactan from maple sap. *Can. J. Chem. 38*:2380.
3. Albersheim, P., W. D. Bauer, K. Keegstra, and K. W.
      Talmadge. 1973. The structure of the wall of suspension-
      cultured sycamore cells. *In* Biogenesis of Plant Cell
      Wall Polysaccharides (F. Loewus, ed.), pp. 117-148.
      Academic Press, New York & London.

4. Bailey, R. W. and H. Kauss. 1974. Extraction of
    hydroxyproline-containing proteins and pectic sub-
    stances from cell walls of growing and non-growing
    mung bean hypocotyl segments. *Planta (Berl.) 119*:233.
5. Berg, R. A., Y. Kishida, Y. Kobayashi, K. Inouye, A. E.
    Tonelli, S. Sakakibara, and D. J. Prockop. 1973.
    Model for the triple-helical structure of (Pro-Hyp-
    Gly)$_{10}$ involving a cis peptide bond and inter-chain
    hydrogen-bonding to the hydroxyl group of hydroxy-
    proline. *Biochim. Biophys. Acta 328*:553.
6. Brysk, M. M. and M. J. Chrispeels. 1972. Isolation
    and partial characterization of a hydroxyproline-rich
    cell wall glycoprotein and its cytoplasmic precursor.
    Biochim. Biophys. Acta 257:421.
7. Butler, W. T. and L. W. Cunningham. 1966. Evidence for
    the linkage of a disaccharide to hydroxylysine in tro-
    pocollagen. *J. Biol. Chem. 241*:3882.
8. Chang, L. M. S. 1976. Phylogeny of DNA polymerase-β.
    Science 191:1183.
9. Clark, C. C. and N. A. Kefalides. 1976. Carbohydrate
    moieties of procollagen: Incorporation of isotopically
    labeled mannose and glucosamine into propeptides of
    procollagen secreted by matrix-free chick embryo tendon
    cells. *Proc. Nat. Acad. Sci. 73*:34.
10. Dashek, W. V. 1970. Synthesis and transport of hydroxy-
    proline-rich components in suspension cultures of
    sycamore. *Plant Physiol. 46*:831.
11. Davies, D. R. and K. Roberts. 1976. Genetics of cell
    wall synthesis in *Chlamydomonas reinhardtii*. *In* The
    Genetics of Algae (R. Lewin, ed.), pp. 63-68. Black-
    wells, London.
12. Downs, F., A. Herp, J. Moschera, and W. Pigman. 1973.
    β-Elimination and reduction reactions and some appli-
    cations of dimethylsulfoxide on submaxillary glyco-
    proteins. *Biochim. Biophys. Acta 328*:182.
13. Eylar, E. H. 1965. On the biological role of glyco-
    proteins. *J. Theoret. Biol. 10*:89.
14. Fincher, G. B., W. H. Sawyer, and B. A. Stone. 1973.
    Chemical and physical properties of an arabinogalactan-
    peptide from wheat endosperm. *Biochem. J. 139*:535.
15. Gardiner, M. and M. J. Chrispeels. 1975. Involvement
    of the Golgi apparatus in the synthesis and secretion
    of hydroxyproline-rich cell wall glycoproteins. *Plant
    Physiol. 55*:536.

16. Garrone, R., A. Huc, and S. Junqua. 1975. Fine struc-
    ture and physicochemical studies on the collagen of
    the marine sponge *Chondrosia reniformis* Nardo. *J.
    Ultrastructure Res. 52:*261.
17. Haeckel, E. 1874. Die Gastraatheorie. *Jenaer Zeit-
    schrift.*
18. Heath, M. F. and D. H. Northcote. 1971. Glycoprotein
    of the wall of sycamore tissue-culture cells. *Biochem.
    J. 125:*953.
19. Hills, G. J., J. M. Phillips, M. R. Gay, and K. Roberts.
    1975. Structure, composition, and morphogenesis of
    the cell wall of *Chlamydomonas reinhardi*. III. Self-
    assembly of a plant cell wall *in vitro*. *J. Mol. Biol.
    96:*431.
20. Jermyn, M. A. and Y. M. Yeow. 1975. A class of lectins
    present in the tissues of seed plants. *Aust. J. Plant
    Physiol. 2:*501.
21. Karr, J. L., Jr. 1972. Isolation of an enzyme system
    which will catalyze the glycosylation of extensin.
    *Plant Physiol. 50:*275.
22. Keegstra, K., K. W. Talmadge, W. D. Bauer, and P.
    Albersheim. 1973. The structure of plant cell walls.
    *Plant Physiol. 51:*188.
23. Lamport, D. T. A. 1967. Hydroxyproline-O-glycosidic
    linkage of the plant cell wall glycoprotein extensin.
    *Nature 216:*1322.
24. Lamport, D. T. A. 1969. The isolation and partial
    characterization of hydroxyproline-rich glycopeptides
    obtained by enzymic degradation of primary cell walls.
    *Biochemistry 8:*1155.
25. Lamport, D. T. A. 1973. The glycopeptide linkages of
    extensin: *O*-D-Galactosyl serine and *O*-L-arabinosyl
    hydroxyproline. *In* Biogenesis of Plant Cell Wall
    Polysaccharides (F. Loewus, ed.), pp. 149–164. Aca-
    demic Press, New York.
26. Lamport, D. T. A., L. Katona, and S. Roerig. 1973.
    Galactosylserine in extensin. *Biochem. J. 133:*125.
27. Lamport, D. T. A. and D. H. Miller. 1971. Hydroxy-
    proline arabinosides in the plant kingdom. *Plant
    Physiol. 48:*454.
28. Lenard, J. 1969. Reactions of proteins, carbohydrates,
    and related substances in liquid hydrogen fluoride.
    *Chem. Revs. 69:*625.

29. Loewus, F. and Labarca. 1973. Pistil secretion product
    and pollen tube wall formation. *In* Biogenesis of
    Plant Cell Wall Polysaccharides (F. Loewus, ed.), pp.
    175-194. Academic Press, New York.
30. Margulis, L. 1970. Origin of Eukaryotic Cells. Yale
    Univ. Press, New Haven.
31. Mescher, M. F. and J. L. Strominger. 1976. Purifica-
    tion and characterization of a prokaryotic glycopro-
    tein from the cell envelope of *Halobacterium salinarium*.
    *J. Biol. Chem. 251:*2005.
32. Miller, D. H., D. T. A. Lamport, and M. Miller. 1972.
    Hydroxyproline heterooligosaccharides in *Chlamydomonas*.
    *Science 176:*918.
33. Miller, D. H., I. S. Mellman, D. T. A. Lamport, and M.
    Miller. 1974. The chemical composition of the cell
    wall of *Chlamydomonas gymnogama* and the concept of a
    plant cell wall protein. *J. Cell Biology 63:*420.
34. Mort, A. M. and D. T. A. Lamport. 1976. Specific de-
    glycosylation *via* anhydrous hydrogen fluoride. *Plant
    Physiol. 57* Suppl.:Abst. No. 297.
35. Muir, L. and Y. C. Lee. 1969. Structures of the D-
    galactose oligosaccharides from earthworm cuticle
    collagen. *J. Biol. Chem. 244:*2343.
36. Nakamura, K., D. F. Bray, and E. B. Wagenaar. 1975.
    Ultrastructure of *Chlamydomonas eugametos* palmelloids
    induced by chloroplatinic acid treatment. *J. Bacteriol.
    121:*338.
37. Nes, W. R. 1974. Role of sterols in membranes. *Lipids
    9:*596.
38. Nordwig, A. and U. Hayduk. 1969. Invertebrate collagens:
    Isolation, characterization and phylogenetic aspects.
    *J. Mol. Biol. 44:*161.
39. Oohira, A., A. Kusakabe, and S. Suzuki. 1975. Isolation
    of a large glycopeptide from cartilage procollagen by
    collagenase digestion and evidence indicating the pre-
    sence of glucose, galactose and mannose in the peptide.
    *Biochem. Biophys. Res. Commun. 67:*1086.
40. Pope, D. G. 1974. Hydroxyproline-rich material secreted
    by cultured *Acer pseudoplatanus* cells: Evidence for
    polysaccharide attached to hydroxyproline. *Plant
    Physiol. 54* (Suppl.):Abst. No. 82.
41. Pope, D. G. 1977. Secreted hydroxyproline-glycoproteins.
    Plant Physiol. (in press).
42. Roberts, K. 1974. Crystalline glycoprotein cell walls
    of algae: Their structure, composition and assembly.
    *Phil. Trans. R. Soc. Lond. B 268:*129.

43. Roberts, K. and D. H. Northcote. 1972. Hydroxyproline: Observations on its chemical and autoradiographical localization in plant cell wall protein. *Planta 107:* 43.

44. Sadava, D. and M. J. Chrispeels. 1971. Hydroxyproline biosynthesis in plant cells peptidyl proline hydroxylase from carrot disks. *Biochim. Biophys. Acta 227:* 278.

45. Sinohara, H. 1972. Molecular evolution of glycoproteins. *Tohoku J. Exp. Med. 106:*93.

46. Talmadge, K. W., K. Keegstra, W. D. Bauer, and P. Albersheim. 1973. The structure of plant cell walls. I. The macromolecular components of the walls of suspension cultured sycamore cells with a detailed analysis of the pectic polysaccharides. *Plant Physiol. 51:* 158.

47. Thompson, E. W. and R. D. Preston. 1968. Proteins in the cell walls of some green algae. *Nature 213:*684.

48. Waechter, C. J. and W. J. Lennarz. 1976. The role of polyprenol-linked sugars in glycoprotein biosynthesis. *Annu. Rev. Biochem. 45:*95.

49. Weathers, P. J., M. Jost, and D. T. A. Lamport. 1977. The gas vacuole membrane of *Microcystis aeruginosa.* A partial amino acid sequence and proposed functional model. *Arch. Biochem. Biophys.* (in press).

50. Winterburn, P. J. and C. F. Phelps. 1972. The significance of glycosylated proteins. *Nature 236:*147.

Chapter Four

DEGRADATION PRODUCTS OF PROTOLIGNIN AND THE STRUCTURE OF
LIGNIN

AKIRA SAKAKIBARA

Department of Forest Products
Faculty of Agriculture, Hokkaido University
Sapporo, Japan

INTRODUCTION

A clue to the chemical structure of lignin was first
given by the studies of K. Freudenberg and coworkers on the
enzymatic dehydrogenation of coniferyl alcohol, in which
several important di-, tri- and oligolignols were isolated
and their structures were elucidated.[11] A series of his
studies threw light not only on the biogenesis but on the
linkage types of lignin macromolecules. However, while precise information about lignin structure may be derived from
various fragments of protolignins, it is also necessary to
confirm the radical coupling mechanism for bond formation
assumed by Freudenberg for these degradation products.

Direct proofs of structure of native lignin have been
very few, because lignin involves very complex irreversible
linkages. Nevertheless, the studies on mild degradation of
protolignins carried out by Sakakibara and coworkers and by
Nimz have yielded several valuable results. The degradation
products from native lignins have given further support to
the concept that lignin is a polymeric product arising from
an enzyme-initiated dehydrogenative polymerization of cinnamic alcohols. Moreover, they have yielded more information

on the linkages in lignin molecules than those from the
dehydrogenation products of coniferyl alcohol.

The oxidation of lignin with permanganate after methylation led to many interesting suggestions about structure.[7,8]
The procedure of a permanganate oxidation was later modified
and yields of products increased remarkably.[3,4] The products
mainly gave information about "condensed type" and diphenyl
ether units which showed that extensive side chain displacement occurred during the formation of linkages through radical coupling. These oxidation products failed to provide
information concerning arrangement of side chains.

This chapter reports on the products of hydrolysis with
dioxane and water, and the catalytic hydrogenolysis of protolignins. These degradation products appear to be almost as
intact in their structure as in the native state, and may be
obtained by cleavage of the α-aryl-alkyl-ether structure
which has been suggested by model experiments.[47] In hydrogenolysis, the cleavage of ether linkages and the reduction
of side chains carrying hydroxyl groups occur.

HYDROLYSIS OF PROTOLIGNIN WITH
DIOXANE AND WATER

Extractive-free wood meal of softwood or hardwood was
hydrolyzed with dioxane-water (1:1) in an autoclave at
180 C.[45,46] By cooking Ezomatsu (*Picea jezoensis* Carr.) and
Buna (*Fagus crenata* Blum.) wood at 180 C for 120 min, 40 to
60% of lignin and 8 to 26% of carbohydrates were dissolved
in the solvent mixture.[46] Cinnamic alcohols and aldehydes
as well as benzoic acid and benzaldehyde derivatives were
isolated as monomeric hydrolysis products. Such compounds
can only be obtained from virtually unchanged lignins such
as protolignins or milled wood lignins. The method may be
used as a criterion to judge to what extent a lignin preparation has been modified during its isolation process. The
hydrolysis products, coniferyl alcohol and sinapyl alcohol,
were assumed also to be obtained through cleavage of β-alkyl-
aryl ethers but were found only in small quantities.[47] A
veratryl model compound, however, gave no cinnamic alcohols.

The important hydrolysis products obtained in our laboratory are summarized in Figure 1. For purposes of reference,
compounds obtained by Nimz are also shown.

Arylglycerols (I-III) with up to three pendant groups were isolated in our laboratory[40],[48] and by Nimz.[32] The glycerol side chain was not only observed in hydrolysis products but also in those from hydrogenolysis[54] and reaction with metallic sodium in liquid ammonia.[55] Arylglycerols have not been isolated from dehydrogenation products of coniferyl alcohol as yet. Whether the glycerol side chain is an artifact formed during reactions or not is uncertain. A model compound, guaiacylglycerol-β-guaiacyl ether (IV), however, did not give guaiacylglycerol under the same hydrolysis conditions.[47] The same side chain structure is also seen in compounds VI and VII. Efforts to detect this function in milled wood lignin by oxidation with periodate have so far been unsuccessful (Yasuda and Sakakibara, unpublished). Perhaps this function constitutes only a minor part of lignin.

Compound VII was isolated from an ether soluble fraction of the hydrolysis products of Yachidamo (*Fraxinus mandshurica* Rupr.) wood meal by using column chromatography. The compound is especially interesting in connection with the formation of so-called syringyl lignin. Freudenberg showed that dehydrogenation of sinapyl alcohol led to racemic syringaresinol in yields of 80 to 88%, and gave no polymer.[14] On the other hand, the existence of syringyl lignin was known from UV spectrascopic observations on sections of birch xylem.[5] In the case of syringyl lignin, a requirement for polymerization may be β-0-4 coupling, because both *ortho* positions to the phenolic hydroxyl are not available for coupling. A circumstance in which so-called end-wise polymer[50] can be formed would be necessary, for hardwood lignins seem to involve β-0-4 linkage as a predominant bonding. Compound VII was isolated for the first time from hardwood lignin as a unit with such a linkage pattern.

Phenylcoumaran VIII was reported to be obtained from spruce wood by treatment with methanolic hydrochloric acid at 20 C.[13] This compound, however, could not be detected by us even in trace amounts on chromatograms. However, compound IX, a coupling product of sinapyl alcohol and coniferyl aldehyde radicals was very recently isolated (Aoyama and Sakakibara, unpublished). Aldehyde type phenylcoumarans and guaiacylglycerol-β-coniferyl aldehyde ether (V) were isolated early as dilignols in the DHP reaction mixture.[10] The latter aldehyde V was later detected among the mild hydrolysis products of spruce lignin.[35] The cinnamic aldehyde function in lignin molecules is responsible for almost all the

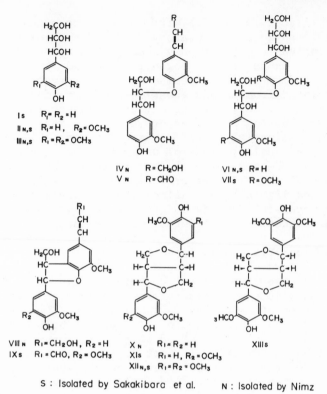

S : Isolated by Sakakibara et al.   N : Isolated by Nimz

Figure 1.   Hydrolysis products of lignin.

characteristic color reactions of lignin, for example, the red violet color in phloroglucinol-HCl solution, the red color with dimethyl-p-phenylenediamine, and the green color with furfuran.

It was reported that pinoresinol (X) also exists in the hydrolysis products with methanolic hydrochloric acid at 20 C,[13] but we could not detect this compound in the hydrolysis

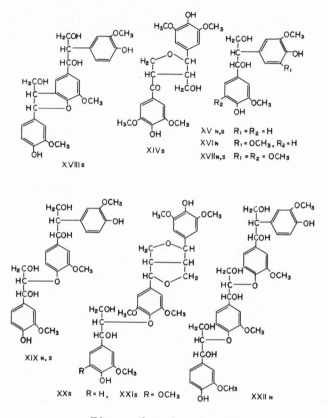

Figure 1.   Continued.

mixture of dioxane and water, even in trace amounts.   The
results obtained by us are in accordance with the report,
following ^{13}C-NMR studies, that the pinoresinol unit may
exist only in small quantities in conifer lignin.[36]

On the other hand, syringaresinol (XII) was isolated
easily in crystalline form from hardwood lignins.[33,39,51]
This compound was obtained from hardwood lignin either by

hydrolysis or by hydrogenolysis in good yield. This indicates that the syringaresinol unit is bonded in lignin molecules only by ether linkages, while pinoresinol can be
connected in addition through C-C linkage at the 5-position
with another phenylpropane unit. DL-Episyringaresinol (XIII)
was also isolated accompanied by DL-syringaresinol.[30,41] It
is very probable that episyringaresinol units may exist originally in the hardwood lignin, as it was found that DL-
epipinoresinol accompanied DL-pinoresinol in the dehydrogenation products of coniferyl alcohol.[10] However, it was
shown that about 10% of DL-syringaresinol was isomerized to
DL-episyringaresinol when subjected to hydrolysis in dioxane-
water (1:1) under the same conditions. Acidolysis may iso-
merize DL-syringaresinol more effectively. At present, the
question is still open to debate.

Earlier, the third lignan type compound XI had been
isolated by us as a hydrolysis product. It has both guaiacyl and syringyl groups and was designated as medioresinol.
The compound has intermediate patterns between pinoresinol
and syringaresinol in IR, NMR and mass spectra and chromatographic behavior.[41] The dimethyl ether of this compound
(magnolin) has been isolated from the extractives of the
flower buds of *Magnolia fargesii*.[20] The lignol XI with both
guaiacyl and syringyl units may constitute a part of copolymer moieties of lignin molecules, together with VII, IX,
XVI, and XX.

Compounds XV, XVI and XVII were first obtained in the
products extracted with water at 100 C.[33] We also have isolated XV and XVII.[40,48] The compounds represent unusual
lignols having $C_6-C_3-C_6$ skeletons. For the formation of this
structure, two mechanisms were proposed,[23,34] as shown in
Figure 2. It is considered that mechanism 1 may be more
favorable. The β-1 coupling, thus, may cause the displacement of a side chain, leaving a substituted glyceraldehyde
residue which may be the source of unconjugated carbonyl
groups in lignin molecules.[24] This diarylpropane unit is
also seen in trilignols XVIII and XIX, and a tetralignol
XXII. A new lignan type dilignol having an α-carbonyl, XIV,
was isolated from Yachidamo wood protolignin. It has been
proposed that the α-carbonyl groups in lignin amount to about
0.06 per methoxyl.[25] Dilignol XIV represents one of these
units.

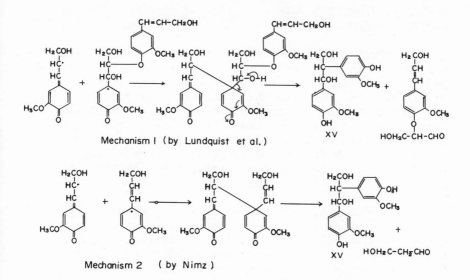

Figure 2.   Proposed mechanisms for the formation of diarylpropanes.

The phenylcoumaran type dilignol, XIII, could not be detected in the reaction products of dioxane-water hydrolysis as mentioned above, but recently, trilignol XVIII was isolated from softwood lignin.[49] Compound XVIII has combined linkages of β-5 and β-1. Trilignol XIX and tetralignol XXII obtained from conifer protolignins have combinations of β-1 and β-O-4 linkages.[33,48]

Trilignol XX was found in the hydrolysate of hardwood protolignin[42] and most recently, a syringyl type trilignol, XXI, was isolated (Omori and Sakakibara, unpublished). Earlier, the homologous trilignols, guaiacylglycerol-β-pinoresinol and guaiacylglycerol-β-epipinoresinol ether had been found among the dehydrogenation products of coniferyl alcohol.[12]

All these compounds obtained from mild hydrolysis processes can be explained by the mechanism of enzymatic dehydrogenation followed by radical coupling. Their side chain functions remained almost unchanged, which illustrates the advantages of mild hydrolysis procedures in elucidating lignin structure.

CATALYTIC HYDROGENOLYSIS
OF PROTOLIGNIN

Catalytic hydrogenolysis mainly cleaves ether linkages
and reduces in part the hydroxyls on side chains. The hydrogenolysis of lignin was reported in many early papers,[2,16,17,1]
in which it was reported that monomeric phenylpropane derivatives were obtained in good yields and it was established that
lignin might be built up from C_6-C_3 units. Later, hydrogenolysis of lignin was investigated as a new process for pulping.[1]
Red spruce was hydrogenated on cobalt carbonyl with hydrogen
and carbon monoxide, and 4,4'-dipropyl-6,6'-biguaiacol was
isolated in 2% yield of lignin.[29]

Hydrogenolysis with the objective of gaining information
about the linkage pattern in lignin molecules has been carried
out by us since 1969. The reaction conditions applied are as
follows: Pre-extracted wood meal of soft- (*Picea jezoensis*
Carr.) and hard- (*Fraxinus mandshurica* Rupr.) woods is subjected to hydrogenolysis in a solvent mixture of dioxane and
water (1:1, 9:1) on copper chromite as a catalyst at 80 kg/cm^2
initial hydrogen pressure and 200 to 220 C for one hour. Under
these conditions, almost all lignin is dissolved.[54]

In Figure 3, the hydrogenolysis products of lignin are
summarized. For reference, the compounds derived by cleavage
of beech wood lignin upon thioacetic acid treatment followed
by the reduction with Raney nickel are also shown.[38]

As stated before, hydrogenolysis cleaves most of the ether
linkages, but a small part of aryl-alkyl ether remains as seen
in compounds XXIII and XXIV.[52] The latter has a hydroxyl at
the α-position. In general, an α-carbon bearing a hydroxyl
can be easily reduced to a methylene group. Presumably, the
α-hydroxyl in compound XXIV might be formed through the reduction of an α-carbonyl in a unit bearing a phenolic hydroxyl at
the para-position since this is known to have stability against
reduction.

Compound XXV which was obtained together with other homo-
logs with different states of reduction in the side chains[26,27]
can be considered to be formed after reductive cleavage of the
phenylcoumaran ring, and this linkage pattern is observed often
among the hydrogenolysis products as, for example, in XXXVI,
XXXVII and XXXVIII.

The structures of compounds XXVI[53] and XXVII[57] throw
light on the origin of metahemipinic acid (XLIV), the well
known acid obtained by permanganate oxidation of methylated
lignin. Although compound XXVII was obtained from compression
wood lignin of Karamatsu (*Larix leptolepis* Gord.), its occurrence in normal wood lignin is very probable. At one time,
it was suspected that metahemipinic acid might be derived as
an artifact of rearrangements occurring in the oxidation process. The problem, however, has been resolved by the isolation of these two compounds. Whether or not compound XXVI can
be derived from XXVII by reductive cleavage of the phenyl-
isochroman ring is open to speculation, but a model experiment
suggests that this does not occur. Thus, hydrogenolysis of
pinoresinol under the same conditions demonstrated that the
alkyl-alkyl ether of the tetrahydrofuran ring is very stable
toward hydrogenolysis, and the starting material, pinoresinol,
was almost completely recovered.[27] This suggests that the
linkage patterns of compounds XXVI and XXVII may exist differently in lignin molecules. A possible mechanism for the
formation of these compounds is shown in Figure 4.

From hardwood lignin hydrolysates, further lignan type
compounds have been obtained. Though it is undeniable that
compounds XXVIII and XXX could be derived after reductive
opening of tetrahydrofuran ring, the model experiment mentioned above and the existence of compound XIV among the
hydrolysis products suggests another mechanism. It seems
likely that these linkages originally occur as different units
in the lignin molecule.

So-called cyclolignans such as compound XXXI[38] were proposed as a source of benzene polycarboxylic acids in the
permanganate oxidation products of lignin.[11] The existence
of the diphenyl ether XXXIII[38] was predicted on the grounds of
the detection of 2,5',6'-trimethoxydiphenyl ether-4,3'-
dicarboxylic acid among the permanganate oxidation products.

Compound XXXIV was isolated very recently from the hydrogenolysis products of hardwood protolignin (*Fraxinus mandshurica* Rupr.) (Sudo and Sakakibara, unpublished). The α-O-γ-
ether pattern in lignin has not been reported as yet. The
compound indicates that the alkyl-alkyl ether is fairly stable
against the reductive cleavage, as demonstrated in the case of
pinoresinol. An aryl ether at the α-position of arylpropane
is known in guaiacylglycerol-α,β-bis-coniferyl ether, a dehydrogenation intermediate product of coniferyl alcohol.[9]

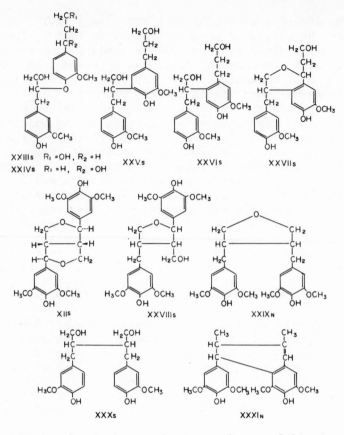

Figure 3.   Hydrogenolysis products of lignin.

The diphenyl-type linkage as in compounds XXXV and XXXVII is most frequently observed in the hydrogenolysis products of lignin.   Compound XXXVI may be formed from compound XVIII unit through reductive cleavage of the phenylcoumaran ring.   Compound XXXVIII may originate from the dehydrotriconiferyl alcohol unit which has double

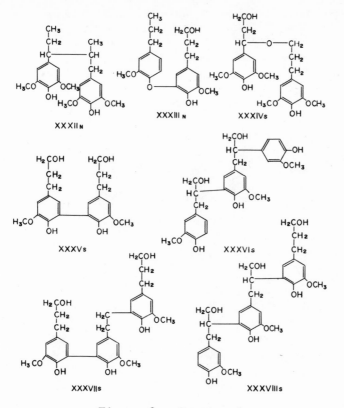

Figure 3.   Continued.

phenylcoumaran rings and was isolated from the dehydrogenation products of coniferyl alcohol.[31]

As stated above, most of the hydrogenolysis products
have carbon to carbon linkages and reduced side chains such
as n-propyl or n-ω-propanol.   The reaction of hydrogenolysis

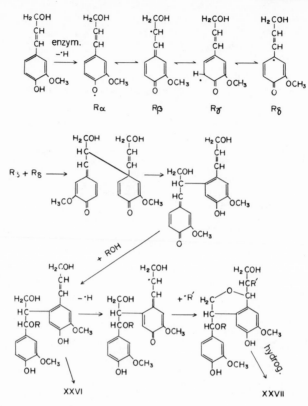

Figure 4.   A proposed mechanism for the formation of
β-6 dilignols.

mainly cleaves alkyl–aryl and aryl–aryl ethers, with little
effect on alkyl–alkyl ethers. The formation of these products may also be explained by the radical coupling mechanism
of the dehydrogenation of cinnamic alcohols assumed in lignin
formation.

COMPRESSION WOOD PROTOLIGNIN

General Chemical Properties. It has long been known
that compression wood has a much higher content of lignin
which contains a low proportion of methoxyl groups. Systematic studies on the chemical composition of lignin in compression wood of Todomatsu (*Abies sachalinensis* Mast.) were
carried out by Morohoshi and Sakakibara.[28] From the results,
it has been shown by means of nitrobenzene-oxidation,
permanganate-oxidation, alcoholysis, and the analysis of NMR
spectra that compression wood lignin involves a much higher
degree of condensed units (79% in compression wood lignin
and 48% in normal wood lignin) and p-hydroxyphenyl units than
normal wood lignin does.

Recently, we have studied the compression wood lignin
of Karamatsu (*Larix leptolepis* Gord.).[56,57] Klason lignin
of the compression wood and of the normal (side) wood (Fig. 5)
amounts to 39.27% and 28.05%, respectively. Results of the
analysis of milled wood lignins from the compression and the
normal woods are summarized in Table 1, and show that the
methoxyl content of milled wood lignin from the compression
wood is lower than that of normal wood.

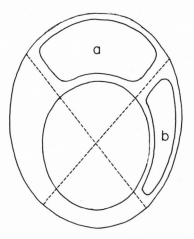

Figure 5. Location of zones in stem containing compression wood (sapwood). Compression wood, a. Side (normal)
wood, b.

As seen from Table 2, more "condensed units" are
involved in compression wood lignin than in normal wood lignin. These structures are shown in Figure 6. Only veratric
acid (XLIII) and metahemipinic acid (XLIV) are present in
greater amount in normal wood lignin. Since p–hydroxyphenyl
units also exist in larger quantity in the compression wood,
it may be presumed that p–hydroxyphenyl units have greater
probability of coupling with other phenylpropanes to produce
"condensed type" units, because both positions *ortho* to the
phenolic hydroxyl are available. Permanganate oxidation
products from the compression wood lignin, XL, XLI, XLII, and
XLVI appear to support the above assumption (Table 2). But
it should not be overlooked that "noncondensed type" p–
hydroxyphenyl units also exist in greater amount in the compression wood lignin as shown by the anisic acid yield. The
results of nitrobenzene oxidation (Table 3), ethanolysis
(Table 4) and hydrogenolysis (Table 5) also show the greater
amount of p–hydroxyphenyl nuclei in the compression wood
lignin. The products of ethanolysis are shown in Figure 7.

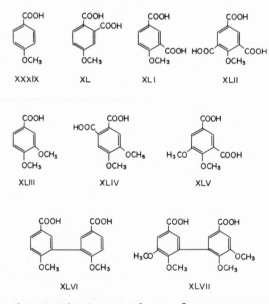

Figure 6. Oxidation products from compression wood
lignin by permanganate.

Table 1. Analysis of milled wood lignins (MWL) from compression and normal woods (*Larix leptolepis* Gord.).

Sample	OCH_3 in MWL (%)	Analysis of MWL C (%)	Analysis of MWL H (%)	Empirical Formula
Compression wood	12.64	62.97	5.91	$C_9H_{7.51}O_{1.67}(OCH_3)_{0.76}(OH)_{1.21}$
Normal wood	15.19	62.12	5.92	$C_9H_{7.64}O_{1.68}(OCH_3)_{0.94}(OH)_{1.20}$

Table 2. Yields of $KMnO_4$ oxidation products (% of Klason lignin, as methyl esters).

Sample	XXXIX	XL	XLI	XLII	XLIII	XLIV	XLV	XLVI	XLVII
Compression wood	2.47	0.84	0.27	0.21	10.17	0.12	2.41	+	0.14
Normal wood	0.91	+	0.11	–	12.14	0.20	1.07	–	0.10

XXXIX, anisic acid; XL, 4-methoxy-o-phthalic acid; XLI, 4-methoxy-isophthalic acid; XLII, methoxytrimesic acid; XLIII, veratric acid; XLIV, metahemipinic acid; XLV, isohemipinic acid; XLVI, dehydrodianisic acid; XLVII, dehydrodiveratric acid.

Table 3. Yields of nitrobenzene oxidation products (% of Klason lignin).

Oxidation Product	Compression Wood	Normal Wood
p-Hydroxybenzoic acid	0.2	0.1
p-Hydroxybenzaldehyde	2.6	0.8
Vanillic acid	2.1	2.6
Vanillin	15.1	22.6
Total	20.0	26.0

Table 4. Yields of ethanolysis products (% of Klason lignin).

Sample	XLVIII	XLIX	L	LI	LII	LIII	LIV	LV	Total
Compression wood	0.31	+	0.33	1.01	1.67	2.52	1.11	3.11	10.06
Normal wood	+	+	+	+	2.44	3.01	1.79	4.78	12.05

XLVIII, 4-hydroxyphenylacetone; XLIX, 4-hydroxybenzoyl acetyl; L, 1-ethoxy-1-(4-hydroxyphenyl)-2-propanone; LI, 4-hydroxy-α-ethoxypropiophenone; LII, guaiacylacetone; LIII, vanilloyl acetyl; LIV, 1-ethoxy-1-(4-hydroxy-3-methoxyphenyl)-2-propanone; LV, α-ethoxypropiovanillone.

Table 5. Yields of hydrogenolysis products (% of Klason lignin).

Sample	LVI	LVII	LVIII	LIX	LX	Total
Compression wood	0.36	0.44	1.10	3.81	20.56	26.27
Normal wood	0.26	0.10	1.80	1.57	27.46	31.19

LVI, p-cresol; LVII, p-hydroxyphenylpropane; LVIII, dihydroeugenol; LIX, p-hydroxyphenylpropan-3-ol; LX, guaiacylpropan-3-ol.

Hydrogenolysis of Compression Wood Lignin. Compression wood meal was subjected to hydrogenolysis under the same conditions as applied to normal wood, to gain more detailed information on the structure of this lignin. The compounds isolated to date from the reaction mixture are summarized in Figure 8. The "condensed units" involving β-5, β-6 and biphenyl linkages are thought to be a common occurrence in compression wood lignin. Compound LXIII is presumed to be produced by a process other than successive α-β cleavage of a three-carbon side chain, because the breakdown of the aliphatic side chain rarely occurs except by β-γ cleavage as in p-cresol, LVI. Isolation of guaiacylglycerol-β-vanillin ether from mild hydrolysis products[35] suggests this

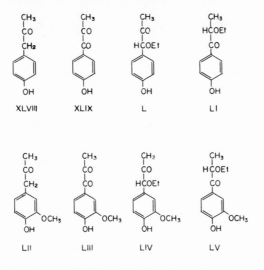

Figure 7. Ethanolysis products from compression wood lignin.

possibility. These "condensed type" compounds involve p-hydroxyphenyl units extensively, as seen in compounds LXV to LXVII and LXIX (Yasuda and Sakakibara, unpublished).

In the family *Gramineae*, a major part of the p-hydroxy-phenyl groups present stem from the associated ester groups in the lignin molecule.[19] It is clear, however, that p-hydroxyphenyl groups in compression wood lignin normally participate in the linkages of so-called "core lignin" as the result of radical couplings of dehydrogenated p-coumaryl alcohol.

Thus compression wood contains much lignin which mainly consists of "condensed units." It may be deduced that the structural features of compression wood lignin reflect the high resistance to compression stress.

CONCLUSIONS

The mild hydrolysis and the catalytic hydrogenolysis products from lignins described above represent almost all the linkage patterns which exist in the enzymatic dehydro-genation products of coniferyl alcohol. Moreover, many new

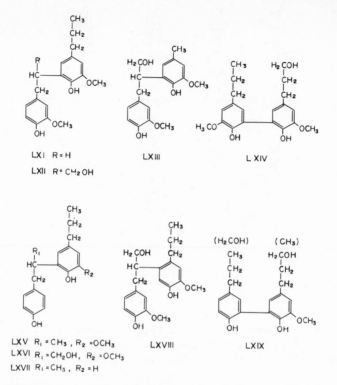

Figure 8. Hydrogenolysis products from compression wood lignin.

types of linkages between phenylpropane units have been found in these products.

Oxidation products from methylated lignin by permanganate give complementary information on the linkages in lignin macromolecules. For example, various kinds of biphenyls and diphenyl ethers are found among them. The formation of some of them requires side chain displacement.[21,22,43,44]

The isolation of compounds from the complex reaction mixtures obtained is very troublesome. However, it is difficult to establish the chemical structure of lignin without them. The quantitative estimation of the various linkages is also necessary. The development of new effective reactions is desirable for this line of research to progress.

So far, a number of schematic formulas for the structure of lignin have been proposed. They are, for example, based on the intermediate products of the dehydrogenation of coniferyl alcohol,[11] gel filtration behavior of lignosulfonic acids from softwood lignin,[6] or on the degradation products for hardwood lignin.[36,37] Glasser built up a lignin model by computer.[15] However, these formulas are only temporary, and they will have to be altered when further information is added. Before constructing a lignin model structure, it will be necessary to accumulate more information about linkages. Nevertheless, the chemical structure of lignin will never be distinctly established like the structures of some other natural macromolecules, but will only be defined as a statistical model from information about linkage types between the phenylpropane present.

REFERENCES

1. Bhaskaran, T. A. and C. Schuerch. 1969. A study of non-volatile hydrogenated maple wood lignin. *Tappi* 52:1948.
2. Brewer, C. D., L. M. Cooke, and H. Hibbert. 1948. Studies on lignin and related compounds. LXXXIV. The high pressure hydrogenation of maple wood: Hydrol lignin. *J. Am. Chem. Soc. 70*:57.
3. Erickson, M., S. Larsson, and G. E. Miksche. 1973. Gaschromatographische analyse von ligninoxydationsprodukten. VII. Ein verbessertes verfahren zur characterisierung von ligninen durch methylierung und oxydativen abbau. *Acta Chem. Scand. 27*:127.
4. Erickson, M., S. Larsson, and G. E. Miksche. 1973. Gaschromatographische analyse von ligninoxydationsprodukten. VIII. Zur struktur des lignins der fichte. *Acta Chem. Scand. 27*:903.
5. Fergus, B. J. and D. A. I. Goring. 1968. The location of guaiacyl and syringyl lignins in birch xylem tissue. Post-Graduate Research Reports. P.G.R. No. 12. Pointe Claire, P.Q. Canada.
6. Forss, K., K.-E. Fremer, and B. Stenlung. 1966. Spruce lignin and its reactions in sulfite cooking. I. The structure of lignin. *Paper and Timber 48*:565.
7. Freudenberg, K. and C. L. Chen. 1960. Methylierte phenolcarbonsäuren aus lignin. *Chem. Ber. 93*:2533.
8. Freudenberg, K. and C. L. Chen. 1967. Weitere oxydationsprodukte des fichtenlignins. *Chem. Ber. 100*:3683.

9. Freudenberg, K. and M. Friedmann. 1960. Oligomere zwischenprodukte der ligninbildung. *Chem. Ber. 93*: 2138.

10. Freudenberg, K. and B. Lehmann. 1960. Aldehydische zwischenprodukte der ligninbildung. *Chem. Ber. 93*: 1354.

11. Freudenberg, K. and A. C. Neish. 1968. The Constitution and Biosynthesis of Lignin. Springer-Verlag, New York, Inc., p. 47.

12. Freudenberg, K. and H. Nimz. 1962. Guaiacylglycerin-β-pinoresinoläther, ein dehydrierungsprodukte des coniferylalkohols. *Chem. Ber. 95*:2057.

13. Freudenberg, K., C. L. Chen, J. M. Harkin, H. Nimz, and H. Renner. 1965. Observations on lignin. *Chem. Commun.*, 224.

14. Freudenberg, K., J. M. Harkin, M. Reichert, and T. Fukuzumi. 1958. Die an der verholzung beteiligten enzyme. Die dehydrierung des sinapinalkohols. *Chem. Ber. 91*:581.

15. Glasser, W. and H. R. Glasser. 1974. Simulation of reactions with lignin by computer (simrel). II. A model for softwood lignin. *Holzforschung 28*:5.

16. Hachihama, Y. and A. Jodai. 1941. Studies on catalytic hydrogenation of lignin. *Kogyo Kagaku Zasshi 44*: 773,775.

17. Hachihama, Y. and A. Jodai. 1943. Studies on catalytic hydrogenation of lignin. *Kogyo Kagaku Zasshi 46*: 132.

18. Harris, E. E., J. D'Ianni, and H. Adkins. 1932. Reaction of hardwood lignin with hydrogen. *J. Am. Chem. Soc. 60*:1467.

19. Higuchi, T., Y. Ito, and I. Kawamura. 1967. p-Hydroxyphenylpropane component of grass lignin and role of tyrosine-ammonia lyase in its formation. *Phytochemistry 6*:875.

20. Kakisawa, H. and T. Kusumi. 1970. Structures of lignans of *Magnolia fargesii*. *Bull. Chem. Soc. Japan 43*:3631.

21. Larsson, S. and G. E. Miksche. 1969. Gaschromatographische analyse von ligninoxydationsprodukten III. Oxydativer abbau von methyliertem Björkmanlignin (fichte). *Acta Chem. Scand. 23*:3337.

22. Larsson, S. and G. E. Miksche. 1972. Gaschromatographische analyse von ligninoxydationsprodukten VI. 4,4-Bisaryloxy-cyclohexa-2,5-dienonstruktur im lignin. *Acta Chem. Scand. 26*:2031.

23. Lundquist, K. and G. E. Miksche. 1965. Nachweis eines neuen verknupfungsprinzips von guajacylpropaneinheiten im fichtenlignin. *Tetrahedron Lett.*, 2131.
24. Lundquist, K. and G. E. Miksche, L. Ericsson, and L. Berndtson. 1967. Über das vorkommen von glyceraldehyd-2-arylätherstrukturen im lignin. *Tetrahedron Lett.*, 4587.
25. Marton, J. and E. Adler. 1959. Zur kenntnis der carbonylgruppen im lignin. I. *Acta Chem. Scand.* 13: 75.
26. Matsukura, M. and A. Sakakibara. 1973a. Hydrogenolysis of protolignin. V. Isolation of some dimeric compounds with carbon to carbon linkage. *Mokuzai Gakkaishi* 17:131.
27. Matsukura, M. and A. Sakakibara. 1973b. Hydrogenolysis of protolignin. VIII. Isolation of a dimer with $C\beta$-$C\beta$ linkage and a biphenyl. *Mokuzai Gakkaishi* 19:171.
28. Morohoshi, N. and A. Sakakibara. 1971. The chemical composition of reaction wood. I. *Mokuzai Gakkaishi* 17:393.
29. Nahum, L. 1965. Deligninification of wood by hydrogenation in presence of dicobalt octacarbonyl. *Ind. Eng. Chem.* 4:71.
30. Nakatsubo, F. and T. Higuchi. 1972. Acidolysis of bamboo lignin. II. Isolation and identification of acidolysis products. *Wood Research* 53:9.
31. Nimz, H. 1963. Dehydro-triconiferylalkohol, ein zwischenprodukt der ligninbildung. *Chem. Ber.* 96:478.
32. Nimz, H. 1965. Über die milde hydrolyse des buchenlignins. I. Isolierung des dimethylpyrogallolglycerins. *Chem. Ber.* 98:3153.
33. Nimz, H. 1966a. Der abbau des lignins durch schonende hydrolyse. *Holzforschung* 20:105.
34. Nimz, H. 1966b. Isolierung von zwei weiteren abbauphenolen mit einer 1,2-diarylpropan-struktur. *Chem. Ber.* 99:469.
35. Nimz, H. 1967. Über zwei aldehydische dilignole aus fichtenlignin. *Chem. Ber.* 100:2633.
36. Nimz, H. 1974. ^{13}C-Kernresonanzspektren von ligninen. 3. Vergleich von fichtenlignin mit künstlichem lignin nach Freudenberg. *Makromol. Chem.* 175:2563.
37. Nimz, H. 1974. Das lignin der buche--Entwurf eines konstitutionsschemas. *Angew. Chem.* 86:336.
38. Nimz, H. and K. Das. 1971. Niedermolekulare spaltprodukte des lignins. 2. Durch abbau mit thioessigsäure erhaltene dimere abbauphenole des buchenlignins. *Chem. Ber.* 104:2359.

39. Omori, S. and A. Sakakibara. 1971. Hydrolysis of lig-
     nin with dioxane and water. IX. Isolation of DL-
     syringaresinol from hydrolysis products of yachidamo
     (*Fraxinus mandshurica*). *Mokuzai Gakkaishi 17*:464.
40. Omori, S. and A. Sakakibara. 1972. Hydrolysis of lig-
     nin with dioxane and water. X. Isolation of aryl-
     glycerols and diarylpropanediol from yachidamo
     (*Fraxinus mandshurica*). *Mokuzai Gakkaishi 18*:355.
41. Omori, S. and A. Sakakibara. 1974. Hydrolysis of lig-
     nin with dioxane and water. XI. Isolation of aryl-
     glycerol-β-aryl ether- and lignan-type compounds from
     hardwood lignin. *Mokuzai Gakkaishi 20*:388.
42. Omori, S. and A. Sakakibara. 1975. Hydrolysis of lig-
     nin with dioxane and water. XII. Isolation of dimeric
     and trimeric compounds with lignan type linkages and
     diguaiacylpropanediol from hardwood lignin. *Mokuzai
     Gakkaishi 21*:170.
43. Pew, J. C. and W. J. Connors. 1967. New structures
     from the dehydrogenation of model compounds related
     to lignin. *Nature 215*:623.
44. Pew, J. C. and W. J. Connors. 1969. New structures
     from the enzymic dehydrogenation of lignin model p-
     hydroxy-α-carbinols. *J. Org. Chem. 34*:580.
45. Sakakibara, A. and N. Nakayama. 1961. Hydrolysis of
     lignin with dioxane and water. I. Formation of
     cinnamic alcohols and aldehydes. *Mokuzai Gakkaishi
     7*:13.
46. Sakakibara, A. and N. Nakayama. 1962. Hydrolysis of
     lignin with dioxane and water. III. Hydrolysis pro-
     ducts of various woods and lignin preparations.
     Mokuzai Gakkaishi 8:157.
47. Sakakibara, A., H. Takeyama, and N. Morohoshi. 1966.
     Untersuchungen über die hydrolyse von lignin mit dioxan
     und wasser. *Holzforschung 20*:45.
48. Sano, Y. and A. Sakakibara. 1970. Hydrolysis of lignin
     with dioxane and water. VII. Isolation of dimeric and
     trimeric compounds with 1,2-diarylpropane structure.
     Mokuzai Gakkaishi 16:121.
49. Sano, Y. and A. Sakakibara. 1975. Hydrolysis of lignin
     with dioxane and water. XIV. Isolation of a new tri-
     meric compound. *Mokuzai Gakkaishi 21*:461.
50. Sarkanen, K. V. and C. H. Ludwig. 1971. Lignins,
     Occurrence, Formation, Structure and Reactions.
     John Wiley & Sons, Inc., pp. 150.

51. Sudo, K. and A. Sakakibara. 1973. Hydrogenolysis of
     protolignin. VII. Isolation of DL-syringaresinol,
     biphenyl and diarylpropane from hardwood lignin.
     Mokuzai Gakkaishi 19:165.
52. Sudo, K. and A. Sakakibara. 1974a. Hydrogenolysis of
     protolignin. X. Isolation of two dimeric compounds
     with β-aryl ether linkage. *Mokuzai Gakkaishi 20*:331.
53. Sudo, K. and A. Sakakibara. 1974b. Hydrogenolysis of
     protolignin. XI. Isolation of a dimer with C_β-C_6 and
     a trimer with two C_β-C_5 linkages. *Mokuzai Gakkaishi
     20*:396.
54. Wada, I. and A. Sakakibara. 1969. Hydrogenolysis of
     protolignin. I. Hydrogenolysis products under various
     reaction conditions. *Mokuzai Gakkaishi 15*:214.
55. Yamaguchi, A. 1973. Degradation of lignin with metallic
     sodium in liquid ammonia. II. Degradation of MWL and
     LCC and their GLC analyses. *Mokuzai Gakkaishi 19*:29.
56. Yasuda, S. and A. Sakakibara. 1975a. Hydrogenolysis of
     protolignin in compression wood. I. Isolation of two
     dimers with C_β-C_5 and C_β-C_3 composed of p-hydroxyphenyl
     and guaiacyl nuclei and two p-hydroxyphenyl nuclei,
     respectively. *Mokuzai Gakkaishi 21*:370.
57. Yasuda, S. and A. Sakakibara. 1975b. Isolation of a
     new dimeric "condensed type" compound from hydro-
     genolysis products of compression wood lignin. *Mokuzai
     Gakkaishi 21*:639.

Chapter Five

BIOSYNTHESIS OF LIGNIN AND RELATED MONOMERS

GEORG G. GROSS

*Lehrstuhl für Pflanzenphysiologie
Ruhr-Universität, 4630 Bochum, FRG*

INTRODUCTION

   After Payen's fundamental discovery in 1838,[103] that wood does not represent a uniform substance but rather con- sists, in addition to cellulose, of a "true lignous material," more than a century of intensive research was required to establish the aromatic nature of this natural product and to elaborate the principles of its constitution. As a result of these investigations we know now that lignin is a complex polymer originating from the random oxidative polymerization of hydroxylated cinnamyl alcohols (Fig. 1) as principal con- stituents, as discussed in the previous chapter. About 20 years ago, when it became increasingly evident that lignin in fact is derived from substituted cinnamyl alcohols, interest was focussed in a variety of laboratories on the reactions leading to lignin and its precursors. The results obtained by a multitude of tracer experiments revealed the principal pathways from CO_2 to lignin,[8,36,65] thus providing an ample basis for our understanding of the lignification process. However, there remained many questions which could not be answered satisfactorily by these techniques.

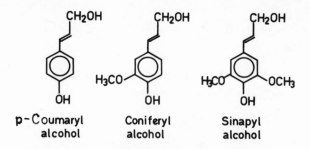

Figure 1.   Structure of lignin monomers.   The cinnamyl alcohol derivatives are characterized by a free OH-group in the 4-position.   The corresponding aromatic residues found in lignin are the *p*-hydroxy, vanillyl and syringyl group, respectively.

More recently, many of these gaps have been filled by the extensive application of enzymatic studies.   The successful isolation and characterization of a variety of enzymes has considerably augmented our knowledge of the reactions and reaction mechanisms involved in lignification.

As a result of both tracer experiments and enzymatic studies we are now able to draw a fairly detailed picture of the pathways leading to the lignin macromolecule.   As shown in Figure 2, this sequence comprises a series of individual minor pathways.   The first among these, the shikimate pathway, leads to the synthesis of aromatic amino acids. There is no doubt that this pathway, which was originally detected in bacteria, operates also in higher plants.[161]   As a constituent of general metabolism, however, it will not be discussed further in this chapter.

The branching point between general metabolism and the true lignification sequence lies in the deamination of phenylalanine to cinnamate.   This latter compound is then substituted in a sequence of hydroxylation and methylation steps.   The resulting cinnamate derivatives are subsequently reduced to the corresponding alcohols *via* intermediate CoA esters.   Finally, these alcohols are polymerized to yield the lignin polymer.

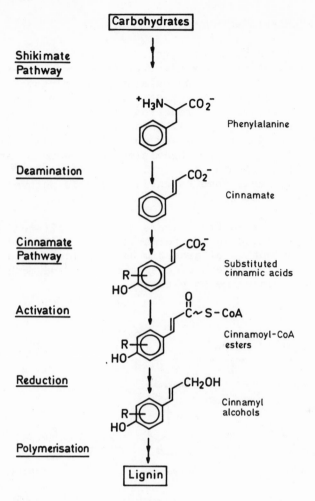

Figure 2.   Schematic representation of the principal steps in lignin biosynthesis.

During the past few years, the most important contributions to the field have been achieved by enzymatic studies. Consequently, emphasis will be put mainly on the enzymology of the lignification process.

THE PHENYLALANINE–CINNAMATE PATHWAY

The biosynthesis of lignin as one of the most abundant secondary plant products diverges from general metabolism by the deamination of L-phenylalanine to cinnamate. This compound is further converted to various substituted cinnamate derivatives in a reaction sequence known as the cinnamate pathway (Fig. 3), which has been summarized in a variety of recent reviews and monographs dealing with the reactions of the entire phenylalanine–cinnamate pathway or special topics within it.[8,12,15,36,65,126,134,136,137,166] Consequently, only some essentials of this pathway are briefly mentioned here.

As can be seen from Figure 3, three different reaction types are involved in the formation of substituted cinnamic acids: deamination of aromatic amino acids, hydroxylations of the aromatic nucleus of cinnamic acids and methylation reactions yielding methoxy groups.

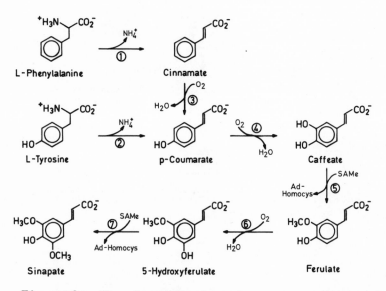

Figure 3.   The phenylalanine–cinnamate pathway.   (1) Phenylalanine ammonia lyase; (2) tyrosine ammonia lyase; (3) cinnamate 4-hydroxylase; (4) p-coumarate 3-hydroxylase; (5, 7) catechol O-methyltransferase; (6) ferulate 5–hydroxylase (hypothetical).

Ammonia Lyases. The deamination of phenylalanine is catalyzed by phenylalanine ammonia lyase (PAL);[97] an analogous enzyme activity (tyrosine ammonia lyase)[79] appears to be confined mainly to grasses. During these reactions, the elements of ammonia are eliminated in an *anti*-periplanar fashion, leading directly to *trans*-cinnamate[60,69] and *trans*-p-coumarate,[26] respectively. According to this mechanism, previous theories on the intermediacy of the corresponding phenylpyruvate and lactate derivatives in this conversion[9] have been ruled out.

The active site of PAL from various sources contains a derivative of dehydroalanine (2-aminoacrylate); recently, a dehydroalanine imine was proposed as the prosthetic group.[63]

PAL operates at an important metabolic branching point. As expected, this key enzyme of phenolic metabolism is controlled by numerous internal and external factors, *e.g.* substrates, products, inhibitors, hormones, light, carbohydrate levels, wounding, or infection, which affect both synthesis and activity of the lyase.[12] With PAL from wheat seedlings, a detailed kinetic analysis of subunit interactions in response to the binding of inhibitors has been reported.[96] Recently, evidence has been presented for the presence of an non-proteinous PAL-inactivator in gherkin seedlings, which is presumably associated with a high molecular compound.[33] Further control over PAL-activity may be exerted through the existence of isoenzymes, as shown for *Quercus*[6] or *Aesculus*.[13]

Hydroxylases. The hydroxylation reactions of the cinnamate pathway comprise the formation of p-coumarate, caffeate and 5-hydroxyferulate (reactions 3, 4 and 6 in Fig. 3). The latter reaction has not yet been found in cell-free systems, but was postulated from tracer experiments. A fourth possible hydroxylation, the formation of tyrosine from phenylalanine, which is known from animal tissues, appears not to occur in higher plants.[62]

The two hydroxylases catalyzing the sequence cinnamate → p-coumarate → caffeate have been studied in detail. Both enzymes are mixed function oxidases or mono-oxygenases (Fig. 4). The first of these, cinnamate 4-hydroxylase, has been identified in extracts from various plants (cf. 68, 143 and literature cited therein). This enzyme is membrane associated, generally requires NADPH as electron donor and possibly a lipid cofactor[10] and is of the cytochrome P_{450} type.[68,104,136]

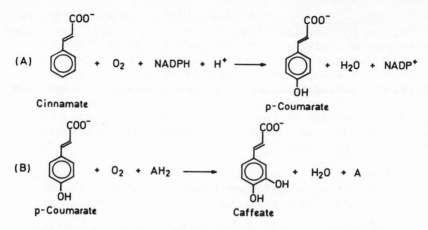

Figure 4. Hydroxylases of the cinnamate pathway. (A)
Cinnamate 4-hydroxylase; (B) *p*-coumarate 3-hydroxylase
(phenolase). $NADPH_2$, ascorbate or reduced pterdines can
serve as the electron donor AH_2.

Recently, a similar cinnamate 2-hydroxylase has been
isolated from chloroplasts of *Melilotus alba*[39] which seems
to be involved in coumarin biosynthesis.

The subsequent hydroxylase, *p*-coumarate 3-hydroxylase
(phenolase) which has recently been purified to apparent
homogeneity,[148] appears to be associated with membranes,
especially with chloroplast lamellae.[3,102] In contrast to
the above enzymes, not only NADPH but also ascorbate or
reduced pteridines can be utilized as cofactors.[136] The
incorporation of molecular oxygen into *p*-coumarate has been
demonstrated with $^{18}O_2$.[37]

In recent kinetic studies on spinach-beet phenolase,[92]
a "ping-pong" mechanism was proposed involving an intermediary oxidized stable enzyme form. Evidence was presented
that the oxidized enzyme is subsequently reduced by catalytic
amounts of caffeate. The resultant caffeoyl-*o*-quinone is
recycled by external electron donors such as the above mentioned cofactors. These results were confirmed in further
experiments where the enzyme was treated with H_2O_2.[149]

However, other mechanisms including the superoxide
radical $O_2^{\cdot-}$ and the hydroxyl radical $HO^{\cdot}$ have also been proposed for this reaction.[58,59]

Methyl Transferases. The third reaction type in the
cinnamate pathway comprises two methylation reactions yielding ferulate and sinapate, respectively (reactions 5 and 7
in Fig. 3). Enzymes catalyzing these reactions have been
isolated from many different plants (literature cited in 83,
105). In most cases, they were found to be *meta*-specific,
but also the methylation of *para* or both *para* and *meta*-hydroxy
groups has been observed (for literature see 156). Recently,
an *O*-methyltransferase has been found in the basidiomycete
Lentinus lepideus.[156] It specifically acts on the *para*-
position. That these enzymes also discriminate between different classes of compounds is demonstrated by the separation
of two methyltransferases from soybean cell cultures which
either act on cinnamate or on flavonoid substrates.[107]

The methylating enzymes of the cinnamate pathway may be
classified as *S*-adenosyl-L-methionine: 3,4-dihydric phenol
3-*O*-methyltransferase or catechol *O*-methyltransferase (OMT).
The reaction catalyzed is illustrated in Figure 5. In experiments on such an enzyme from spinach beet,[105] an enzyme hydrolyzing the *S*-adenosylhomocysteine formed in this reaction was
also detected,[105,106] and it was thought that this degradation might be important in relation to the preceeding methylation reaction.

In investigations on the substrate specificities of OMT's
from gymnosperms and angiosperms,[130] it was found that the
enzyme from *Pinus* preferably methylates caffeate and is almost

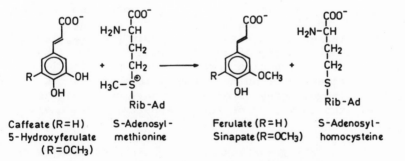

Caffeate (R=H)   S-Adenosyl-   Ferulate (R=H)   S-Adenosyl-
5-Hydroxyferulate   methionine   Sinapate (R=OCH₃)   homocysteine
   (R=OCH₃)

Figure 5. Reaction catalyzed by catechol *O*-methyltrans-
ferase in the biosynthesis of substituted cinnamic acids.

inactive with 5-hydroxyferulate.[83,127] In contrast, the
analogous enzyme from bamboo methylates both these substrates
at approximately equal rates.[129] These two enzyme activities
were called *mono-* and *di-*function OMT, respectively. It was
concluded that these different substrate specificities might
represent at least one biochemical explanation for the known
significant differences between the lignin from gymnosperms
and angiosperms (cf. 66).

Enzyme Complexes. Finally, it should be mentioned that
it has recently been proposed that the enzymes of the entire
phenylalanine-cinnamate pathway might form a membrane-
associated multi-enzyme complex.[137] Similar conclusions have
also been drawn from experiments on the formation of hydroxy-
cinnamic acids from phenylalanine.[17,18]

REDUCTION OF CINNAMIC ACIDS
TO ALCOHOLS

General Considerations. Many details of the cinnamate
pathway were discovered without any knowledge of the enzymes
involved in this sequence. In contrast, investigations on
the subsequent steps in lignin biosynthesis clearly demonstrated the limitations of tracer studies. In spite of the
early finding that radioactive ferulate administered to wheat
plants was reduced to coniferyl alcohol, with the probable
participation of intermediate coniferyl aldehyde,[67] no information on the mechanism of this conversion could be obtained
by this technique. This problem has been clarified only
recently by the application of enzymatic studies.

Any consideration concerning the mechanism of the reduction of carboxylic acids must take into account that such a
reaction represents a highly endergonic process, involving a
change in free energy of about 13 kcal mol^{-1}.[70] Thus, the
enzymatic reduction of a carboxyl group principally requires
the participation of an activated intermediate. To our present knowledge, three alternatives have been realized in
nature: the formation of acylphosphates, acyladenylates and
thioesters (*e.g.* CoA esters).[165]

Whereas the reduction of acylphosphates is a rather
common reaction in numerous metabolic sequences, the participation of acyladenylates appears to occur only sporadically.

This latter reaction has been found in yeast for the reduction of 2-amino-adipate[123,132] and in *Neurospora* for the reduction of benzoic and cinnamic acids.[44]

Aldehyde formation *via* the reduction of aliphatic acyl-CoA derivatives, in contrast, is obviously of especial importance. Enzymes catalyzing this reaction have been isolated from bacteria,[11,108,152] animal tissues,[71,135] algae,[77] and higher plants.[78] A special case is represented by the reduction of 3-hydroxy-3-methylglutaryl-CoA. This reaction leads directly to the alcohol (mevalonate) *via* a hemithioacetal[111] with the consumption of 2 moles of NADPH and without the liberation of aldehyde. That other thiols than CoA can also be utilized in reactions of this type is demonstrated by the reduction of S-formylglutathione to formaldehyde.[146]

Concerning the reduction of cinnamic acids, CoA esters have long been postulated as the activated intermediates.[49,98,99,163] This assumption was mainly based on investigations of the degradation of cinnamic acids which yielded benzoic acids, aldehydes and alcohols with the simultaneous liberation of acetate units.[74,151,163] By analogy to β-oxidation of fatty acids, these transformations were assumed to proceed *via* acyl-CoA derivatives which were reduced to the corresponding aldehydes.

In this context, it should be mentioned that cinnamoyl-CoA esters have also been suggested as intermediates in the biosynthesis of flavonoids,[41] stilbenes,[49,73] various amides and esters,[49,98] and simple phenols.[163] Experimental proof of these hypotheses has been achieved by enzymatic studies in several cases. It was shown that the syntheses of the flavanone naringenin,[80,81] styrylpyrones,[82] p-coumaroyl-quinate,[119] and chlorogenic acid[119,141] involve cinnamoyl-CoA esters. Participation of these activated compounds has also been indicated in the β-oxidation of cinnamic acids.[2] Other authors working with membrane-bound enzymes were unable to find a CoA dependence for this reaction;[17,18,52,74,86] it was considered, however, that the enzyme preparations might contain endogenous pools of this coenzyme.[74,86] Thus, the previously suggested central role of cinnamoyl-CoA esters in plant phenolic metabolism[164,165] is now established unequivocally.

Considering the above results, it is not surprising that the reduction of cinnamic acids was recently found to proceed

via CoA esters, too. This was demonstrated for the first
time by Mansell *et al.*[90] with cell-free extracts from *Salix*
which catalyzed the reduction of ferulate to coniferyl alcohol only in the presence of CoA. This result has been confirmed by several reports on the CoA-dependent reduction of
p-coumarate or ferulate to the corresponding alcohols with
enzyme preparations from cell suspension cutures of *Glycine*,[22]
lignifying tissues of *Forsythia*[140] and root tissue of
Brassica.[116]

The probable intermediary function of cinnamoyl-CoA
esters was further substantiated by the observation that
cinnamyl alcohol formation also occurred when both the cinnamate substrate and the cofactors ATP and CoA were substituted by authentic *p*-coumaroyl-CoA[22,140] or feruloyl-CoA.[23,48]

Apart from *p*-coumaryl or coniferyl alcohol as main products, some formation of the corresponding aldehydes was also
observed on several occasions.[89,116,140] These results indicated that, in accordance with earlier *in vivo* experiments,[67]
the biosynthesis of cinnamyl alcohols involves two consecutive
reductive steps.

The results of these enzymatic studies suggested a biosynthetic pathway along which the combined action of a
cinnamate-activating CoA ligase, a cinnamoyl-CoA reductase
and an alcohol dehydrogenase gives rise to the formation of
cinnamyl alcohols. The existence of such a reaction sequence
was demonstrated for the first time by the isolation of three
enzymes from lignifying tissues of *Forsythia*[48] (Fig. 6). As
discussed above, the activity of the first enzyme of this
sequence, an acyl-CoA ligase, is not necessarily restricted
to the synthesis of lignin monomers. However, since the inter
mediary formation of energy-rich acyl derivatives is an
essential step in the reduction of cinnamic acids, it appears
rational in this context to include the activating reaction
into a common pathway. Some properties of the enzymes involve
in this metabolic sequence will be discussed in the following
sections.

Cinnamate:CoA Ligases. Cinnamate:CoA ligases have
been identified in extracts from a wide variety of different
plants.[47] More specific reports on enzymes of this type are
summarized in Table 1. An analogous enzyme was recently
detected in the basidiomycete *Polyporus hispidus*.[147] Several
of these ligases have been characterized in detail. From

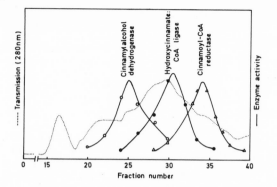

Figure 6. Separation of enzymes from *Forsythia* involved in the activation and reduction of cinnamic acids by gel filtration on Sephadex G-200. Transmission and enzyme activities in arbitrary units.

these investigations, it becomes evident that their basic properties are fairly uniform. A molecular weight of about 55,000 has been reported for both isoenzymes from soybean[75,84] and the *Forsythia* enzyme.[51] A strict dependence on ATP has been observed throughout; other nucleoside triphosphates cannot substitute for this coenzyme.[51,75,115] Mg^{2+} is generally required as cofactor, but can be replaced by Mn^{2+} or CO^{2+} without loss of activity.[75,115]

Inconsistency exists concerning the effect of thiol blocking reagents. Whereas the ligases from soybean remained unaffected,[75] complete inhibition of the enzyme from swede root was observed with 1 mM N-ethylmaleimide or p-chloromercuribenzoate.[115] A stimulatory effect of added thiols was observed with the ligase from *Forsythia*.[51]

The substrate specificity of cinnamate:CoA ligases from several plant species has been studied in detail. As shown in Table 2, the enzymes isolated from *Forsythia* and *Brassica*, as well as the isoenzyme 2 from soybean exhibit an almost identical affinity towards differently substituted cinnamic acids. With one exception (cinnamate in the case of *Glycine*), CoA

Table 1. Cinnamate:CoA ligases from higher plants

Species	Tissue Examined	Remarks	References
Acer saccharinum, sugar maple	Young xylem	---	47
Apium graveolens, celery	Vascular bundles	Not detected in parenchyma	47
Beta vulgaris, spinach beet	Leaves	Activity dependent on age and growth conditions	153,154
Brassica napo-brassica, swede root	Roots	Activated by aging or ethylene	112,115,118
Brassica oleracea, red cabbage	Seedlings	Unaffected upon illumination	91
Fagopyrum esculentum, buckwheat	Seedlings	Unaffected upon illumination	91
Forsythia suspensa	Young xylem	---	47,48,51
Glycine max, soybean	Cell suspension cultures	Two isoenzymes; activity increased upon illumination	84
Lilium longiflorum	Anthers	Activity correlated with flavonol synthesis	23,56,75,76
Lycopersicum esculentum, tomato	Fruits	---	142
Narcissus pseudonarcissus	Anthers	Activity correlated with flavonol synthesis	119
Petroselinum hortense, parsley	Cell suspension cultures	Activity increases upon illumination or change in culture conditions	53-55,57
Petunia hybrida	Callus cultures	Two isoenzymes; activity dependent on age and medium	109,110
Raphanus sativus, radish	Seedlings	Unaffected upon illumination	91
Sinapis alba, white mustard	Seedlings	Unaffected upon illumination	91
Solanum tuberosum, potato	Tubers	Only present in illuminated, aged disks	117
Taxus baccata, yew	Young xylem	---	47
Tulipa sp., tulip	Anthers	---	142

esters are formed only from hydroxylated cinnamic acids,
whereas cinnamate and its fully methoxylated derivatives are
inactive. Among the substrates of importance with respect to
lignification, p-coumarate and ferulate show comparable high
reaction rates. On the other hand, no or only minimal activation occurred with sinapic acid. Identical results were
obtained with enzyme preparations from *Acer* and *Taxus*.[47]
Similar specificities were found for the ligases from potato
tubers[117] or tomato fruits,[119] but in these cases some activity with mono- or dimethoxycinnamic acids was also observed.

In contrast to the above enzymes, the isoenzyme 1 from
soybean exhibits a quite different substrate specificity.
This enzyme preferentially acts on p-coumarate, but sinapic
and methoxycinnamic acids are also converted with high reaction rates.

In further experiments it was shown that the ligase from
Forsythia is inactive with a multitude of various phenyl-
acetic, benzoic and aliphatic mono- or dicarboxylic acids, as
well as with aliphatic or aromatic amino acids. As expected,
glucoferulic acid, lacking a free OH-group, was not activated.[51] The same exclusive specificity towards cinnamate
substrates was found with extracts from *Acer* and *Taxus*[47] and
also seems to apply to the enzymes from *Glycine*.[75] With the
latter enzyme it was further shown that only *trans-p*-coumarate,
but not the *cis*-isomer is activated.[75]

During the formation of CoA esters, ATP undergoes cleavage
to AMP and pyrophosphate.[51,75] Stoichiometric investigations
revealed that the consumption of ferulate and ATP, and the
formation of feruloyl-CoA, AMP and pyrophosphate occur in
equimolar ratios.[51] It thus appears most likely that the synthesis of cinnamoyl-CoA esters occurs in an analogous manner
to the common mechanism of fatty acid activation. This reaction is known to proceed *via* intermediary acyladenylates
which further react with CoA to form the thioesters.

The existence of such a reaction mechanism can easily be
demonstrated with the aid of the ^{32}P-pyrophosphate-ATP exchange
reaction. After preliminary experiments,[154] unequivocal evidence was presented in recent detailed studies[51] that this
mechanism, in fact, also applies to the synthesis of cinnamoyl-
CoA esters. This indirect proof of an intermediary acyladenylate was further substantiated in experiments with authentic feruloyl-AMP.[51] Upon incubation of this compound together

Table 2.   Substrate specificity of cinnamate:CoA ligases from various plants

Substrate	Relative activity (%)*			
	Glycine[75]		Brassica[115]	Forsythia[47]
	Isoenzyme 1	Isoenzyme 2		
Cinnamate	5	24	0	0
o-Coumarate	–	–	15	26
m-Coumarate	–	–	21	54
p-Coumarate	177	104	137	104
Caffeate	100	90	93	64
Ferulate	100	100	100	100
Isoferulate	74	104	83	87
3,4,5-Trihydroxycinnamate	–	–	–	43
Sinapate	81	0	2	0
p-Methoxycinnamate	44	0	0	0
3,4-Dimethoxycinnamate	159	0	0	0
3,4,5-Trimethoxycinnamate	81	0	0	0

*Data refer to the activity observed with ferulate = 100%.

with CoA, the enzymatic formation of feruloyl-CoA was
observed. After the addition of pyrophosphate, a reverse
reaction leading to ATP and ferulate occurred. As expected,
the addition of CoA drastically inhibited this latter reaction by competition of pyrophosphate and CoA for the acyladenylate.

Summarizing the above results, the activation of cinnamic
acids in higher plants can be formulated as shown in Figure 7
for ferulate. According to their substrate specificities,
most of the enzymes catalyzing this reaction may be classified as hydroxycinnamate: CoA ligases (AMP) (EC 6.2.1.-) or,
as proposed especially for the ligases from soybean,[75] as
trans-p-coumarate: CoA ligases.

Cinnamoyl-CoA Reductases. Enzymes catalyzing the reduction of cinnamoyl-CoA esters have been reported from cell
suspension cultures of soybean,[22] lignifying tissues of
Forsythia[48,140] and aged swede root tissue.[112,118] In recent
more detailed studies it was shown that the enzymes from
Forsythia[46] and *Glycine*[157] are similar with respect to their
molecular weights (*ca.* 40,000 and 38,000, respectively) and
their high specificities for NADPH as cofactor. Further,
almost identical affinities towards various cinnamoyl-CoA
esters have been determined for these enzymes (Table 3).
Feruloyl-CoA is by far the most active substrate, whereas the
CoA esters of *p*-coumarate and sinapate, which also are of
interest with respect to lignification, are reduced at

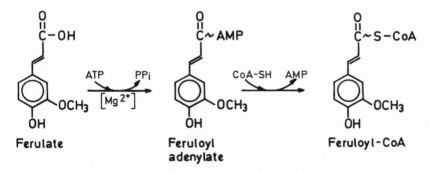

Figure 7. Activation of cinnamic acids by hydroxycinnamate: CoA ligase. Reaction shown with feruloyl-CoA
as substrate.

comparatively low rates. An analogous enzyme which has been
purified from extracts of *Acer pseudoplatanus* (Gross, unpublished) shows an even more pronounced specificity towards
feruloyl-CoA. In this case, only a slight reaction occurred
with sinapoyl-CoA and none with *p*-coumaroyl-CoA.

By analogy to the ligases described above, the reductases appear to act exclusively on cinnamyl substrates; no
or only minimal activity was observed with the CoA esters of
several benzoic of aliphatic acids.[46,157]

In further investigations, coniferyl aldehyde was demonstrated unequivocally as the reaction product formed from
feruloyl-CoA.[46,157] Thus the results from previous *in vivo*[67]
and *in vitro* experiments[48,90,140] have been confirmed. This
also excludes the occasionally discussed possibility[22,42]
that, by analogy to mevalonate biosynthesis, the CoA esters
might be reduced directly to the alcohol level.

Based on these results, the systematic designation
cinnamoyl-CoA:NADP oxidoreductase, or, according to the
best substrate, feruloyl-CoA:NADP oxidoreductase was proposed.[46] Other authors[157] have stressed that, due to the
undemonstrated reversibility of this reaction, the name

Table 3. Substrate specificity of cinnamoyl-CoA
reductases from higher plants

Substrate	Relative Activity (%)[*]		
	Forsythia[46]	*Glycine*[157]	*Acer*[**]
Cinnamoyl-CoA	10	10	0
p-Coumaroyl-CoA	20	13	0
p-Methoxycinnamoyl-CoA	25	–	–
Caffeoyl-CoA	10	20	4
Feruloyl-CoA	100	100	100
3,4-Dimethoxycinnamoyl-CoA	40	–	34
5-Hydroxyferuloyl-CoA	20	25	–
Sinapoyl-CoA	20	30	8

[*]Data refer to the activity observed with feruloyl-
CoA = 100%.
[**]Gross, unpublished.

cinnamoyl-CoA:NADPH reductase would be more appropriate.
Doubtless, the term "reductase" represents an illustrative
trivial name, but enzymes of this type are generally classified as oxidoreductases. The reaction catalyzed by this
enzyme is illustrated in Figure 8.

In experiments with specifically ^{3}H-labeled A and B forms
of NADPH it was found that the B-hydrogen atom (H_s) is transferred to the cinnamyl moiety.[46] Thus, cinnamoyl-CoA reductase belongs to the B-group of NAD(P)-specific dehydrogenases
(Fig. 9).

Cinnamyl Alcohol Dehydrogenases. As the final step in
cinnamyl alcohol biosynthesis, reduction of the corresponding
aldehydes is mediated by an alcohol dehydrogenase. Compared
to the multitude of such enzymes known to catalyze the reversible oxidation of aliphatic alcohols, little is known about
alcohol dehydrogenases acting on aromatic substrates. Several
of these enzymes have been isolated from bacteria, fungi and
animal tissues (literature cited in 20, 89). Only recently,
the existence of aromatic alcohol dehydrogenases in a wide
variety of higher plants has been demonstrated.[88,89] Further
reports on such enzymes are summarized in Table 4.

The most detailed studies have been reported on the
enzymes from *Forsythia*[89] and *Glycine*.[159] The dehydrogenase
from *Forsythia* has a molecular weight of *ca.* 80,000. From
Glycine, two isoenzymes were isolated with molecular weights
of *ca.* 43,000 and 69,000, respectively. All these enzymes

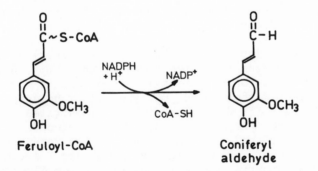

Feruloyl-CoA   Coniferyl
                                          aldehyde

Figure 8. Reduction of cinnamoyl-CoA esters by
cinnamoyl-CoA reductase. Reaction shown with feruloyl-CoA
as substrate.

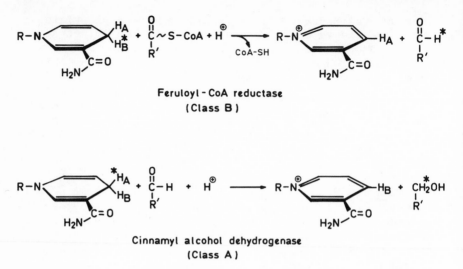

Feruloyl - CoA reductase
(Class B)

Cinnamyl alcohol dehydrogenase
(Class A)

Figure 9.   Stereospecificity of hydrogen transfer catalyzed by the two reductases involved in cinnamyl alcohol
biosynthesis.

were found to be exclusively dependent on NADP(H).   The
Forsythia enzyme was shown to belong to class A of NAD(P)-
specific oxidoreductases (Fig. 9).   As mentioned above, the
preceding cinnamoyl-CoA reductase is of class B.   Analogous
results were observed for the reduction of aromatic acids in
Neurospora, where the aromatic acid reductase is of class B[50]
and the subsequent alcohol dehydrogenase belongs to class A
(Gross, unpublished).   These results are contradictory to the
assumption[19] that consecutive dehydrogenases in a metabolic
pathway should have the same stereospecificity.

Determination of the substrate specificity of the alcohol
dehydrogenase from *Forsythia* revealed that this enzyme catalyzes the oxidation of cinnamyl alcohol and of a variety of
hydroxylated, methoxylated, methylated, or halogenated phenyl-
allyl alcohols.[89]   Fully methoxylated substrates were, in
contrast to the enzymes discussed in the preceding sections,
converted most readily.   Analogous results were observed for
the reduction of various cinnamyl aldehydes.   A similar specificity was reported for the isoenzyme 2 from *Glycine*.   In
contrast, the isoenzyme 1 was found to catalyze exclusively
the oxidation of coniferyl alcohol[159] (Table 5).

Table 4. Cinnamyl alcohol dehydrogenases from higher plants

Species	Tissue Examined	Remarks	References
Apium graveolens, celery	Vascular bundles	Predominantly in xylem	47
Brassica napo-brassica, swede root	Roots	Two (3 ?) isoenzymes, specificity unknown	112, 118
Cryptomeria japonica	Callus culture	---	94
Fagopyrum esculentum, buckwheat	Seedlings	---	Gross, unpubl.
Forsythia suspensa	Lignifying tissue	Specific for cinnamyl substrates; tissues contain also 2 unspecific enzymes	48, 89, 140
Glycine max, soybean	Cell suspension cultures	Two isoenzymes; specific for cinnamyl substrates	159
Petunia hybrida	Callus culture	Activity dependent on age and benzyladenine	109
Populus nigra, poplar	Shoots	---	94
Solanum tuberosum, potato	Tubers	Unspecific; also a wide variety of benzaldehydes is reduced	20

Table 5. Substrate specificity of cinnamyl alcohol dehydrogenases from higher plants

Substrate	Relative Activity (%)*		
	Forsythia[89]	_Glycine_[159]	
		Isoenzyme 1	Isoenzyme 2
Cinnamyl alcohol	39	0	55
p–Coumaryl alcohol	112	0	1
p–Methoxycinnamyl alcohol	290	–	–
Coniferyl alcohol	100	100	100
3,4–Dimethoxycinnamyl alcohol	384	0	44
3,4–Methylenedioxycinnamyl alcohol	383	–	–
Sinapyl alcohol	(*)**	0	95
4–Methylcinnamyl alcohol	128	–	–
2,4–Dimethylcinnamyl alcohol	90	–	–
4–Chlorocinnamyl alcohol	55	–	–
2,4–Dichlorocinnamyl alcohol	13	–	–
2,6–Dichlorocinnamyl alcohol	6	–	–
4–Bromocinnamyl alcohol	51	–	–

*Data refer to the activity observed with coniferyl alcohol = 100%.
**Sinapyl alcohol is converted rapidly; color development of the aldehyde formed prevents quantitative determination under the assay conditions employed.

Considering the frequently encountered nonspecificity of many alcohol dehydrogenases, it was surprising to find that the enzyme from *Forsythia* was absolutely inactive towards a wide variety of aliphatic aldehydes or alcohols, as well as benzaldehydes or alcohols. Also no reaction was observed with coniferin (glucoconiferyl alcohol) or hydroconiferyl and homovanillyl alcohols.[89] Thus, this alcohol dehydrogenase seems virtually specific for substrates having both an aromatic ring and a propene side chain. The same situation appears to apply for the enzymes from *Glycine*.[159]

According to the results discussed above, the reversible reduction of cinnamyl aldehydes occurs as depicted in Figure 10 for coniferyl aldehyde. The equilibrium of this reaction,

$$(K = \frac{[\text{coniferyl aldehyde}]\ [\text{NADPH}]\ [\text{H}^+]}{[\text{coniferyl alcohol}]\ [\text{NADP+}]} = 2.8 \cdot 10^{-9}),[159]$$

highly favours the formation of cinnamyl alcohols as the immediate lignin precursors.

POLYMERIZATION OF CINNAMYL ALCOHOLS TO LIGNIN

The Polymerization Process. Polymerization of cinnamyl alcohols, the final step in the lignification sequence, is obviously initiated by oxidation of their phenolic hydroxyl groups. It is now widely accepted that a peroxidase functions as the natural catalyst of this reaction, since this

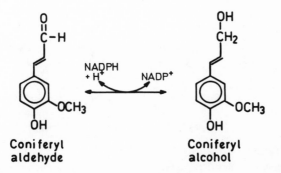

Figure 10. Reversible reduction of cinnamyl aldehyde by cinnamyl alcohol dehydrogenase. Reaction shown with coniferyl aldehyde as substrate.

enzyme is widely distributed in higher plants and has unequivocally been demonstrated in lignifying tissues.[61] This
assumption is further strengthened by the well documented
existence of cell wall-bound[16,38,139,158] and even cell wall-
specific isoperoxidases.[85,87] For recent reviews on the role
and distribution of peroxidase see 136 and 137 in literature
cited.

  The attack of peroxidase causes the formation of phenoxyradicals which have long ago been theorized to be involved in
lignification.[28,29] These radicals exist in several mesomeric
forms (Fig. 11), thus being stabilized. Even in aqueous
media, comparatively long half-lives have been measured for
the free radicals.[35,134,138] Dimerization especially of
radicals a, b and d, followed by further dehydrogenation and
polymerization reactions, finally leads, *via* oligomeric intermediates, to the lignin macromolecule. Especially the work
of Freudenberg on the constitution of this complex macromolecule has led to the proposal of a structural scheme which
is now generally accepted.[36,126]

  According to this radical mechanism, lignification is
only initiated enzymatically and thus differs from the formation of all other biopolymers which are synthesized under
strict enzymatic control. Consequently, no valid conclusions
can be drawn from polymerization experiments with peroxidase
preparations from special tissues, as has been reported occasionally.[94,101] Further, it cannot be expected to find a
correlation between peroxidase levels and lignification as
long as peroxidase is present in the tissues examined (cf. 7,
21, 40).

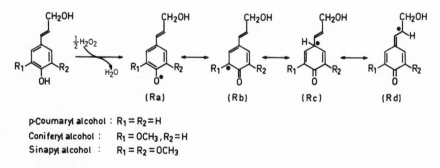

p-Coumaryl alcohol : $R_1 = R_2 = H$
Coniferyl alcohol : $R_1 = OCH_3, R_2 = H$
Sinapyl alcohol : $R_1 = R_2 = OCH_3$

Figure 11.   Formation of phenoxyradicals by peroxidase.

Origin of Hydrogen Peroxide. H_2O_2 (required as substrate in the peroxidative oxidation of cinnamyl alcohols) has recently been detected in high concentrations in xylem and bark of *Populus gelrica.*[122] In spite of the existence of several systems generating H_2O_2 in the cytoplasm (cf. 136, 137), its natural origin in the cell wall remained obscure. An interesting approach to this question was recently reported by Elstner and Heupel.[27] These authors were able to demonstrate the formation of H_2O_2 by isolated cell walls from horseradish at the expense of NAD(P)H. This complex reaction involves the superoxide free radical ion $(O_2{\cdot}^-)$ as an intermediate, requires Mn^{2+} and it's stimulated by monophenols. This reaction was ascribed to the activity of a cell wall-bound peroxidase[27] and has also been reported for isolated soluble peroxidases.[1,72]

The problem now arises: by which mechanism(s) is the NAD(P)H required as electron donor provided to the bound peroxidase? In recent investigations on this question,[45] it was found that horseradish cell walls appear to contain bound malate dehydrogenase. Under the catalysis of this enzyme, sufficient amounts of NADH are formed to allow the subsequent synthesis of H_2O_2 by the peroxidase.

Considering the known malate-oxalacetate shuttles involved in the transport of reducing equivalents through the membranes of chloroplasts or mitochondria,[64,93,144] one might also visualize such a mechanism across the plasmalemma. By this means, cytoplasmic reducing equivalents would be transported as malate through the outer cell membrane into the cell wall.

The reactions discussed above are illustrated in Figure 12. It should be noted that, apart from the enzymatic reactions which are thought to initiate these reactions by the formation of $O_2{\cdot}^-$, non-enzymatic reactions must also take place. This was concluded from the fact that superoxide dismutase does not, as expected, stimulate the formation of H_2O_2, but rather partially inhibits this reaction.[27] Thus, also other mechanisms utilizing $O_2{\cdot}^-$ must be involved.

Such a scheme would offer several advantages. For instance, no transport of toxic H_2O_2 from the cytoplasm to the cell wall is required, since this compound is synthesized directly at the proper site. Further, the lignification process could easily be regulated, either by producing H_2O_2 only

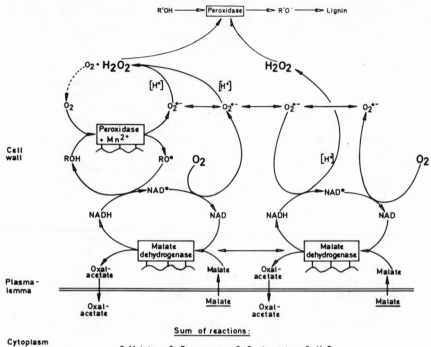

Figure 12. Hypothetical scheme of reactions generating hydrogen peroxide in cell walls. The sequence of reactions is thought to be initiated by the action of peroxidase in a reaction stimulated by Mn^{2+} and monophenols. Most likely, also the subsequent non-enzymatic steps (right part of the scheme) involve participation of these cofactors (Elstner and Gross, in preparation).

within the lignifying areas or by destroying H_2O_2 in the non-lignifying parts through the action of a wall-bound catalase which has also been found in horseradish cell walls.[27] One also could suppose that both ROH and ROH in this scheme represent hydroxycinnamyl alcohols. In this case, the entire reaction sequence would depend on the availability of these lignin precursors.

At present, the proposed scheme is still hypothetical. It should be noted, however, that recent preliminary investigations have revealed the presence of both malate dehydrogenase and peroxidase in isolated cell walls from actively lignifying tissues of *Forsythia*.

TISSUE SPECIFICITY OF ENZYMES
INVOLVED IN LIGNIFICATION

In the preceding sections, the properties of a variety
of enzymes which catalyze the formation of monomeric lignin
units were discussed. In most cases, these enzymes were
extracted from actively lignifying tissues, and their marked
specificity towards phenylpropanoid substrates suggested an
intimate relation to xylogenesis. These conclusions have
been further substantiated in experiments on the tissue specificity of the enzymes involved in the lignification sequence.

It has previously been shown that PAL is predominantly
located in the xylem of several plants, and that phenylalanine or cinnamate are preferably incorporated into this
tissue rather than into parenchyma or cambia.[121] Similarly,
PAL activity from peripheral xylem layers of poplar was found
to be about 30-times higher than that in phloem (cf. 47). It
was further shown that these extracts from lignifying tissue
also catalyzed the hydroxylation of p-coumarate to caffeate
as well as the methylation of the latter compound to ferulate.

Moreover, in experiments with celery (*Apium graveolens*)
it was found that hydroxycinnamate:CoA ligase was confined
exclusively to extracts from isolated vascular bundles; no
enzyme activity was observed in the parenchyma.[47] In further
investigations, xylem and phloem of vascular bundles were processed separately. As shown in Table 6, the activities of
the enzymes involved in the activation and reduction of cinnamic acids were found to be predominantly located in the
xylem. The least difference between these tissues was observed
in the case of alcohol dehydrogenase. However, in analogous
experiments with *Heracleum*,[89] an approximately 10-fold concentration of this enzyme was found in the xylem.

From these results it becomes obvious that lignifying
tissues possess the potential for the entire reaction sequence
leading from phenylalanine to cinnamyl alcohols as the immediate lignin precursors. Similar conclusions have also been
drawn by several other authors.[4,121] Thus, the suggestion of
Freudenberg[34] is no longer tenable that cinnamyl alcohols
should be synthesized in the cambium or phloem and transported
in the form of glucosides into the xylem where they, after the
attack of β-glucosidase, are polymerized to lignin. This
hypothesis has also been questioned by Brown[8] considering the
only limited occurrence of these glucosides (*e.g.* coniferin)
in higher plants.

Table 6. Distribution of enzymes involved in the reduction of cinnamic acids in vascular bundles of celery (*Apium graveolens*)

Enzyme	Substrate	Specific Activity (mU/mg protein)			Total Activity (mU/g tissue)		
		X*	P*	X/P	X	P	X/P
Hydroxycinnamate: CoA ligase[47]	*p*-Coumarate	2.0	0.18	11.1	1.9	0.18	10.6
	Ferulate	2.5	0.14	17.8	2.4	0.14	17.1
Cinnamoyl-CoA reductase**	Feruloyl-CoA	1.4	~0	–	1.3	~0	–
Cinnamyl alcohol dehydrogenase**	Coniferyl alcohol	26.0	13.9	1.9	25.0	14.5	1.7

*X = xylem; P = phloem.
**Gross, unpublished.

TAXONOMIC CONSIDERATIONS

It is well known that lignin from various taxonomic
classes differs with respect to its composition. Angiosperm
lignin is characterized by a high portion of syringyl residues. For instance, analysis of beech lignin showed a ratio
of about 4:56:40 for p-hydroxybenzyl, guaiacyl and syringyl
residues, respectively.[100] In contrast, gymnosperm lignin
is almost devoid of the latter group; for spruce lignin a
ratio of about 14:80:6 has been determined.[36] Similar data
have also been reported for pteridophyta.[30] Previous assump-
tions on the existence of a "primitive" moss lignin being
mainly derived from p-coumaryl units[5] have recently been
seriously questioned.[31,32] One must assume that, with the
eventual exception of some giant mosses,[131] this group contains other phenolic cell wall constituents, perhaps like the
recently identified sphagnum acid[145] or some dimeric compounds.[31,32]

How do the above reported enzyme activities agree with
these data? It was found that formation of feruloyl-CoA was
mediated by extracts from a wide variety of gymnosperms and
angiosperms, whilst no activation of ferulate occurred with
Sphagnum.[47]

Feruloyl-CoA reductase has been detected, though occasionally at only low levels, in lignifying tissues from
gymnosperms (4 species) and dicotyledons (7 species) (Gross,
unpublished).

Detailed studies have been carried out on the taxonomic
distribution of cinnamyl alcohol dehydrogenase. This enzyme
has been found in a wide variety of species from bryophyta,
pteridophyta, gymnosperms, monocotyledons, and dicotyledons.[89]
In further studies, cell-free extracts from numerous species
from these taxonomic groups were subjected to starch gel
electrophoresis.[88] Apart from multiple nonspecific alcohol
dehydrogenases, only a single NADP-specific cinnamyl alcohol
dehydrogenase was found in most plants. Multiple forms of
this enzyme were observed only within the genus *Salix*.

Thus, the existence of these enzymes appears to be a
common feature of lignifying plants. Moreover, on summarizing
the data on the distribution of CoA ligase[47] and alcohol
dehydrogenase[88,89] it becomes evident that there exists a
close correlation between enzyme levels and the degree of

lignification. In general, enzyme activities from woody gymnosperms and dicotyledons are appreciably higher than those from the usually herbaceous members of the other groups. Tissue cultures showed enzyme levels of the same order as found for monocotyledons. It is interesting to note in this context that an analogous correlation to woodiness was reported for the contents of shikimate and quinate in a wide variety of higher plants.[162]

In contrast to these obvious correspondences, the following point demands a closer discussion. As mentioned above, angiosperm lignin contains a considerable amount of syringyl units derived from sinapyl alcohol. Sinapic acid, however, is not or only slightly activated by most of the reported ligases. On the other hand, participation of these enzymes in the lignification process must be concluded from their close association with lignifying tissues. Similar considerations also apply to cinnamoyl-CoA reductase, which converts feruloyl-CoA much better than the CoA esters of p-coumarate or sinapate. Only with alcohol dehydrogenase have such limitations of the substrate specificity not been observed.

These enzyme activities allow the biosynthesis of the monomers usually found in gymnosperm lignins. It was found, in fact, that tissue slices from *Pinus* or *Gingko* are capable of reducing ferulate, but not sinapate.[94] On the other hand, alcohol dehydrogenase isolated from gymnosperms catalyzed, like the analogous enzymes from angiosperms, the reduction of both coniferyl and sinapyl aldehyde. The typical composition of gymnosperm lignins thus may be attributed to several factors: the possible absence of ferulate 5-hydroxylase,[128] the poor affinity of OMT towards 5-hydroxyferulate[127] and the lack of activation and/or reduction of sinapate.

However, how can the high proportion of syringyl residues characteristic of angiosperm lignins be brought in line with the observed enzyme specificities? This question was answered by investigations[94] which showed that sinapate was reduced only by tissues of fully differentiated angiosperms, but not by less differentiated tissues or cell cultures. Accordingly, almost no syringyl lignin was found in the latter tissues.

These results agree well with previous reports that the methoxyl content of angiosperm lignins increases during maturation. The characteristic high amount of syringyl units occurs only in the wood of fully differentiated plants. It

was further reported that lignin from the youngest annual
increments next to the cambium has a significantly lower
methoxyl content (literature cited in 125). In preliminary
experiments (Gross, unpublished), it was found that Björkman
lignin from cambial regions of *Forsythia* has only about one-
half of methoxy groups as compared to lignin from several-
year-old stems.

These results indicate that sinapate is not activated or
reduced during the early stages of lignification. Accordingly,
most of the above reported enzymes which have been isolated
from undifferentiated tissues or very young tissues just initiating lignification, have none or only little affinity for
this substrate. Thus it will be of interest to see whether
enzymes can be extracted from fully differentiated woody
tissues which are capable of reducing sinapate.

CONCLUDING REMARKS

As discussed in the preceding sections, there is ample
evidence that lignifying tissues contain all the enzymes
necessary to mediate the biosynthesis of this important
natural product. Except for the still hypothetical ferulate
5-hydroxylase, the enzymes catalyzing the individual steps in
this sequence have been isolated and characterized.

Accordingly, it is very unlikely that the "metabolically
active insoluble esters" of cinnamic acids found in wheat
plants are directly involved in the biosynthesis of lignin or
its precursors, as has occasionally been assumed.[8,24,25,165]
These insoluble esters probably represent the *p*-coumarate and
ferulate groups associated with grass lignins.[124] An interesting alternative was recently discussed by Stafford.[137] She
proposed that these esters might be multienzyme complexes of
cinnamate metabolism being overloaded with exogeneous substrates.

Whereas we are now able to draw a fairly detailed picture
of the biosynthetic events involved in lignification, no conclusive statements can be made concerning the metabolic regulation of this pathway, albeit a variety of endogenous or
exogenous factors is known which directly or indirectly affect
this process.[8,155]

Hormones, for example, appear to be of importance in
xylogenesis. Gibberellin was found to promote both lignin

synthesis and PAL activity.[14] Kinetin stimulates lignification,[4],[150] possibly by a general activation of phenylpropanoid
metabolism.[4] Lignin formation in tissue cultures grown in
the presence of this hormone was paralleled by enhanced activity of PAL[120] and O-methyltransferase.[160] Benzyladenine
retards lignification, eventually by decreasing peroxidase
activity, since free coniferyl alcohol was detected in callus
cultures from *Castanea* in the presence of this growth factor.[15]
A stimulatory effect of ethylene on lignin synthesis was
observed in swede roots,[113],[114] paralleled by a dramatic
increase of the activities of PAL, cinnamate hydroxylase and
cinnamate: CoA ligase. A moderate stimulation was found of
shikimate dehydrogenase, O-methyltransferase and feruloyl-CoA
reductase, whereas the activities of cinnamyl alcohol dehydrogenase and peroxidase remained unaffected.[112]

The effect of light on lignification is difficult to
assess. In tissue cultures, illumination concomitantly
increases the activities of PAL, cinnamate hydroxylase and
isoenzyme 1 of cinnamate: CoA ligase,[23],[43],[55] but these
enzymes are also involved in flavonoid biosynthesis.[133] A
similar light effect has been observed along the pathway leading to hispidine in a basidiomycete.[95] On the other hand, no
stimulation by light was observed with respect to the levels
of isoenzyme 2 of CoA ligase (the substrate specificity of
this enzyme closely resembles those isolated from xylem tissues), O-methyltransferases or the enzymes catalyzing the two
reductive steps yielding cinnamyl alcohols.[23] Furthermore,
in several dicotyledonous seedlings only PAL activity, but not
the activity of CoA ligase[91] or of cinnamyl alcohol dehydrogenase (Gross, unpublished), was raised upon illumination.
It thus appears most likely that light affects lignification
only indirectly,[155] presumably by the increased synthesis of
cinnamic acids.

In conclusion, it can be stated that enzymatic studies
have considerably augmented our knowledge of the individual
reactions involved in the lignification process. It is hoped
that this will facilitate further investigations, especially
on the regulatory phenomena controlling the biosynthesis of
this important natural product.

ACKNOWLEDGMENT

The work from the Bochum laboratory referred to in this
article was generously supported by grants from the Deutsche
Forschungsgemeinschaft to Professor M. H. Zenk.

REFERENCES

1. Akazawa, T. and E. E. Conn. 1958. The oxidation of
    reduced pyridine nucleotides by peroxidase. *J. Biol.
    Chem. 232:*403–415.
2. Alibert, G. and R. Ranjeva. 1971. Recherches sur les
    enzymes catalysant la biosynthese des acides pheno-
    liques chez *Quercus pedunculata* (Ehrh.): I. Formation
    des premiers termes des series cinnamique et benzoique.
    *FEBS Lett. 19:*11–14.
3. Bartlett, D. J., J. E. Poulton, and V. S. Butt. 1972.
    Hydroxylation of *p*-coumaric acid by illuminated chloro-
    plasts from spinach beet leaves. *FEBS Lett. 23:*265–
    267.
4. Bergmann, L. 1964. Der einfluss von kinetin auf die
    ligninbildung und differenzierung in gewebekulturen
    von *Nicotiana tabacum. Planta 62:*221–254.
5. Bland, D. E., A Logan, M. Menshun, and S. Sternhell.
    1968. The lignin of *Sphagnum. Phytochemistry 7:*
    1373–1377.
6. Boudet, A., R. Ranjeva, and P. Gadal. 1971. Propriétés
    allostériques spécifiques des deux isoenzymes de la
    phénylalanine-ammoniaque lyase chez *Quercus peduncu-
    lata. Phytochemistry 10:*997–1005.
7. Bowling, A. C. and R. K. Crowden. 1973. Peroxidase
    activity and lignification in the pod membrane of
    Pisum sativum L. *Aust. J. Biol. Sci. 26:*679–684.
8. Brown, S. A. 1966. Lignins. *Ann. Rev. Plant Physiol.
    17:*223–244.
9. Brown, S. A., D. Wright, and A. C. Neish. 1959. Stu-
    dies of lignin biosynthesis using isotopic carbon.
    VII. The role of *p*-hydroxyphenylpyruvic acid. *Can.
    J. Biochem. Physiol. 37:*25–34.
10. Büche, T. and H. Sandermann. 1973. Lipid dependence of
    plant microsomal cinnamic acid 4-hydroxylase. *Arch.
    Biochem. Biophys. 158:*445–447.
11. Burton, R. M. and E. R. Stadtman. 1953. The oxidation
    of acetaldehyde to acetyl coenzyme A. *J. Biol. Chem.
    202:*873–890.
12. Camm, E. L. and G. H. N. Towers. 1973. Phenylalanine
    ammonia lyase. *Phytochemistry 12:*961–973.
13. Charriere-Ladreix, Y. 1975. Sur la presence de deux
    isoenzymes de la phenylalanine-ammoniac-lyase chez
    *Aesculus hippocastanum. Phytochemistry 14:*1727–1731.

14. Cheng, C. K. C. and H. V. Marsh. 1968. Gibberellic
     acid-promoted lignification and phenylalanine ammonia-
     lyase activity in a dwarf pea (*Pisum sativum*). *Plant
     Physiol*. *43*:1755-1759.

15. Creasy, L. L. and M. Zucker. 1974. Phenylalanine
     ammonia-lyase and phenolic metabolism. *Recent Adv.
     Phytochem*. *8*:1-19.

16. Curtis, C. R. and N. M. Barnett. 1974. Isoelectric
     focusing of peroxidases released from soybean hypo-
     cotyl cell walls by *Sclerotium rolfsii* culture fil-
     trate. *Can. J. Bot*. *52*:2037-2040.

17. Czichi, U. and H. Kindl. 1975. A model of closely
     assembled consecutive enzymes on membranes: Formation
     of hydroxycinnamic acids from L-phenylalanine on tyla-
     koids of *Dunaliella marina*. *Z. Physiol. Chem*. *356*:
     475-485.

18. Czichi, U. and H. Kindl. 1975. Formation of *p*-coumaric
     acid and *o*-coumaric acid from L-phenylalanine by
     microsomal membrane fractions from potato: Evidence
     of membrane-bound enzyme complexes. *Planta 125*:115-125

19. Davies, D. D., A. Texeira, and P. Kenworthy. 1972. The
     stereospecificity of nicotinamide-adenine dinucleotide-
     dependent oxidoreductases from plants. *Biochem. J.
     127*:335-343.

20. Davies, D. D., E. N. Ugochukwu, K. D. Patil, and G. H.
     N. Towers. 1973. Aromatic alcohol dehydrogenase from
     potato tubers. *Phytochemistry 12*:531-536.

21. DeJong, D. W. 1967. An investigation of the role of
     plant peroxidase in cell wall development by the histo-
     chemical method. *J. Histochem. Cytochem. 15*:335-346.

22. Ebel, J. and H. Grisebach. 1973. Reduction of cinnamic
     acids to cinnamyl alcohols with an enzyme preparation
     from cell suspension cultures of soybean (*Glycine max*).
     FEBS Lett. 30:141-143.

23. Ebel, J., B. Schaller-Hekeler, K. H. Knobloch, E.
     Wellmann, H. Grisebach, and K. Hahlbrock. 1974.
     Coordinated changes in enzyme activities of phenyl-
     propanoid metabolism during the growth of soybean cell
     suspension cultures. *Biochim. Biophys. Acta 362*:417-
     424.

24. El-Basyouni, S. Z., A. C. Neish, and G. H. N. Towers.
     1964. The phenolic acids in wheat. III. Insoluble
     derivatives of phenolic cinnamic acids as natural
     intermediates in lignin biosynthesis. *Phytochemistry
     3*:627-639.

25. El-Basyouni, S. Z. and A. C. Neish. 1966. Occurrence
    of metabolically-active bound forms of cinnamic acid
    and its phenolic derivatives in acetone powders of
    wheat and barley plants. *Phytochemistry* 5:685-691.
26. Ellis, B. E., M. H. Zenk, G. W. Kirby, J. Michael, and
    H. G. Floss. 1973. Steric course of the tyrosine
    ammonia-lyase reaction. *Phytochemistry* 12:1057-1058.
27. Elstner, E. F. and A. Heupel. 1976. Formation of
    hydrogen peroxide by isolated cell walls from horse-
    radish (*Armoracia lapathifolia* Gilib.). *Planta 130:*
    175-180.
28. Erdtman, H. 1933. Dehydrierungen in der coniferyl-
    reihe. I. Dehydrodi-eugenol und dehydrodiiso-eugenol.
    *Biochem. Z. 258:*172-180.
29. Erdtman, H. 1933. Dehydrierungen in der coniferyl-
    reihe. II. Dehydrodi-isoeugenol. *Liebigs Ann. Chem.
    503:*283-294.
30. Erickson, M. and G. E. Miksche. 1974. Charakterisierung
    der lignine von pteridophyten durch oxidativen abbau.
    *Holzforschung 28:*157-159.
31. Erickson, M. and G. E. Miksche. 1974. Two dibenzofurans
    obtained on oxidative degradation of the moss *Poly-
    trichum commune* Hedw. *Acta Chem. Scand. B 28:*109-113.
32. Erickson, M. and G. E. Miksche. 1974. On the occurrence
    of lignin or polyphenols in some mosses and liverworts.
    *Phytochemistry 13:*2295-2299.
33. French, C. J. and H. Smith. 1975. An inactivator of
    phenylalanine ammonia-lyase from gherkin hypocotyls.
    *Phytochemistry 14:*963-966.
34. Freudenberg, K. 1965. Lignin: Its constitution and
    formation from *p*-hydroxycinnamyl alcohols. *Science
    148:*595-600.
35. Freudenberg, K., C. L. Chen, J. M. Harkin, H. Nimz, and
    H. Renner. 1965. Observations on lignin. *Chem.
    Commun.* :224-225.
36. Freudenberg, K. and A. C. Neish. 1968. Constitution
    and Biosynthesis of Lignin. Springer, Berlin.
37. Fritz, G. J., R. W. King, and R. C. Dougherty. 1974.
    Incorporation of molecular oxygen into caffeic acid
    by green *Helianthus annuus. Phytochemistry 13:*1473-
    1475.
38. Gardiner, M. G. and R. Cleland. 1974. Peroxidase iso-
    enzymes of the *Avena* coleoptile. *Phytochemistry 13:*
    1707-1711.
39. Gestetner, B. and E. E. Conn. 1974. The 2-hydroxyla-
    tion of *trans*-cinnamic acid by chloroplasts from
    Melilotus alba Desr. *Arch. Biochem. Biophys. 163:*617-624.

40. Goff, C. W. 1975. A light and electron microscopic study of peroxidase localization in the onion root tip. *Amer. J. Bot. 62:*280-291.

41. Grisebach, H. 1962. Die biosynthese der flavonoide. *Planta Medica 10:*385-397.

42. Grisebach, H. 1973. Comparative biosynthetic pathways in higher plants. *Pure Appl. Chem. 34:*487-513.

43. Grisebach, H. and K. Hahlbrock. 1974. Enzymology and regulation of flavonoid and lignin biosynthesis in plants and plant cell suspension cultures. *Recent Adv. Phytochem. 8:*21-52.

44. Gross, G. G. 1972. Formation and reduction of intermediate acyladenylate by aryl-aldehyde: NADP oxidoreductase from *Neurospora crassa. Eur. J. Biochem. 31:*585-592.

45. Gross, G. G. 1976. Cell wall-bound malate dehydrogenase from horseradish. *Phytochemistry*, submitted for publication.

46. Gross, G. G. and W. Kreiten. 1975. Reduction of coenzyme A thioesters of cinnamic acids with an enzyme preparation from lignifying tissue of *Forsythia. FEBS Lett. 54:*259-262.

47. Gross, G. G., R. L. Mansell, and M. H. Zenk. 1975. Hydroxycinnamate:coenzyme A ligase from lignifying tissue of higher plants. Some properties and taxonomic distribution. *Biochem. Physiol. Pflanzen 168:*41-51.

48. Gross, G. G., J. Stöckigt, R. L. Mansell, and M. H. Zenk. 1973. Three novel enzymes involved in the reduction of ferulic acid to coniferyl alcohol in higher plants: ferulate:CoA ligase, feruloyl-CoA reductase and coniferyl alcohol oxidoreductase. *FEBS Lett. 31:*283-28

49. Gross, G. G. and M. H. Zenk. 1966. Darstellung und eigenschaften von coenzyme A-thiolestern substituierter zimtsäuren. *Z. Naturforsch. 21b:*683-690.

50. Gross, G. G. and M. H. Zenk. 1969. Reduktion aromatischer säuren zu aldehyden und alkoholen im zellfreien system. I. Reinigung und eigenschaften von arylaldehyd:NADP-oxidoreductase aus *Neurospora crassa. Eur. J. Biochem. 8:*413-419.

51. Gross, G. G. and M. H. Zenk. 1974. Isolation and properties of hydroxycinnamate:CoA ligase from lignifying tissue of *Forsythia. Eur. J. Biochem. 42:*453-459

52. Hagel, P. and H. Kindl. 1975. *p*-Hydroxybenzoate synthase: A complex associated with mitochondrial membranes of roots of *Cucumis sativus. FEBS Lett. 59:*123-124.

BIOSYNTHESIS OF LIGNIN

BIOSYNTHESIS OF LIGNIN

BIOSYNTHESIS OF LIGNIN ... 175

53. Hahlbrock, K., J. Ebel, R. Ortmann, A. Sutter, E. Wellmann, and H. Grisebach. 1971. Regulation of enzyme activities related to the biosynthesis of flavone glycosides in cell suspension cultures of parsley (*Petroselinum hortense*). *Biochim. Biophys. Acta 244:* 7–15.

54. Hahlbrock, K. and H. Grisebach. 1970. Formation of coenzyme A esters of cinnamic acids with an enzyme preparation from cell suspension cultures of parsley. *FEBS Lett. 11:*62–64.

55. Hahlbrock, K., K. H. Knobloch, F. Kreuzaler, J. R. M. Potts, and E. Wellmann. 1976. Coordinated induction and subsequent activity changes of two groups of metabolically interrelated enzymes. Light-induced synthesis of flavonoid glycosides in cell suspension cultures of *Petroselinum hortense*. *Eur. J. Biochem. 61:*199–206.

56. Hahlbrock, K., E. Kuhlen, and T. Lindl. 1971. Änderungen von enzymaktivitäten während des wachstums von zell-suspensionskulturen von *Glycine max:* Phenylalanin ammonium-lyase und *p*-cumarat:CoA ligase. *Planta 99:* 311–318.

57. Hahlbrock, K. and J. Schröder. 1975. Specific effects on enzyme activities upon dilution of *Petroselinum hortense* cell cultures into water. *Arch. Biochem. Biophys. 171:*500–506.

58. Halliwell, B. 1975. Hydroxylation of *p*-coumaric acid by illuminated chloroplasts. The role of superoxide. *Eur. J. Biochem. 55:*355–360.

59. Halliwell, B. and S. Ahluwalia. 1976. Hydroxylation of *p*-coumaric acid by horseradish peroxidase. The role of superoxide and hydroxyl radicals. *Biochem. J. 153:*513–518.

60. Hanson, K. R., R. H. Wightman, J. Staunton, and A. R. Battersby. 1971. Stereochemical course of the deamination catalyzed by L-phenylalanine ammonia-lyase and the configuration of 2-benzamidocinnamic azlactone. *Chem. Commun.* 185–186.

61. Harkin, J. M. and J. R. Obst. 1973. Lignification in trees: Indication of exclusive peroxidase participation. *Science 180:*296–298.

62. Hasegawa, S. and V. P. Maier. 1972. Cinnamate hydroxylation and the enzymes leading from phenylpyruvate to *p*-coumarate synthesis in grapefruit tissues. *Phytochemistry 11:*1365–1370.

63. Havir, E. A. and K. R. Hanson. 1975. L-Phenylalanine ammonia lyase (maize, potato and *Rhodotorula glutinis*).

Studies of the prosthetic group with nitromethane.
*Biochemistry 14:*1620-1626.

64. Heber, U. 1974. Metabolite exchange between chloroplasts and cytoplasm. *Ann. Rev. Plant Physiol. 25:*
393-421.

65. Higuchi, T. 1971. Formation and biological degradation
of lignins. *Adv. Enzymol. 34:*207-283.

66. Higuchi, T. 1976. Biochemical aspects of lignification
and heartwood formation. *Wood Research 59/60:*180-199.

67. Higuchi, T. and S. A. Brown. 1963. Studies of lignin
biosynthesis using isotopic carbon. VIII. The phenylpropanoid system in lignification. *Can. J. Biochem.
Physiol. 41:*621-628.

68. Hill, A. C. and M. J. C. Rhodes. 1975. The properties
of cinnamic acid 4-hydroxylase of aged swede root
disks. *Phytochemistry 14:*2387-2391.

69. Ife, R. and E. Haslam. 1971. The shikimate pathway.
Part III. The stereochemical course of the L-phenylalanine ammonia lyase reaction. *J. Chem. Soc. C:*2818-
2821.

70. Johnson, M. J. 1960. Enzyme equilibria and thermo-
dynamics. *In* The Enzymes (P. P. Boyer, H. Lardy, and
K. Myrbäck, eds.) Vol. 3, pp. 407-441. Academic Press,
New York.

71. Johnson, R. C. and J. R. Gilbertson. 1972. Isolation,
characterization, and partial purification of a fatty
acyl coenzyme A reductase from bovine cardiac muscle.
*J. Biol. Chem. 247:*6991-6998.

72. Kalyanaraman, V. S., S. A. Kumar, and S. Mahadevan.
1975. Oxidase-peroxidase enzymes of *Datura innoxia*.
Oxidation of reduced nicotinamide-adenine dinucleotide
in the presence of formylphenylacetic acid ethyl ester.
*Biochem. J. 149:*577-584.

73. Kindl, H. 1971. Aromatische aminosäuren im stoffwechsel
höherer pflanzen. *Naturwissenschaften 58:*554-563.

74. Kindl, H. and H. Ruis. 1971. Subcellular distribution
of *p*-hydroxybenzoic acid formation in castor bean endosperm. *Z. Naturforsch. 26b:*1379-1380.

75. Knobloch, K. H. and K. Hahlbrock. 1975. Isoenzymes of
p-coumarate:CoA ligase from cell suspension cultures
of *Glycine max*. *Eur. J. Biochem. 52:*311-320.

76. Knobloch, K. H. and K. Hahlbrock. 1975. Isoenzyme der
p-cumarat: CoA ligase aus zellsuspensionskulturen von
Glycine max. *Planta Medica*, Suppl., pp. 102-106.

77. Kolattukudy, P. E. 1970. Reduction of fatty acids to alcohols by cell-free preparations of *Euglena gracilis*. *Biochemistry* 9:1095-1102.
78. Kolattukudy, P. E. 1971. Enzymatic synthesis of fatty alcohols in *Brassica oleracea*. *Arch. Biochem. Biophys.* 142:701-709.
79. Koukol, J. and E. E. Conn. 1961. Purification and properties of the phenylalanine deaminase of *Hordeum vulgare*. *J. Biol. Chem.* 236:2692-2698.
80. Kreuzaler, F. and K. Hahlbrock. 1972. Enzymatic synthesis of aromatic compounds in higher plants: Formation of naringenin (5,7,4'-trihydroxyflavanone) from p-coumaryl coenzyme A and malonyl coenzyme A. *FEBS Lett.* 28:69-72.
81. Kreuzaler, F. and K. Hahlbrock. 1975. Enzymatic synthesis of an aromatic ring from acetate units. Partial purification and some properties of flavanone synthase from cell-syspension cultures of *Petroselinum hortense*. *Eur. J. Biochem.* 56:205-213.
82. Kreuzaler, F. and K. Hahlbrock. 1975. Enzymatic synthesis of aromatic compounds in higher plants. Formation of bis-noryangonin (4-hydroxy-6[4-hydroxystyryl]-2-pyrone) from p-coumaroyl-CoA and malonyl-CoA. *Arch. Biochem. Biophys.* 169:84-90.
83. Kuroda, H., M. Shimada, and T. Higuchi. 1975. Purification and properties of O-methyltransferase involved in the biosynthesis of gymnosperm lignin. *Phytochemistry* 14:1759-1763.
84. Lindl, T., F. Kreuzaler, and K. Hahlbrock. 1973. Synthesis of p-coumaroyl coenzyme A with a partially purified p-coumarate:CoA ligase from cell suspension cultures of soybean (*Glycine max*). *Biochim. Biophys. Acta* 302:457-464.
85. Liu, E. H. and D. T. A. Lamport. 1974. An accounting of horseradish peroxidase isoenzymes associated with the cell wall and evidence that peroxidase does not contain hydroxyproline. *Plant Physiol.* 54:870-876.
86. Löffelhardt, W. and H. Kindl. 1975. The conversion of L-phenylalanine into benzoic acid on the thylakoid membrane of higher plants. *H-S. Z. Physiol. Chem.* 356:487-493.
87. Mäder, M., Y. Meyer, and M. Bopp. 1975. Lokalisation der peroxidase-isoenzyme in protoplasten und zellwänden von *Nicotiana tabacum* L. *Planta* 122:259-268.

88. Mansell, R. L., G. R. Babbel, and M. H. Zenk. 1976.
     Multiple forms and specificity of coniferyl alcohol
     dehydrogenase from cambial regions of higher plants.
     Phytochemistry, in press.
89. Mansell, R. L., G. G. Gross, J. Stöckigt, H. Franke,
     and M. H. Zenk. 1974. Purification and properties
     of cinnamyl alcohol dehydrogenase from higher plants
     involved in lignin biosynthesis. *Phytochemistry 13:*
     2427-2435.
90. Mansell, R. L., J. Stöckigt, and M. H. Zenk. 1972.
     Reduction of ferulic acid to coniferyl alcohol in a
     cell free system from a higher plant. *Z. Pflanzen-
     physiol. 68:*286-288.
91. McClure, J. W. and G. G. Gross. 1975. Diverse photo-
     induction characteristics of hydroxycinnamate:co-
     enzyme A ligase and phenylalanine ammonia lyase in
     dicotyledonous seedlings. *Z. Pflanzenphysiol. 76:*
     51-55.
92. McIntyre, R. J. and P. F. T. Vaughan. 1975. Kinetic
     studies on the hydroxylation of *p*-coumaric acid to
     caffeic acid by spinach-beet phenolase. *Biochem. J.
     149:*447-461.
93. Meijer, A. J. and K. van Dam. 1974. The metabolic
     significance of anion transport in mitochondria.
     *Biochim. Biophys. Acta 346:*213-244.
94. Nakamura, Y., H. Fushiki, and T. Higuchi. 1974. Meta-
     bolic differences between gymnosperms and angiosperms
     in the formation of syringyl lignin. *Phytochemistry
     13:*1777-1784.
95. Nambudiri, A. M. D., C. P. Vance, and G. H. N. Towers.
     1973. Effect of light on enzymes of phenylpropanoid
     metabolism and hispidin biosynthesis in *Polyporus
     hispidus. Biochem. J. 134:*891-897.
96. Nari, J., C. Mouttet, F. Fouchier, and J. Ricard. 1974.
     Subunit interactions in enzyme catalysis. Kinetic
     analysis of subunit interactions in the enzyme L-
     phenylalanine ammonia-lyase. *Eur. J. Biochem. 41:*499-
     515.
97. Neish, A. C. 1961. Formation of *m*- and *p*-coumaric acids
     by enzymatic desamination of the corresponding isomers
     of tyrosine. *Phytochemistry 1:*1-24.
98. Neish, A. C. 1964. Major pathways of biosynthesis of
     phenols. *In* Biochemistry of Phenolic Compounds (J. B.
     Harborne, ed.), pp. 295-359. Academic Press, London.

99. Neish, A. C. 1964. Cinnamic acid derivatives as inter-
     mediates in the biosynthesis of lignin and related
     compounds. *In* The Formation of Wood in Forest Trees
     (M. H. Zimmermann, ed.), pp. 219-239. Academic Press,
     New York.
100. Nimz, H. 1974. Beech lignin-proposal of a constitu-
     tional scheme. *Angew. Chem. Internat. Edit. 13:*313-
     320.
101. Nozu, Y. 1967. Studies on the biosynthesis of lignin.
     III. Dehydrogenative polymerization of coniferyl alco-
     hol by peroxidase. *J. Biochem. 62:*519-530.
102. Parish, R. W. 1972. The intracellular location of
     phenol oxidases in the leaves of spinach beet (*Beta
     vulgaris* L. subspecies *vulgaris*). *Eur. J. Biochem.
     31:*446-455.
103. Payen, A. 1838. Mémoire sur la composition du tissu
     propre des plantes et du ligneux. *Compt. Rend. 7:*
     1052-1056.
104. Potts, J. R. M., R. Weklych, and E. E. Conn. 1974. The
     4-hydroxylation of cinnamic acid by sorghum microsomes
     and the requirement for cytochrome P-450. *J. Biol.
     Chem. 249:*5019-5026.
105. Poulton, J. E. and V. S. Butt. 1975. Purification and
     properties of *S*-adenosyl-L-methionine:caffeic acid
     O-methyltransferase from leaves of spinach beet (*Beta
     vulgaris* L.). *Biochim. Biophys. Acta 403:*301-314.
106. Poulton, J. E. and V. S. Butt. 1976. Purification and
     properties of *S*-adenosyl-L-homocysteine hydrolase from
     leaves of spinach beet. *Arch. Biochem. Biophys. 172:*
     135-142.
107. Poulton, J. E., H. Grisebach, J. Ebel, B. Schaller-
     Hekeler, and K. Hahlbrock. 1976. Two distinct *S*-
     adenosyl-L-methionine:3,4-dihydric phenol 3-*O*-methyl-
     transferases of phenylpropanoid metabolism in soybean
     cell suspension cultures. *Arch. Biochem. Biophys. 173:*
     301-305.
108. Quayle, J. R. 1966. Glyoxalate dehydrogenase. *Methods
     Enzymol. 9:*342-346.
109. Ranjeva, R., A. M. Boudet, H. Harada, and G. Marigo.
     1975. Phenolic metabolism in *Petunia* tissues. I. Cha-
     racteristic responses of enzymes involved in different
     steps of polyphenol synthesis to different hormonal
     influences. *Biochim. Biophys. Acta 399:*23-30.

110. Ranjeva, R., R. Faggion, and A. M. Boudet. 1975.
      Metabolisme des composés phénoliques des tissus de
      Pétunia. II. Étude de système enzymatique d`acti-
      vation des acides cinnamiques. Mise en évidence de
      deux formes de la "cinnamoyl-coenzyme A ligase."
      Physiol. Veg. 13:725-734.
111. Rétey, J., E. von Stetten, U. Coy, and F. Lynen. 1970.
      A probable intermediate in the enzymatic reduction
      of 3-hydroxy-3-methylglutaryl coenzyme A. *Eur. J.
      Biochem. 15*:72-76.
112. Rhodes, M. J. C., A. C. R. Hill, and L. S. C. Wooltorton
      1976. Activity of enzymes involved in lignin biosyn-
      thesis in swede root disks. *Phytochemistry 15*:707-710
113. Rhodes, M. J. C. and L. S. C. Wooltorton. 1973. Stimu-
      lation of phenolic acid and lignin biosynthesis in
      swede root tissue by ethylene. *Phytochemistry 12*:107-
      118.
114. Rhodes, M. J. C. and L. S. C. Wooltorton. 1973. Change
      in phenolic acid and lignin biosynthesis in response
      to treatment of root tissue of the Swedish turnip
      (*Brassica napo-brassica*) with ethylene. *Qualitas
      Plant. 23*:145-155.
115. Rhodes, M. J. C. and L. S. C. Wooltorton. 1973. Forma-
      tion of CoA esters of cinnamic acid derivatives by
      extracts of *Brassica napo-brassica* root tissue.
      Phytochemistry 12:2381-2387.
116. Rhodes, M. J. C. and L. S. C. Wooltorton. 1974. Reduc-
      tion of the CoA thioesters of *p*-coumaric and ferulic
      acids by extracts of aged *Brassica napo-brassica* root
      tissue. *Phytochemistry 13*:107-110.
117. Rhodes, M. J. C. and L. S. C. Wooltorton. 1975. *p*-
      Coumaryl-CoA ligase of potato tubers. *Phytochemistry
      14*:2161-2164.
118. Rhodes, M. J. C. and L. S. C. Wooltorton. 1975. Enzyme
      involved in the reduction of ferulic acid to coniferyl
      alcohol during the aging of disks of swede root tissue
      Phytochemistry 14:1235-1240.
119. Rhodes, M. J. C. and L. S. C. Wooltorton. 1976. The
      enzymic conversion of hydroxycinnamic acids to *p*-
      coumaroylquinic and chlorogenic acids in tomato fruits
      Phytochemistry 15, in press.
120. Rubery, P. H. and D. E. Fosket. 1969. Changes in pheny
      alanine ammonia-lyase activity during xylem differen-
      tiation in *Coleus* and soybean. *Planta 87*:54-62.

121. Rubery, P. H. and D. H. Northcote. 1968. Site of
     phenylalanine ammonia-lyase activity and synthesis
     of lignin during xylem differentiation. *Nature 219:*
     1230-1234.
122. Sagisaka, S. 1976. The occurrence of peroxide in a
     perennial plant, *Populus gelrica*. *Plant Physiol.*
     *57:*308-309.
123. Sagisaka, S. and K. Shimura. 1962. Studies in lysine
     biosynthesis. IV. Mechanism of activation and reduc-
     tion of α-aminoadipic acid. *J. Biochem.* (Tokyo) *52:*
     155-161.
124. Sarkanen, K. V. 1971. Precursors and their poly-
     merization. *In* Lignins (K. V. Sarkanen and C. H.
     Ludwig, eds.), pp. 95-163. Wiley-Interscience, New
     York.
125. Sarkanen, K. V. and H. L. Hergert. 1971. Classifica-
     tion and distribution. *In* Lignins (K. V. Sarkanen
     and C. H. Ludwig, eds.), pp. 43-94. Wiley-Inter-
     science, New York.
126. Sarkanen, K. V. and C. H. Ludwig. 1971. Lignins.
     Occurrence, Formation, Structure and Reactions.
     Wiley-Interscience, New York, London, Sydney, Toronto.
127. Shimada, M., H. Fushiki, and T. Higuchi. 1972. *O*-Methyl-
     transferase activity from Japanese black pine. *Phyto-*
     *chemistry 11:*2657-2662.
128. Shimada, M., H. Fushiki, and T. Higuchi. 1973. Mechan-
     ism of biochemical formation of the methoxyl groups
     in softwood and hardwood lignins. *Mokuzai Gakkaishi*
     *19:*13-21.
129. Shimada, M., H. Kuroda, and T. Higuchi. 1973. Evi-
     dence for the formation of methoxyl groups of ferulic
     and sinapic acids in *Bambusa* by the same *O*-methyl-
     transferase. *Phytochemistry 12:*2873-2875.
130. Shimada, M., H. Ohashi, and T. Higuchi. 1970. *O*-Methyl-
     transferase involved in the biosynthesis of lignins.
     *Phytochemistry 9:*2463-2470.
131. Siegel, S. M. 1969. Evidence for the presence of lig-
     nin in moss gametophytes. *Amer. J. Bot 56:*175-179.
132. Sinha, A. K. and J. K. Bhattacharjee. 1971. Lysine
     biosynthesis in *Saccharomyces*. Conversion of α-
     aminoadipate into α-aminoadipic σ-semialdehyde.
     *Biochem. J. 125:*743-749.
133. Smith, H. 1973. Regulatory mechanisms in the photo-
     control of flavonoid biosynthesis. *In* Biosynthesis
     and its Control in Plants (B. V. Milborrow, ed.), pp.
     303-321. Academic Press, London, New York.

134. Smith, G. J. and I. J. Miller. 1975. Free radicals
      of coniferyl alcohol and isoeugenol. *Aust. J. Chem.*
      28:193-196.
135. Snyder, F. and B. Malone. 1970. Enzymic interconversion
      of fatty alcohols and fatty acids. *Biochem. Biophys.*
      Res. Commun. 41:1382-1387.
136. Stafford, H. A. 1974. The metabolism of aromatic com-
      pounds. *Ann. Rev. Plant Physiol. 25*:459-486.
137. Stafford, H. A. 1974. Possible multienzyme complexes
      regulating the formation of C_6-C_3 phenolic compounds
      and lignins in higher plants. *Recent Adv. Phytochem.*
      8:53-79.
138. Steelink, C. 1972. Biological oxidation of lignin
      phenols. *Recent Adv. Phytochem. 4*:239-271.
139. Stephens, G. J. and R. K. S. Wood. 1974. Release of
      enzymes from cell walls by an endopectate-trans-
      eliminase. *Nature 251*:358.
140. Stöckigt, J., R. L. Mansell, G. G. Gross, and M. H. Zenk.
      1973. Enzymic reduction of *p*-coumaric acid *via p*-
      coumaroyl-CoA to *p*-coumaryl alcohol by a cell-free
      system from *Forsythia* sp. *Z. Pflanzenphysiol. 70*:305-
      307.
141. Stöckigt, J. and M. H. Zenk. 1974. Enzymatic synthesis
      of chlorogenic acid from caffeoyl coenzyme A and
      quinic acid. *FEBS Lett. 42*:131-134.
142. Sütfeld, R. and R. Wiermann. 1974. Die bildung von
      coenzym A-thiolestern verschieden substituierter
      zimtsäuren durch enzymextrakte aus antheren. *Z.*
      Pflanzenphysiol. 72:163-171.
143. Tanaka, Y., M. Kojima, and I. Uritani. 1974. Proper-
      ties, development and cellular-localization of cinna-
      mic acid 4-hydroxylase in cut-injured sweet potato.
      Plant Cell Physiol. 15:843-854.
144. Tedeschi, H. 1976. Mitochondria: Structure, Biogenesis
      and Transducing Functions. Cell Biology Monographs,
      Vol. 4, p. 102. Springer, Wien, New York.
145. Tutschek, R. 1975. Isolierung und charakterisierung
      der *p*-hydroxy-β-(carboxymethyl)-zimtsäure (sphagnum-
      säure) aus der zellwand von *Sphagnum magellanicum*
      Brid. *Z. Pflanzenphysiol. 76*:353-365.
146. Uotila, L. and M. Koivusalo. 1974. Formaldehyde dehydro-
      genase from human liver. Purification, properties,
      and evidence for the formation of glutathione thiol-
      esters by the enzyme. *J. Biol. Chem. 249*:7653-7663.

147.   Vance, C. P., A. M. D. Nambudiri, C. K. Wat, and G. H.
       N. Towers. 1975. Isolation and properties of hydroxy-
       cinnamate:CoA ligase from *Polyporus hispidus*. *Phyto-
       chemistry 14:*967-969.
148.   Vaughan, P. F. T., R. Eason, J. Y. Paton, and G. A.
       Ritchie. 1975. Molecular weight and amino acid com-
       position of purified spinach beet phenolase. *Phyto-
       chemistry 14:*2383-2386.
149.   Vaughan, P. F. T. and R. J. McIntyre. 1975. The action
       of hydroxygen peroxide on the hydroxylation of *p*-
       coumaric acid by spinach-beet phenolase. *Biochem. J.
       151:*759-762.
150.   Vieitez, A. M., A. Ballester, and E. Vieitez. 1975.
       Coniferyl alcohol from callus of *Castanea sativa* cul-
       tured *in vitro*. *Experientia 31:*1163.
151.   Vollmer, K. O., H. J. Reisener, and H. Grisebach. 1965.
       The formation of acetic acid from *p*-hydroxycinnamic
       acid during its degradation to *p*-hydroxybenzoic acid
       in wheat shoots. *Biochem. Biophys. Res. Commun. 21:*
       221-225.
152.   Wang, L., K. Takayama, D. S. Goldman, and H. K. Schnoes.
       1972. Synthesis of alcohol and wax ester by a cell-
       free system in *Mycobacterium tuberculosis*. *Biochim.
       Biophys. Acta 260:*41-48.
153.   Walton, E. and V. S. Butt. 1970. The activation of
       cinnamate by an enzyme from leaves of spinach beet
       (*Beta vulgaris* L. spp. *vulgaris*). *J. Exptl. Bot 21:*
       887-891.
154.   Walton, E. and V. S. Butt. 1971. The demonstration of
       cinnamyl-CoA synthetase activity in leaf extracts.
       *Phytochemistry 10:*295-304.
155.   Wardrop, A. B. 1971. Occurrence and formation in
       plants. *In* Lignins (K. V. Sarkanen and C. H. Ludwig,
       eds.), pp. 19-41. Wiley-Interscience, New York.
156.   Wat, C. K. and G. H. N. Towers. 1975. Phenolic *O*-
       methyltransferase from *Lentinus lepideus* Basidio-
       mycete). *Phytochemistry 14:*663-666.
157.   Wengenmayer, H., J. Ebel, and H. Grisebach. 1976.
       Enzymic synthesis of lignin precursors. Purification
       and properties of a cinnamoyl-CoA:NADPH reductase
       from cell suspension cultures of soybean (*Glycine max*).
       *Eur. J. Biochem. 65:*529-536.
158.   Whitmore, F. W. 1976. Binding of ferulic acid to cell
       walls by peroxidases of *Pinus elliottii*. *Phytochemis-
       try 15:*375-378.

159. Wyrambik, D. and H. Grisebach. 1975. Purification
     and properties of isoenzymes of cinnamyl-alcohol
     dehydrogenase from soybean-cell-suspension cultures.
     *Eur. J. Biochem. 59:*9–15.

160. Yamada, Y. and T. Kuboi. 1976. Significance of caffeic
     acid *O*-methyltransferase in lignification of cultured
     tobacco cells. *Phytochemistry 15:*395–396.

161. Yoshida, S. 1969. Biosynthesis and conversion of aro-
     matic amino acids in plants. *Ann. Rev. Plant Physiol.
     20:*40–62.

162. Yoshida, S., K. Tazaki, and T. Minamikawa. 1975.
     Occurrence of shikimic and quinic acids in angio-
     sperms. *Phytochemistry 14:*195–197.

163. Zenk, M. H. 1966. Biosynthesis of C_6–C_1-compounds.
     In Biosynthesis of Aromatic Compounds (G. Billek, ed.),
     pp. 45–60. Pergamon, Oxford.

164. Zenk, M. H. 1971. Metabolism of prearomatic and aro-
     matic compounds in plants. *In* Pharmacognosy and
     Phytochemistry (H. Wagner and L. Hörhammer, eds.),
     pp. 314–346. Springer, Berlin, Heidelberg, New York.

165. Zenk, M. H. and G. G. Gross. 1972. The enzymic reduc-
     tion of cinnamic acids. *Recent Adv. Phytochem. 4:*87–
     106.

166. Zucker, M. 1972. Light and enzymes. *Ann. Rev. Plant
     Physiol. 23:*133–156.

Chapter Six

LIPID POLYMERS AND ASSOCIATED PHENOLS, THEIR CHEMISTRY,
BIOSYNTHESIS, AND ROLE IN PATHOGENESIS

PAPPACHAN E. KOLATTUKUDY

Department of Agricultural Chemistry
Washington State University, Pullman, Washington

INTRODUCTION

The presence of aliphatic chains in wood was indicated
in 1797 when La Grange obtained suberic acid by treating
cork with nitric acid. However, until late in the 19th
century occurrence of polymeric materials derived from
fatty acids in plants was not recognized. The chemical
investigations into such recognized polymeric materials
began when Fremy and Urbain[33],[34] isolated two new fatty
acids from the alkaline hydrolysate of an insoluble material
obtained from plant cuticle. This acid-resistant insoluble
material obtained from plant cuticle was called "cutose"
which later became known as cutin. The term suberin originally used by Chevreul in 1815[17] to describe the insoluble
material in bottle cork has come to mean the substance(s)
which make cork tissue or periderm rather impermeable to
water. The chemical similarity between cutin and suberin
was recognized when von Höhnel showed that suberin, like
cutin, gave rise to fatty acids upon alkaline hydrolysis.[116]
As the experimental techniques available to these early
investigators were indeed limited, neither the ultrastructure nor the chemistry of these complex polymeric materials
could be investigated. With the advent of modern instrumentation much progress has been made in this field during
the last quarter of a century. In this chapter the present
state of our knowledge concerning both cutin and suberin
will be summarized with special emphasis on structure,
biosynthesis and their possible role in pathogenesis. A
more detailed account of the historical development of this
field has been published.[86]

The non-carbohydrate polymers associated with wood
might be viewed as belonging to three classes; lignin, the
aromatic polymer, suberin, the mixed polymer of aromatic
and aliphatic monomers, and cutin, the aliphatic polymer
(Fig. 1). It is interesting that even the location of these
polymers in the stem reflects such an admittedly oversimp-
lified classification; the highly lignified wood is
surrounded by the mixed polymer suberin whereas the aliphatic polymer is found in the outermost layer, the cuticle.
Lignin is most highly concentrated in the middle lamella[103],
[119] the suberin lamellae are deposited inside the primary
cell wall in close contact with the cytoplasm[108],[120] (Fig.
2) whereas cutin is an extracellular polymer. Thus the
ultrastructure of these polymers also reflect the above

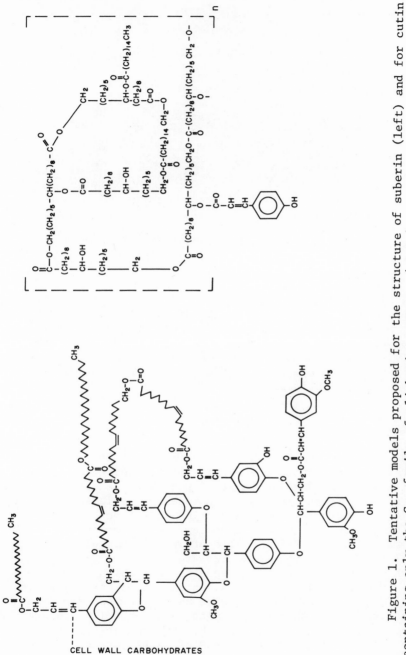

Figure 1. Tentative models proposed for the structure of suberin (left) and for cutin containing only the C_{16} family of aliphatic monomers (right).

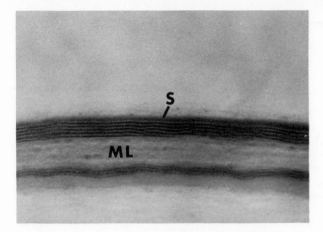

Figure 2.   Electron micrograph of potato suberin.
ML, middle lamellae; S, suberin lamellae.

classification. It must be pointed out that the terms
cutin and suberin as used in this chapter, refer to the
polymeric material and do not include the interlamellar
components such as carbohydrates and/or waxes. As the
trivial names used in the literature are often confusing,
a list of such names and the corresponding systematic
names are given in Table 1.

CHEMISTRY

   Methods. *Isolation of the Polymers*. The cuticular
layer is attached to the outside of the epidermal layer of
cells with an adhesive layer of pectinaceous material
(Fig. 3). Therefore most techniques of isolation of cutin
involves disruption of the pectinaceous layer with chemicals such as oxalic acid-ammonium oxalate or enzymes such
as pectinase.[31,60,87,92] The strips of cuticular layer
collected from such procedures are thoroughly extracted
with organic solvents to remove the wax which embeds cutin.
The resulting material often contains some carbohydrates
and residual wax. Therefore we usually repeatedly treat
the crude cutin preparation with a mixture of cellulase and
pectinase, each time followed by thorough solvent extraction.
The final product is often free of carbohydrates particularly
in the case of cutin preparations from tissues with well

Table 1. Hydroxyacids relevant to cutin and suberin

Systematic name	Trivial name
9,10-Dihydroxy C_{18} dioic acid	Phloionic acid
9,10,18-Trihydroxy C_{18} acid	Phloionolic acid
C_{22}Dioic acid	Phellogenic acid
16-Hydroxyhexadecanoic acid	Juniperic acid
12-Hydroxydodecanoic acid	Sabinic acid
18-Hydroxyoctadeca-*cis*-9-*trans*-11-*trans*-13-trienoic acid	α-Kamlolenic acid
22-Hydroxy C_{22} acid	Phellonic acid
9-Hydroxyoctadeca-*trans*-10-*trans*-12-dienoic acid	Dimorphecolic acid
C_{16} Dioic acid	Thapsic acid
12-Hydroxy oleic acid	Ricinoleic acid
11-Hydroxy C_{14} acid	Convolulinolic acid
11-Hydroxy C_{16} acid	Jalapinolic acid
12-Hydroxy-*cis*-9-*cis*-15-dienoic C_{18} acid	Densipolic acid
14-Hydroxy-*cis*-11-C_{20} acid	Lesquerolic acid
3,11-Dihydroxy C_{14} acid	Ipurolic acid
15,16-Dihydroxy C_{16} acid	Ustilic acid A
2,15,16-Trihydroxy C_{16} acid	Ustilic acid B
13-Hydroxy-*cis*-9-*trans*-11-dienoic C_{18} acid	Coriolic or artemisic acid
8-Hydroxy-9a,11a-17-enoic C_{18} acid*	Isanolic acid
9-Hydroxy-*trans*-10-12a-C_{18} acid*	Helenynolic acid

*a, acetylenic

developed cuticles such as apple fruits. However in many cases contaminating polymeric materials are present in the cutin preparations used thus far.

Suberin from root crops was isolated using similar techniques.[67,73] However, considering the subcellular location of this polymer and its possible attachment to cell wall polymers, it is not surprising that suberin preparations most probably contain varying amounts of other polymeric contaminants.

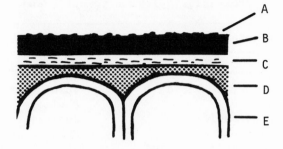

Figure 3. Schematic representation of the structure
of leaf cuticle. A, surface wax; B, cutin embedded in wax;
C, a mixed layer containing some cutin, wax, and carbohydrate polymers possibly with traces of protein; D, pectin;
E, cellulose wall of the epidermal cells.

Depolymerization. As the nature of the linkages
present in cutin and suberin are not fully understood a
method for complete depolymerization of these polymers cannot be chosen at the present time. All of the presently
available methods are directed towards cleaving the ester
linkages, which constitute the major type of linkage, present
in these polymers. Hydrolysis of alcoholic alkali has been
used for over a century to release the acidic monomers.[10,
12,19,31,33,34,53,54,78,80,87,89,104] More recently trans-
esterification with methanol containing sodium methoxide[50,
52,76] or BF_3 [67,73] was used to release the monomer acids as
their methyl esters. This technique avoids the possibility
of reesterification of the monomers during isolation and
storage. Functional groups such as epoxides and aldehydes
are derivatized during the methanolysis procedure and the
derivatives are stable and identifiable. However, it should
be born in mind that prolonged treatment with BF_3 in methanol
could result in the addition of methanol at the double
bonds of monomers.[67,72] In an attempt to cleave the inter-
chain peroxide bridges and ether linkages which might be
present in the polymer, cutin was treated sequentially with
alcoholic alkali, NaI and HI and thus a more complete depolymerization was achieved than that obtained with alkali
alone.[19] However, the products obtained from these treatments were not identified and therefore the nature of the
bonds cleaved by such techniques remain obscure. Exhaustive
hydrogenolysis with $LiAlH_4$ in tetrahydrofuran offers certain

advantages over the other techniques particularly in biosynthetic studies.[60,74,118] This treatment converts
carboxyl groups of the monomers into primary alcohol
functions which are indistinguishable from the naturally
occurring primary hydroxyl groups. Other LiAlH$_4$-susceptible
functions, such as carbonyls and epoxides, are also converted into alcohols. However, LiAlD$_4$ can be used to
detect and locate the presence of such groups in cutin.[64,75,118] This hydrogenolysis technique is not as convenient
as the transesterification methods for structural analysis
of suberin because both the ω-hydroxyacids and the corresponding dicarboxylic acids, which constitute the major
aliphatic components of suberin, are converted into indistinguishable diols.[67,72,73]

   Methods of Analysis of the Monomers. A fairly complex
mixture of monomers are obtained from cutin and suberin,
irrespective of the depolymerization technique used. Therefore it is not surprising that the earlier techniques of
fractional crystallization, counter-current distribution
and reversed-phase partition chromatography, gave way to
the more modern techniques of thin-layer chromatography
(TLC) and combined gas-liquid chromatography and mass
spectrometry (GLC-MS). Usually a single TLC step on silica
gel is quite adequate to fractionate the monomers into
individual classes (Fig. 4). Each fraction can then be
readily separated into individual components by gas chromatography (Fig. 4). If the monomer mixture contains
phenolic components they can be extracted with aqueous base
and recovered from this aqueous phase by extraction with
organic solvents after acidification.[99] As hydroxyl groups
and carboxyl groups are often the only functional groups
generally present in the monomers, for gas chromatography
the trimethyl silyl ethers of the methyl esters can be
easily prepared. As the resolution of these derivatives
are quite good, the depolymerization products can often be
separated by gas chromatography without prior fractionation
by thin-layer chromatography.[118] Since the monomers often
give readily interpretable mass spectra combined GLC-MS of
the depolymerization product can give structure and composition of the monomers in a single analysis although supple-
mental evidence would be required for novel monomers.[29,65]
In any case combined GLC-MS is the method of choice for
analysis of cutin monomers.

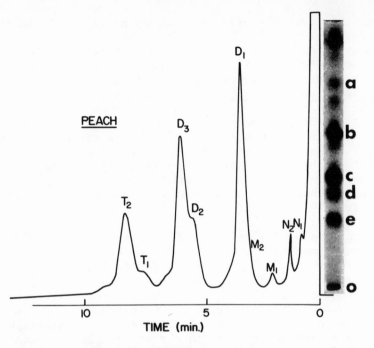

Figure 4. (Right) Thin-layer chromatogram of the hydrogenolysis product of peach fruit cutin. The major spots from top to bottom are: (a) fatty alcohol, (b) alkane diol, (c) C_{18} triol, (d) C_{16} triol, and (e) C_{18} tetral. (Left) Gas-liquid chromatogram of the hydrogenolysate (as Me_3Si ethers) of peach fruit cutin. Peaks were assigned on the basis of mass spectral data; N_1, hexadecanol; N_2, octadecanol; M_1, hexadecanediol; M_2, C_{18} diol; D_1, hexadecanetriol; D_2, octadecenetriol; D_3, octadecanetriol; T_1, octadecenetetraol; T_2, octadecanetetraol.[118]

Structure Determination. The structure of the aliphatic monomers are such that they yield readily interpretable mass spectra.[11,19,31,32,48,49,50,51,52,67,72,73,74,118] Major ions found in the high mass region of spectra of the major monomer acids of cutin and suberin are shown in Table 2. In the case of monomers, which are not in-chain substituted, fragments in the high mass region are the most characteristic ions. In-chain hydroxylated monomers give, in addition to the characteristic ions at the high mass

Table 2.   Major ions in the mass spectra of the aliphatic
monomers of cutin and suberin.

Compound	Major ions at the high mass region
Alkane-α,ω-dioic acid dimethyl ester	M^+ (weak), M-31, M-73, M-92, M-105, M-123
Alkene-α,ω-dioic acid dimethyl ester	M^+, M-32, M-50, M-64, M-92
Methyl ω-hydroxyalkanoate tri-methylsilyl ether	M^+ (small), M-15, M-31 and M-47 Metastable for [M-15] → [M-47]
Methyl ω-hydroxyalkenoate tri-methylsilyl ether	M^+ (moderate), M-15, M-31, M-47, Metastable for [M-15] → [M-47]
Methyl 16,x-dihydroxyhexade-canoate trimethylsilyl ether	M^+, M-15, M-31, M-47, M-47-32, M-15-90, M-31-90, M-47-90
Methyl 16-hydroxy-10-oxohexa-decanoate trimethylsilyl ether	M-15, M-31, M-47
Methyl 18-hydroxy-9(or 10)-methoxy-10(or 9),18-dihydroxy-octadecanoate trimethylsilyl ether	M^+, M-15, M-31, M-47, M-90-31
Methyl 9,10,18-trihydroxyocta-decanoate trimethylsilyl ether	M^+ (not significant), M-15, M-31, and M-15-90
Alkane-α,ω-diol trimethylsilyl ether	M^+ (extremely weak), M-15, M-31, M-90, M-15-90, M-31-90, [M-30/2 is a diagnostic ion]
Alkene-α,ω-diol	M^+ (strong), M-15, M-31, M-90, M-15-90, M-31-90, [M-30/2 is a dianostic ion]
Alkane triol and alkane tetraol	M^+, M-15, M-90, M-15-90

region, intense ions representing α-cleavage on either side
of the in-chain functional group, thus revealing the position of the in-chain substituent. As trimethylsiloxy
substituents are particularly prone to such α-cleavage
trimethylsilyl ether is the most preferred derivative for
the GLC-MS analysis of the monomers. Even mixtures of
positional isomers can be analysed by this technique as
illustrated by the spectrum of the dihydroxypalmitic acid
which is probably the most abundant monomer acid in cutin
and is also present in small quantities in suberin. The
mass spectrum of this component from potato suberin (as the
trimethylsilyl ether of the methyl ester) gives in addition
to the ions listed in Table 2, three pairs of intense α-
cleavage ions one at m/e 273,275, another at m/e 259,289
and a third at m/e 245,303 (Fig. 5). These ion pairs are
obviously derived from 10,16-, 9,16-, and 8,16-dihydroxy
isomers respectively. The relative intensities of these
ion pairs can be used to estimate the propoartion of each
isomer present in the mixture, although the accuracy of
this method of quantitation has not yet been verified with
synthetic mixtures of isomers.

     The GLC-MS method can be used for structure determina-
tion even when the monomer contains a relatively reactive
functional group such as an epoxide. As the oxirane function is rather labile, particularly to the available
methods of depolymerization of the polymers, the method of
choice for identification of the naturally occurring
epoxides in the polymer is to convert the oxirane into a
characteristic derivative during the depolymerization
process. Reductive cleavage of the polymer with LiAlD$_4$
cleaves the oxirane ring and labels the carbon atoms
involved in the epoxide with deuterium giving rise to
isomeric trideuterated C$_{18}$ triols which give unique even-
mass ions by α-cleavage as shown in Figure 6.[75,118] Depolymerization with BF$_3$ or sodium methoxide in methanol cleaves
the oxirane ring by methanolysis giving rise to methoxy-
hydrins in the place of the oxirane and these derivatives
also give characteristic spectra[52,67] as illustrated by
the α-cleavage pattern shown in Figure 6. Other labile
groups such as aldehydes can be labeled with deuterium by
treating the polymer with NaBD$_4$ prior to depolymerization or
can be derivatized with carbonyl reagents such as hydroxyl-
amine.[64,65] In either case the mass spectra of the final
product can be readily interpreted. Monomers which contain

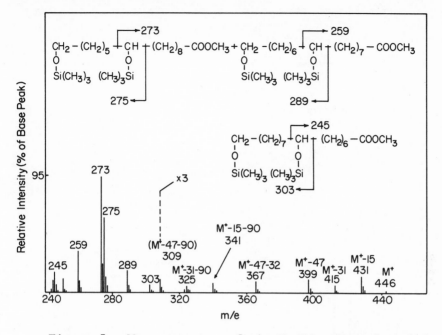

Figure 5. Mass spectrum of the Me_3Si ether of methyl dihydroxy hexadecanoate derived from BF_3-methanol cleavage of potato suberin.[67]

more stable carbonyl groups such as a keto group can be identified without any labeling as keto groups in the products give characteristic cleavage patterns (Fig. 7).[29]

The hydrogenolysis products of cutin also give readily interpretable spectra. The trimethyl siloxy alkanes give the characteristic ions derived by the cleavage of methyl groups and by elimination of trimethylsilanol. The position of the trimethyl siloxy function in the middle of the aliphatic chain can be ascertained from the intense ions produced by α-cleavage on either side of the in-chain substituent. Use of $LiAlD_4$ results in the incorporation of two deuterium atoms at the carbinol originating from the carboxyl carbon of the cutin acid. Consequently the α-cleavage fragment containing this carbinol function is heavier by 2 amu, thus the structure of the cutin acid from

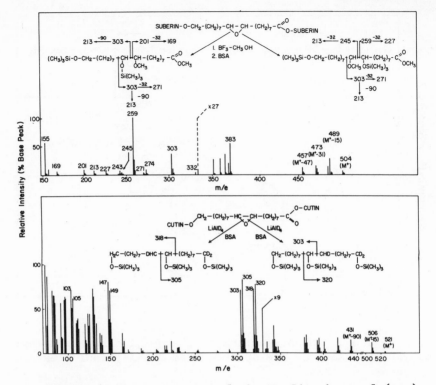

Figure 6. Mass spectra of the Me₃Si ethers of (top)
methyl dihydroxy methoxy octadecanoate derived from BF₃-
methanol cleavage of potato suberin[67] and (bottom) octadecanetriol derived from deuteriolysis of apple cutin.[118]
(BSA, bis-N,O-trimethylsilyl acetamide).

which the polyhydroxy alkane is derived can be easily
deduced.[118] Deuterolysis of epoxy acids of cutin results
in the incorporation of 3 deuterium atoms, and the isomeric
trideuteroalkane triols give characteristic spectra as
indicated above.

The phenolic fraction obtained from the hydrogenolysates
and deuterolysates of cutin and suberin can be analyzed
either as acetates or as trimethylsilyl derivatives (Fig. 8)
by combined GLC-MS.[99] Such analyses of cutin from several
species of plants give rise to two major reduction products,
dihydro-p-coumaryl alcohol and dihydroconiferyl alcohol
(Fig. 9) which can be identified by the major ions in the

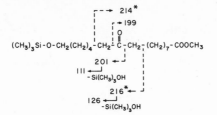

Figure 7. The major diagnostic fragments in the mass spectrum of 16-hydroxy-10-oxohexadecanoate. *Cleavage with H-transfer (McLafferty rearrangement).[29]

mass spectra of their derivatives listed in Figure 10. Reductive cleavage minimizes the possibility of oxidative structural alterations of the phenolic materials during prolonged treatments, which are necessary for depolymerization of the polymer.

Monomer Composition of Cutin and Suberin. *Cutin.* Analysis of the depolymerization products of cutin from a variety of plants by GLC-MS revealed that this polymer is composed of a C_{16}-family and a C_{18}-family of cutin acids (Fig. 11). The former, which appears to be the dominant component of the cutin of fast-growing plants, consists of palmitic acid, 16-hydroxypalmitic acid, and 10,16-dihydroxy-palmitic acid and/or its positional isomers, in which the in-chain hydroxyl group is at C-9, C-8, or C-7 [51,74,118] (Table 3). Monomers in which the alcoholic groups have undergone further oxidation are also found in smaller quantities or in some cases as substantial components. Thus 16-oxo-9- or 10-hydroxypalmitic acid and 16-hydroxy-10-oxo-palmitic acid have been identified as significant components of cutin in embryonic shoots of *Vicia faba*[64,65] and citrus[29] respectively and very small amounts of dicarboxylic acids are also found in virtually all cutin samples thus far examined. Recently hexdecane-1,8,16-triol was reported to be covalently attached to the stem cutin of

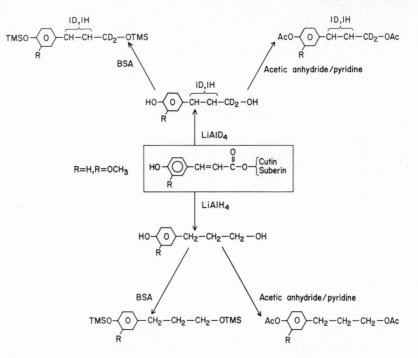

Figure 8. Procedure for the identification of p-coumaric and ferulic acids from cutin and suberin. The phenol function is shown free; however, it could be esterified in cutin and suberin.[99]

$Psilotum$[14] and monohydroxy and dihydroxy acids analogous to those discussed above, but with a hydroxyl group at the penultimate carbon instead of at the ω-carbon, were reported to be present in the cutin of certain liverwort species.[13] However these monomers appear to be of limited occurrence in higher plants. The C_{18}-family of acids, which are usually found as substantial components of cutin only in relatively slow growing plants, consists of 18-hydroxyoleic acid, 18-hydroxy-9,10-epoxy stearic acid, and 9,10,18-trihydroxy stearic acid together with their Δ^{12} monounsaturated analogs (Table 3). Smaller quantities of Δ^{12}, Δ^{15} diunsaturated analogs are also found.[31,50,118] 9,10,18-Trihydroxy-12,13-epoxystearic acid and 9,10,12,13,18-penta-hydroxystearic acid have been identified in the cutin of $Rosmarinus$ $officinalis$[11] and at least small amounts of such acids are likely

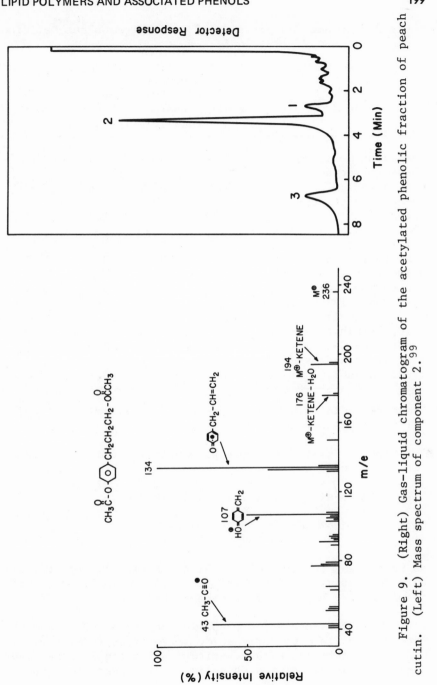

Figure 9. (Right) Gas-liquid chromatogram of the acetylated phenolic fraction of peach cutin. (Left) Mass spectrum of component 2.[99]

Table 3. Monomer composition of cutin from Golden Delicious apple fruit and *V. faba* leaf.

| | Monomers (% of total) | | | |
| Acid | Golden Delicious Apple | | | *V. faba* Leaf[a] |
	Fruit	Leaf (adaxial)	Leaf (abaxial)	
16-Hydroxyhexadecanoic	2.1	1.8	2.1	7.1
18-Hydroxyoctadeca-9,12-dienoic	14.3	---	1.6	---
18-Hydroxyoctadec-9-enoic	2.4	Tr	1.0	---
Monohydroxyhexadecane-1,16-dioic	---	0.9	1.6	---
Dihydroxyhexadecanoic[b]	19.7	32.9	53.9	84.5[c]
9,10-Epoxy-18-hydroxy-octadec-12-enoic	22.5	1.7	5.5	---
9,10-Epoxy-18-hydroxy-octadecanoic	15.3	32.4	21.8	---
9,10,18-Trihydroxy-octadec-12-enoic	4.3	1.1	2.2	---
9,10,18-Trihydroxy-octadecanoic	14.4	13.6	3.7	---

[a]Palmitic acid (3.6%), stearic acid (2.2%), and oleic acid (1%) were also found.

[b]Mixture of positional isomers, mainly 10,16-dihydroxyhexadecanoic acid.

[c]77.8% 10,16-dihydroxy-, 7.1% 9,16-dihydroxy-.

Data taken from references 50 and 74.

to be present in other plants also. Similarly recinoleic acid derivatives, which have been detected in *R. officinalis*[21] might also be present in other plants.

The major esterified phenolic consituent of cutins thus far examined is *p*-coumaric acid. Esterified ferulic acid was also identified as a constituent of cutin in some plants. However these esterified phenolic acids appear to be only

Dihydro-p-coumaryl alcohol (di-TMS derivative): m/e 296($M^{\oplus}$), 281($M^{\oplus}$-CH$_3$), 206 [base peak, $M^{\oplus}$-HOSi(CH$_3$)$_3$], and 191 [$M^{\oplus}$-CH$_3$-HOSi(CH$_3$)$_3$].

Dihydro-p-coumaryl alcohol-d$_3$ (di-TMS derivative): m/e 299 ($M^{\oplus}$), 284 ($M^{\oplus}$-CH$_3$), 209 [base peak, $M^{\oplus}$-HOSi(CH$_3$)$_3$], and 194 [$M^{\oplus}$-CH$_3$-HOSi(CH$_3$)$_3$].

Dihydro-p-coumaryl alcohol (diacetate derivative): m/e 236 ($M^{\oplus}$), 194 ($M^{\oplus}$-ketene), 176 ($M^{\oplus}$-ketene-H$_2$O), 134 (base peak, O=⟨·⊕⟩-CH$_2$-CH=CH$_2$), 107 ($\overset{\oplus}{\underset{H}{O}}$=⟨ ⟩=CH$_2$), 43 (CH$_3$-C≡O$^{\oplus}$).

Dihydro-p-coumaryl alcohol (diacetate derivative-d$_3$): m/e 239 ($M^{\oplus}$), 197 ($M^{\oplus}$-ketene), 179 ($M^{\oplus}$-ketene-H$_2$O), 137 (base peak, O=⟨·⊕⟩-C$_3$H$_2$D$_3$), 108 ($\overset{\oplus}{\underset{H}{O}}$=⟨ ⟩=CHD), 43 (CH$_3$-C≡O$^{\oplus}$).

Dihydroconiferyl alcohol (di-TMS derivative): m/e 326 ($M^{\oplus}$), 311 ($M^{\oplus}$-CH$_3$), 236 [$M^{\oplus}$-HOSi(CH$_3$)$_3$], 206 [base peak, $M^{\oplus}$-2CH$_3$-HOSi(CH$_3$)$_3$].

Dihydroconiferyl alcohol (diacetate derivative): m/e 266 ($M^{\oplus}$), 224 ($M^{\oplus}$-ketene), 164 (base peak, O=⟨·⊕⟩-CH$_2$-CH=CH$_2$), 137 ($\overset{\oplus}{\underset{H}{O}}$=⟨ ⟩=CH$_2$), 43 (CH$_3$-C=O$^{\oplus}$). OCH$_3$, OCH$_3$

Dihydroconiferyl alcohol-d$_3$ (diacetate derivative): m/e 269 ($M^{\oplus}$), 227 ($M^{\oplus}$-ketene), 167 (base peak, O=⟨·⊕⟩-C$_3$H$_2$D$_3$), 138 (H-$\overset{\oplus}{O}$=⟨ ⟩=CHD), 43 (CH$_3$-C≡$\overset{\oplus}{O}$). OCH$_3$, OCH$_3$

Figure 10. Major ion fragments in the mass spectra of the derivatives of dihydro-p-coumaryl alcohol and dihydroconiferyl alcohol.[99]

minor components of cutin as they constituted <1% of the cutin samples analyzed.[99] However as only a very limited number of plants have been examined for the phenolic content of their cutin, it is difficult to assess the significance of the phenolic content of cutin.

Suberin. The major aliphatic components of suberin are ω-hydroxy acids, the corresponding dicarboxylic acids, very long (>C$_{20}$) acids and corresponding alcohols (Table 4). Among the bifunctional molecules monounsaturated C$_{18}$ predominates in many plants, although in some cases saturated C$_{16}$ constitutes an equally dominant component.[8,9,48,49,67,72,73,102] Very long chained components constitute a more significant proportion of the ω-hydroxy acids than of the

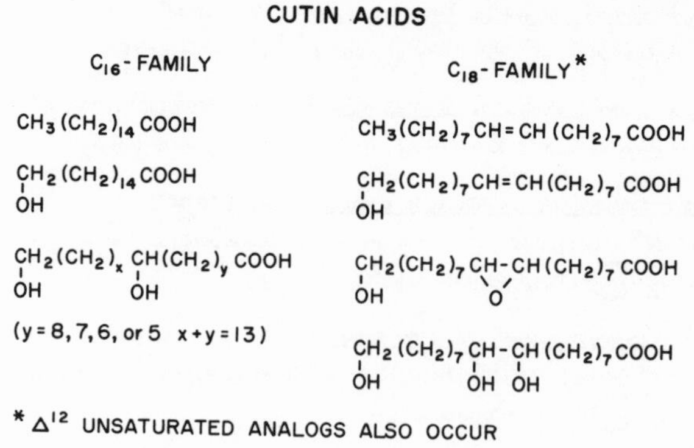

CUTIN ACIDS

C_{16}- FAMILY C_{18}- FAMILY*

$CH_3(CH_2)_{14}COOH$ $CH_3(CH_2)_7CH=CH(CH_2)_7COOH$

$\underset{\text{OH}}{CH_2}(CH_2)_{14}COOH$ $\underset{\text{OH}}{CH_2}(CH_2)_7CH=CH(CH_2)_7COOH$

$\underset{\text{OH}}{CH_2}(CH_2)_x\underset{\text{OH}}{CH}(CH_2)_yCOOH$ $\underset{\text{OH}}{CH_2}(CH_2)_7\underset{\text{O}}{CH-CH}(CH_2)_7COOH$

(y = 8, 7, 6, or 5 x + y = 13)

$\underset{\text{OH}}{CH_2}(CH_2)_7\underset{\text{OH OH}}{CH-CH}(CH_2)_7COOH$

* Δ^{12} UNSATURATED ANALOGS ALSO OCCUR

Figure 11.   Structures of the major components of cutin.

Table 4.   Composition of aliphatic monomers of suberin

Chain Length	Fatty acids			Fatty alcohols		
	Carrot (14%)	Rutabaga (4%)	Sweet Potato (9%)	Carrot (3%)	Rutabaga (5%)	Sweet Potato (4%)
C_{16}	3.0	3.0	0.9	ND	ND	ND
C_{18}	9.4	1.8	1.1	9.4	28.4	68.3
$C_{18:1}$	2.3	3.2	3.9	ND	10.7	ND
C_{20}	17.6	71.4	0.9	37.7	38.8	15.5
C_{22}	39.8	16.4	3.3	52.8	22.4	6.1
C_{24}	23.3	3.9	4.7	D	D	3.0
C_{26}	0.6	ND	16.6	D	D	3.0
C_{28}	ND	ND	20.1	ND	ND	0.6
C_{30}	ND	ND	34.3	ND	ND	0.1

Data taken from reference 73.
D, detectable; ND, not detectable.

dicarboxylic acids. In-chain substituted polyfunctional components similar to those found in cutin occur in suberin only in very small quantities. Depolymerization by hydrogenolysis, trans-esterification with BF_3 in methanol, or alkaline hydrolysis releases only 20 to 50% (by weight) of the crude suberin samples. The intensely colored residue is presumably composed of ligninaceous material which might possibly also contain aliphatic chains attached by bonds not susceptible to the depolymerization techniques mentioned above. However little experimental information concerning the composition of such residues is currently available. A large portion of the soluble materials released by the depolymerization techniques is composed of phenolic materials which probably represents only the esterified phenols. In such soluble materials obtained from suberin preparations from potato, sweet potato, turnip, rutabaga, carrot, and red beet, ferulic acid was identified.[99] That ferulic acid thus identified, was in fact covalently attached to suberin was strongly suggested by the observation that ferulic acid

Table 4. Continued

ω-Hydroxy acids			Dicarboxylic acids		
Carrot (31%)	Rutabaga (36%)	Sweet Potato (36%)	Carrot (24%)	Rutabaga (22%)	Sweet Potato (21%)
23.0	41.5	4.9	26.2	42.7	6.6
6.9	4.0	1.3	14.9	11.2	9.1
53.7	42.0	90.7	54.6	35.9	80.5
7.8	9.1	0.3	3.0	5.4	0.2
7.8	3.4	0.3	0.2	3.3	ND
0.3	ND	0.5	ND	D	ND
ND	ND	0.9	ND	ND	ND
ND	ND	0.7	ND	ND	ND
ND	ND	ND	ND	ND	ND

could be recovered from the residual material remaining
after partial hydrolysis of suberin. Much further work is
necessary before the composition of the phenolic materials
in suberin can be fully understood.

   Structure of the Polymers. Very little information
concerning the structure of cutin is available mainly
because the depolymerization techniques thus far used do
not give soluble oligomers which can be subjected to structural studies. As cutin is a polyester, all linkages are
nearly equally susceptible to cleavage, and therefore chemical depolymerization techniques cannot be used to examine
the structure of this polymer. Recently a little progress
has been made in our understanding of the structure of
this polymer as the amount and the nature of free hydroxyl
groups present in cutin have become known in a few cases.
Of the four methods tested for this purpose, two have given
meaningful values. One method involves oxidation of free
hydroxyl groups with CrO_3-pyridine complex for 18 hr
followed by depolymerization with 0.01 M sodium methoxide
in anhydrous methanol and subsequent analysis of the
monomers.[30] The other method involves treatment of cutin
with methane sulphonyl chloride followed by depolymerization
with $LiAlD_4$, which results in replacement of each free
hydroxyl group with a deuterium.[100] Combined GLC and MS of
the resulting mixture of monomers allows quantitation of
products as well as localization of deuterium. Both methods
have been applied only to cutins which contain only the C_{16}
family of monomers. Both studies lead to the conclusion
that the in-chain hydroxyl group of dihydroxy C_{16} acid
constitute the bulk of the free hydroxyl groups present in
cutin, strongly suggesting that the primary hydroxyl groups
are preferentially esterified. However it is clear that a
good portion of the secondary hydroxyl groups are also
esterified. The mesylation technique showed that tomato
fruit cutin contained 0.38 free hydroxyl groups per monomer,
a value which is only slightly lower than that (0.41)
obtained from the acetylation method (Table 5). On the
other hand, using the CrO_3 oxidation technique, the number
of free in-chain hydroxyl groups were found to be somewhat
higher. Until much more progress is made in this area, it
is difficult to assess the merits of these techniques. In
any case, the conclusion that the primary hydroxyl groups
are preferentially esterified appears valid and the model
of cutin shown on the right in Figure 1 reflects this
conclusion, as well as the finding that small amounts of

Table 5. The amount and nature of free hydroxyl groups in
tomato (Scout) cutin.

Dihydroxy C_{16} acid with free hydroxyl[a]	Moles/mole of monomer
at C-10	0.29
at C-16	0.03
at C-10 and C-16	0.03
None	0.65
Total free hydroxyl	0.38
[1-^{14}C]Acetyl group incorporated[b]	0.41

[a]Determined by methane sulphonyl chloride treatment followed
by depolymerization by $LiAlD_4$ and subsequent analysis of
monomers by combined GLC-MS.

[b]Determined with [1-^{14}C]acetyl chloride assuming an average
monomer molecular weight of 270.

Data taken from reference 100.

esterified p-coumaric acid and ferulic acid occur in cutin.
X-ray powder patterns of crude cutin samples indicated
crystalline regions in the polymer, but removal of residual
waxes embedded in the polymer by repeated solvent extraction
of the sample resulted in loss of the diffraction pattern,
strongly suggesting an amorphous structure for cutin.[4]

     Little information concerning the phenolic constituents
of suberin is available. In most cases, about 50% of the
polymer, i.e. material, remains insoluble after hydrogenolysis with $LiAlH_4$ and this core might be somewhat similar to
lignin. Reduction product of ferulic acid is a significant
component of the monomeric phenols released by the hydrogenolysis, but much of the phenolic constituents released
(one-half to two-thirds) remain unidentified. The major
components of the aliphatic fraction released by hydrogenolysis (one-third to one-half) are ω-hydroxy acids and
dicarboxylic acids. The methane sulphonyl chloride technique for the determination of free hydroxyl groups showed
that the ω-hydroxyl groups of the aliphatic monomers are
all esterified.[100] From such limited structural studies the

working model shown on the left in Figure 1 is proposed for
the structure of suberin. As the bark lignin usually
contains less methoxyl groups than the internal lignin,[112]
it is speculated that the phenolic core of suberin also
contains less methoxyl groups than lignin. The suberin
phenolics may be attached to the cell wall in a manner
similar to that suggested for lignin attachment to carbohydrates,[35,77] and the aliphatic compounds are probably esterified to give a waterproofing layer. The dicarboxylic acids
and ω-hydroxy acids might also be used to cross-link
aromatic oligomers which might be somewhat similar to those
obtained from lignin (see Chapter 5).

BIOSYNTHESIS

Cutin. The old concept[79,93] that cutin is formed by
the spontaneous oxidation of the unsaturated fatty acids,
which reach the plant surface, was seriously in doubt by
the 1950's when the polyester nature of cutin and the structure of cutin monomers became known.[53] In the 1960's,
speculative pathways for the biosynthesis of cutin were
suggested.[7,46] These suggestions, which were based mainly
on observations on wound-healing in leaves, postulated that
some ill-defined oxidases called "stearic acid oxidase" and
"oleic acid oxidase" in addition to lipoxidase were involved
in the so-called "cutin resynthesis," which was assumed to
occur during wound-healing. As none of the products formed
during wound-healing were actually chemically examined,
these investigators did not realize that wound-healing
involves suberization,[26] and the changes they were observing
most probably had no relationship to the biosynthesis of
cutin. In fact, these speculations were not even consistent
with the known structure of cutin monomers and the known
properties of lipoxidases. However in the absence of
reliable biochemical evidence, these postulates were accepted
and widely quoted even as late as 1974.[86,88,115]

Modern techniques of metabolic biochemistry such as the
use of radioactive precursors were not used to elucidate
the biochemical pathways involved in the synthesis of cutin
until 1970.[60,62] The application of modern biochemical
techniques to this problem has yielded results which have
fairly well established the pathways involved in the biosynthesis of cutin. For a discussion of the biosynthesis of

cutin monomers it is convenient to divide the cutin monomers into two families, a C_{16}-family and a C_{18}-family, as indicated in an earlier section.

   Biosynthesis of C_{16}-Family of Cutin Acids. Most rapidly growing plants contain the C_{16}-family of acids as the dominant components of their cutin. Therefore, young rapidly expanding *V. faba* leaves were incubated with the most probable precursor of this family of acids, $[1-^{14}C]$ palmitic acid.[60] After a thorough extraction of the tissue with solvents, the insoluble residue was found to be labeled. Exhaustive hydrogenolysis of this residue with LiAlH$_4$ released the radioactivity into chloroform soluble products. Radio chromatographic analysis and chemical degradation showed that the major part of the label was contained in the C_{16}-triol expected to be produced with 10,16-dihydroxy-palmitic acid, the major monomer of *V. faba* cutin.[74] The only other labeled components found in the labeled cutin derived from exogenous $[1-^{14}C]$palmitic acid was 16-hydroxy-palmitic acid and palmitic acid itself. The time-course of incorporation of the labeled palmitic acid into various cutin components did not reveal any precursor-product relationships. However, exogenous 16-hydroxy$[1-^{14}C]$palmitic acid was directly converted into the dihydroxy C_{16} acid of cutin, whereas labeled 10,16-dihydroxy-C_{16} acid was directly incorporated into cutin. These results strongly suggested that palmitic acid was first hydroxylated at the ω-position and subsequently at C-10. Synthetic 9- or 10-hydroxy C_{16} acid was hydroxylated at the ω-position, raising the possibility of an alternate order of hydroxylation.[68] However, as all of the hydroxy acids of cutin contain a hydroxyl group at the ω-carbon, it appears that ω-hydroxylation occurs prior to C-10 hydroxylation.

   Attempts to demonstrate the postulated hydroxylation reactions in cell-free preparations have been stifled by the occurrence of competing enzymes in such preparations. The major interfering activities are thioesterases and α- and β-hydroxylating enzymes. Therefore, it is often difficult to demonstrate the requirement for thioesters and the hydroxylated products obtained are mixtures of positional isomers. For example, palmitic acid gives rise to 2- and 3-hydroxylated products in addition to the 16-hydroxylated acid. Similarly, 2,16- and 3,16-dihydroxy acids, in addition to 10,16-dihydroxy acids, are generated from 16-hydroxy-palmitic acid in cell-free preparations. Therefore assays for these activities require resolution of such mixtures of positional isomers. Recently techniques involving gas chromatography and high pressure liquid chromatography have been developed for this purpose.

A microsomal preparation from the embryonic shoots of germinating *V. faba* catalyzed ω-hydroxylation of palmitic acid,[110] as shown by a radio gas chromatographic assay (Fig. 12). This hydroxylation showed a pH optimum of 8.0 and required NADPH (not NADH) and molecular oxygen as cofactors. This preparation also catalyzed ω-hydroxylation of myristic acid, oleic acid, and stearic acid, consistent with the previous observations, which also suggested that the ω-hydroxylating system may not show a stringent substrate specificity. The microsomal ω-hydroxylation was inhibited by cytochrome c and metal ion chelators such as phenanthroline. This enzyme was inhibited by CO, but light (450 nm) did not substantially reverse this inhibition. Therefore, if a cytochrome P_{450}-type protein is involved in this hydroxylation, it must have a higher affinity for CO than that observed with cytochrome P_{450} found in other hydroxylases.

A cell-free preparation obtained from the excised epidermis of *V. faba* leaves catalyzed C-10 hydroxylation of 16-hydroxypalmitic acid.[117] This hydroxylation which showed a pH optimum of 7.3, required NADPH (but not NADH) and O_2 as cofactors. In addition, ATP and CoA were required, strongly suggesting that activation of the carboxyl group located eight methylene groups away from the hydroxylation site was necessary for this reaction. More recently a high pressure liquid chromatographic assay for C-10 hydroxylase was developed (Fig. 13), and such an assay demonstrated that a microsomal preparation from the shoots of germinating *V. faba* catalyzed hydroxylation of 16-hydroxypalmitic acid specifically at C-10 (Fig. 13).[111] This system also required NADPH (but not NADH) and O_2, but this preparation did not show any stimulation of hydroxylation with the addition of ATP and CoA. However it was found that this microsomal preparation catalyzed rapid hydrolysis of CoA esters of fatty acids and that the addition of ATP and CoA resulted in great stimulation of the competing reaction, namely β-oxidation. These observations provide an explanation for the lack of stimulation of C-10 hydroxylation by ATP and CoA. It appears probable that *in vivo* it is a thioester such as the CoA ester which undergoes hydroxylation followed by

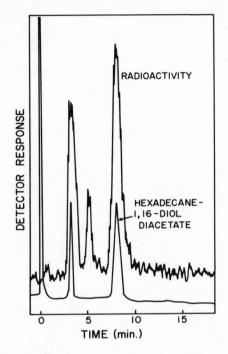

Figure 12. Radio gas–liquid chromatogram of diacetates
prepared from diols, separated by TLC, from the hydrogenolysis product of the soluble lipids obtained from the incubation of [1-^{14}C]palmitic acid and NADPH with a microsomal
preparation from the embryonic shoots of germinating *V. faba*.
Authentic hexadecane-1,2-diol diacetate and hexadecane-1,16-
diol diacetate coinjected with the labeled material are
shown in the lower tracing. The radioactive components
with retention times shorter than that of hexadecane-1,16-
diol diacetate are products of competing reactions such as
α- and β-oxidation.[110]

transfer of the hydroxyacyl group to the growing polymer.
The C-10 hydroxylation catalyzed by the microsomal preparation was severely inhibited by CO, and this inhibition was
reversed by light at 450 nm, suggesting the involvement of
a cytochrome P_{450}-type protein.[111] It is not clear whether
the ω-hydroxylation and C-10 hydroxylation, both of which
are catalyzed by the same microsomal preparation, are

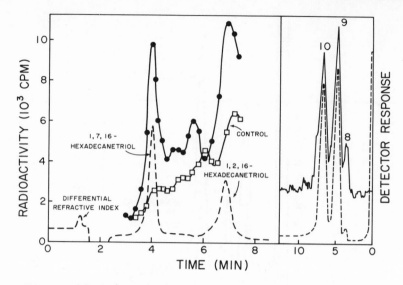

Figure 13.   (Left) High pressure liquid chromatogram of the C_{16}-triols produced by the hydrogenolysis of the dihydroxy C_{16} acids generated by the hydroxylation of ω-hydroxy[G-³H]palmitic acid by a microsomal preparation from *Vicia faba* embryonic shoots.   Control, boiled enzyme. (●——● and □——□ , radioactivity).   (Right) Radio gas-liquid chromatogram of the dimethyl esters of the dicarboxylic acids obtained by CrO₃ oxidation of the labeled 1,7,16-hexadecanetriol fraction.[110]   Number on each peak represents the chain length and dashed line represents flame ionization detector response of coinjected standards.

related as far as the enzymes involved are concerned. Furthermore, it is also not clear why the CO inhibition of ω-hydroxylation is not readily reversed by light (450 nm), whereas C-10 hydroxylation is reversibly inhibited by CO.

Oxidation of the terminal hydroxyl group of cutin monomers can occur as 16-oxo-9-hydroxypalmitic or 16-oxo-10-hydroxypalmitic acid and dicarboxylic acids are found in cutin.[64,65]   Furthermore, in tissue slices of embryonic shoots of *V. faba* exogenous 10,16-dihydroxy[G-³H]palmitic

acid was converted into the 16-oxo derivative.[65] Cell-free
preparations from the excised epidermis of *V. faba* leaves
catalyzed oxidation of 16-hydroxy[G-^{3}H]palmitic acid to the
corresponding dicarboxylic acid with NADP as the preferred
cofactor.[71] This ω-hydroxy acid dehydrogenase was located
mainly in the 100,000g supernatant and it required a pH
optimum near 8.0. The apparent K_m values for 16-hydroxypal-
mitic acid and NADP were 1.3 x 10^{-5} M and 3.6 x 10^{-4} M,
respectively. Thiol reagents such as p-chloromercuriben-
zoate and N-ethylmaleimide inhibited this ω-hydroxy acid
dehydrogenase. Although a 16-oxo intermediate did not
accumulate during the reaction, such an intermediate could
be trapped as the 2,4-dinitrophenylhydrazone. The involvement of such an oxo intermediate in the reaction was also
suggested by the observation that the enzyme preparation also
catalyzed the oxidation of 16-oxopalmitic acid to the
corresponding dicarboxylic acid.

On the basis of the results of the *in vivo* and *in vitro*
experiments, the pathway shown in Figure 14 was proposed for
the biosynthesis of the C_{16}-family of cutin acids. The
exact chemical nature of the substrates remains to be
elucidated, although a CoA ester or a similar thioester
can account for all of the experimental results thus far
obtained. The naturally occurring hexadecane-1,8,16-triol
is probably formed by the reduction of 9,16-dihydroxyhexa-
decanoyl-CoA by a reductase similar to that found in other
plants,[63] and 16-hydroxy-10-oxo-hexadecanoic acid is
produced by the oxidation of the corresponding dihydroxy
acid.

Biosynthesis of C_{18}-Family of Cutin Acids. The presence
of an epoxide or a *vic*-diol function at the 9,10-position of
the C_{18}-family of acids suggested that these monomers are
derived by modification of the Δ^9 double bond of oleic acid
which is known to be present in all cells. In agreement
with this suggestion, [1-^{14}C]oleic acid was readily incorporated specifically into the C_{18}-family of cutin acids,
whereas [1-^{14}C]stearic acid was not incorporated into the
major C_{18} monomers of cutin in apple fruit skin slices and
leaf discs of apple and *S. odoris*.[75],[76] Therefore, it
appeared that ω-hydroxylation of oleic acid followed by
epoxidation and subsequent hydration gave rise to the C_{18}-
family of cutin acids. In agreement with this hypothesis,
exogenous 18-hydroxy[18-^{3}H]oleic acid was directly converted
to 18-hydroxy-9,10-epoxystearic acid of cutin in the skin

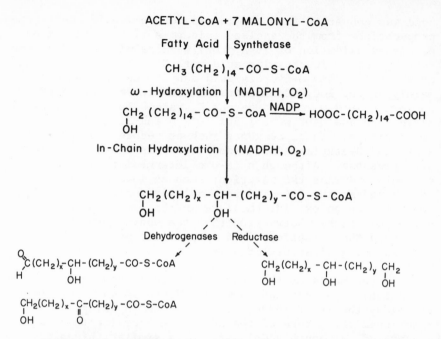

Figure 14. Pathway for the biosynthesis of the C_{16} family of cutin acids. Even though CoA esters are indicated to be the substrates, the nature of the intermediates is yet to be established. Palmitoyl-CoA is probably generated by activation of the free acid, and not directly from fatty acid synthetase.

slices of grape berries, in which this epoxy acid constitutes a major component of cutin.[22] Similarly, exogenous 18-hydroxy-9,10-epoxy[18-³H]stearic acid was converted to 9,10,18-trihydroxystearic acid in the skin slices of young apple fruits. These experimental results strongly supported the above hypothesis concerning the biosynthesis of the C_{18}-family of acids. The Δ^{12} unsaturated acids of cutin could be formed by ω-hydroxylation of linoleic acid, followed by the specific epoxidation at the Δ^9 position, and subsequent hydration of the epoxide. In fact, exogenous [1-¹⁴C]-linoleic acid was converted into 18-hydroxylinoleic acid, 18-hydroxy-9,10-epoxyoctadec-12-enoic acid, and 9,10,18-trihydroxyoctadec-12-enoic acid in the skin slices of apple fruit.[76] Exogenous [1-¹⁴C]linolenic acid was also converted into similar hydroxy and epoxy acids containing one additional double bond. Even though in these cases epoxidation and hydration occur only at Δ^9 double bond, in some plants such as *Rosinarinus officinalis*, similar reactions presumably take place also at Δ^{12} double bond, thus giving rise to 9,10,12,13,18-pentahydroxystearic acid. In young leaves of *R. officinalis*, exogenous [1-^{14}C]linoleic acid was converted into the pentahydroxy acid and into all of the intermediates expected to be involved in this conversion.[21] All of these labeled products were identified by radio chromatographic techniques in conjunction with chemical degradation. Furthermore, combined GLC-MS showed that all of the intermediates in the proposed pathway (Fig. 15) were present in *R. officinalis* cutin.

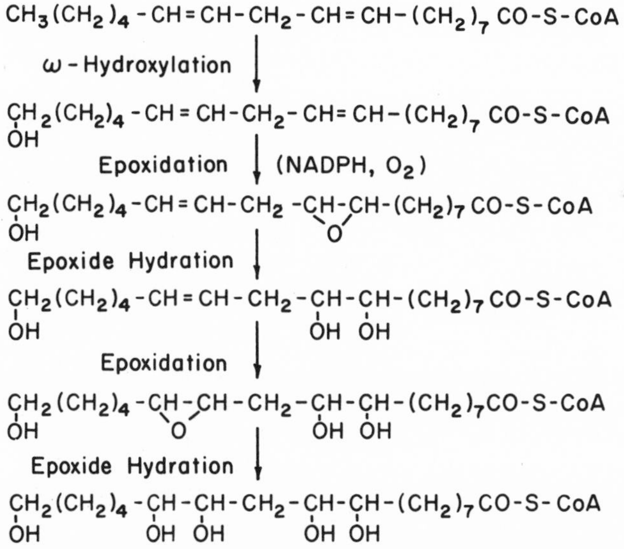

$CH_3(CH_2)_4 - CH=CH-CH_2-CH=CH-(CH_2)_7 CO-S-CoA$

ω - Hydroxylation $\downarrow$

$\underset{OH}{CH_2}(CH_2)_4 - CH=CH-CH_2-CH=CH-(CH_2)_7 CO-S-CoA$

Epoxidation $\downarrow$ (NADPH, O_2)

$\underset{OH}{CH_2}(CH_2)_4-CH=CH-CH_2 -\underset{\diagdown O \diagup}{CH-CH}-(CH_2)_7 CO-S-CoA$

Epoxide Hydration $\downarrow$

$\underset{OH}{CH_2}(CH_2)_4-CH=CH-CH_2-\underset{OH}{CH}-\underset{OH}{CH}-(CH_2)_7 CO-S-CoA$

Epoxidation $\downarrow$

$\underset{OH}{CH_2}(CH_2)_4-\underset{\diagdown O \diagup}{CH-CH}-CH_2-\underset{OH}{CH}-\underset{OH}{CH}-(CH_2)_7CO-S-CoA$

Epoxide Hydration $\downarrow$

$\underset{OH}{CH_2}(CH_2)_4-\underset{OH}{CH}-\underset{OH}{CH}-CH_2-\underset{OH}{CH}-\underset{OH}{CH}-(CH_2)_7 CO-S-CoA$

Figure 15. Pathway for the biosynthesis of the C_{18} family of cutin acids. Linoleic acid is used for illustration but oleic acid and probably to a lesser extent linolenic acid undergo analogous reactions.

From the experimental results discussed above it
became quite clear that the three key enzymes involved in
the biosynthesis of the C_{18}-family of cutin acids must be
ω-hydroxylase, epoxidase, and epoxide hydrase. These three
enzymes were recently demonstrated in extracts of cutin
synthesizing tissues. As discussed in the previous section,
an ω-hydroxylating enzyme was demonstrated in microsomal
preparations from *V. faba*. A rather unique epoxidase was
discovered in a 3,000*g* particulate fraction obtained from
young spinach leaves.[24] This preparation catalyzed the
conversion of 18-hydroxy[18-^{3}H]oleic acid to 18-hydroxy-
cis-9,10-epoxy[18-^{3}H]stearic acid. This conversion, which
occurred optimally at pH 9.0, required O_2 and NADPH (but
not NADH) as cofactors. In addition, ATP and CoA were also
required, suggesting that activation of the carboxyl group,
seven methylenes away from the epoxidation site, was
required for this reaction. Light-reversible inhibition of
the epoxidation by CO suggested the involvement of a cytochrome P_{450}-type protein in this reaction. Chelators, such
as *o*-phenanthroline, inhibited epoxidation and this inhibition was reversed by Fe^{+2}. The apparent Km for 18-hydroxy-
oleic acid was 7.5×10^{-5} M and the enzyme showed a fairly
stringent substrate specificity in that the *cis*-Δ^9 and a
free hydroxyl group at C-18 were required structural features
of the substrate. Hydration of 18-hydroxy-*cis*-9,10-epoxy-
stearic acid to *threo*-9,10,18-trihydroxystearic acid was
catalyzed by a 3,000*g* particulate fraction obtained from
the skin of young apple fruit.[25] This epoxide hydrase,
which showed a pH optimum of 6.5, required no cofactors
and was located exclusively in the 3,000*g* particulate
fraction from the skin. The apparent Km for 18-hydroxy-*cis*-
9,10-epoxystearic acid was 4×10^{-4} M and any modification
of the functional groups of this substrate rendered it a
poor substrate for the hydrase. The substrate specificity
of this biosynthetic epoxide hydrase was also illustrated
by the observation that styrene oxide, a commonly used
substrate for the animal epoxide hydrase,[91] was not hydrated
by the plant enzyme. The epoxide hydrase from apple was
strongly inhibited by thiol-directed reagents but was only
mildly inhibited by trichloropropeneoxide, a potent inhibitor
of the animal epoxide hydrase.

On the basis of the experimental results obtained with
tissue slices and cell-free preparations, the pathway shown
in Figure 15 appears to be involved in the synthesis of the

C_{18}-family of cutin acids. The involvement of CoA ester postulated for all of the reactions is yet to be proven for some of the reactions.

Biosynthesis of the Polymer from Monomers. Since cutin is an extracellular polymer, it is probable that synthesis of this polymer occurs at an extracellular location. In agreement with this hypothesis is the finding that incorporation of labeled monomer acids into an insoluble polymer was catalyzed by a 3,000*g* particulate fraction (containing the cuticular membranes) obtained from the excised epidermis of *V. faba*.[20] This incorporation of hydroxy acids into the insoluble material appears to represent cutin biosynthesis, as similar preparations from mesophyll tissue did not catalyze incorporation of hydroxy acids into insoluble polymers. Conclusive evidence that the incorporation of labeled palmitic acid, 16-hydroxypalmitic acid, and 10,16-dihydroxypalmitic acid into the insoluble material, catalyzed by the particulate preparation, did represent cutin biosynthesis, was provided by enzymatic degradation of the insoluble polymer. Cellulase, pectinase, and pronase treatments failed to release the radioactivity from the insoluble material, while a fungal cutinase (discussed in a later section) released the bulk of the radioactivity contained in the polymeric material. Treatment of the insoluble polymer with $LiAlH_4$, $NaBH_4$, alcoholic KOH, and BF_3 in methanol, and analysis of the soluble products released, showed that the carboxyl group of the hydroxy acid was esterified to a hydroxyl group of cutin (Table 6). In support of this conclusion was the observation that methyl esters and fatty alcohols were not significantly incorporated into cutin, but acetates of hydroxy acids were incorporated at 60 to 70% of the rate observed with the parent acid.[23]

Incorporation of monomer acids into the polymer required ATP and CoA, strongly suggesting that hydroxyacyl-CoA is the substrate for the esterification reaction.[20,23] Two pH optima near 7.0 and 8.5 were observed when free acids were used as the substrates; but palmitoyl-CoA incorporation showed a single optimum pH near 8.5, suggesting that activation of the carboxyl group and transacylation occurred optimally at pH 7.0 and 8.5, respectively. The apparent Km values for palmitic acid, 16-hydroxypalmitic acid, and 10,16-dihydroxypalmitic acid were 2.0×10^{-5} M, 6.7×10^{-5} M, and 1.1×10^{-4} M, respectively. However,

Table 6. Sequential enzyme treatment and chemical depolymerization of the insoluble residue derived from 10,16-
dihydroxy[G-^{3}H]palmitic acid in a cell-free preparation
from *V. faba* epidermis.

Treatment	% Residue solubilized		Identity of Soluble Product
	Weight	Radioactivity	
Cellulase + pectinase	41	21	ND
Pronase	6	4	ND
Cutinase	3	61	10,16-Dihydroxypalmitic acid
LiAlH$_4$	42	95	1,7,16-Hexadecanetriol
LiBH$_4$	35	96	1,7,16-Hexadecanetriol
NaBH$_4$	9	7	ND
KOH in ethanol	47	94	10,16-Dihydroxypalmitic acid
BF$_3$ in methanol	41	95	Methyl 10,16-Dihydroxy-palmitate

ND — not determined.

Data taken from reference 23.

these values may not reflect the true specificity of the
transacylase which catalyzes the synthesis of the polymer,
as the present apparent Km values were determined with free
acids. It is likely that activation of the monomers is a
limiting factor in the cell-free preparation, but *in vivo*,
the hydroxy acids might be synthesized as their CoA esters,
and therefore there might not be a requirement for the
thiokinase. Substrate specificity of the enzyme preparation
did reflect the chain length of the acyl moieties found in
V. faba cutin, as the C_{16}-family of acids was the preferred
substrate, although fatty acids from C_{10} to C_{18}, and C_{18}
hydroxy acids, could also be incorporated into the polymer
by the enzyme preparation. Apparently, other plant tissues

also contain similar enzymes, as we have obtained enzyme preparations capable of esterifying acyl moieties to cutin from *V. faba* flower petals and the epidermis of *Senecio odoris* leaves, but these preparations have not been examined in detail.

The particulate preparation described above did not require exogenous primer, most probably because the preparation contained endogenous primer to which the monomer acids could be esterified.[20,23] In order to determine the nature of the primer, the transacylase had to be dissociated from the primer. Mild ultrasonic treatment dissociated the enzyme from the endogenous primer so that an enzyme preparation which required exogenous primer was obtained. With this preparation the most active primer was purified cutin from very young leaves of *V. faba*, although a variety of chemically distinct cutins from several plant species could also function as primers. Other possible acyl acceptors such as cellulose, glycerol, hexadecanol, cholesterol, and polyethylene glycol were ineffective. The primer-dependent incorporation of monomer acids into cutin showed the same cofactor requirements and substrate specificity as those observed with the particulate preparation. With the use of chemically modified primers it could be demonstrated that the hydroxyl groups of the primer were used in attaching the incoming monomer. For example, acetylation of *V. faba* cutin decreased its primer efficiency by 70%, whereas methylation of the carboxyl groups of cutin had no effect on its efficiency as a primer. Chemical or enzymatic treatment of cutin, which increased the number of hydroxyl groups or opened up the polymer matrix also increased priming efficiency of the cutin. For example, a brief treatment of *V. faba* cutin with fungal cutinase doubled the priming efficiency of the cutin. The more developed cutin from the older leaves was not as good a primer as the cutin from younger leaves. All of the experimental evidence thus far obtained suggests that the monomer acids generated as their CoA esters are transferred to the hydroxyl groups of the polymer by the transacylase which is closely associated with the growing polymer (Fig. 16). In the case of cutin where C_{18} monomers are also involved, a similar strategy is probably used. In addition, at least some of the monomers, such as epoxy acids and the trihydroxy acids, appear to be generated by enzymes present in close association with the polymer.[24,25] Thus, the thioesters of fatty acids generated

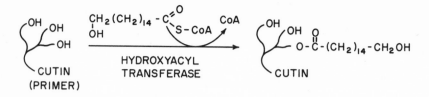

Figure 16.   Schematic representation of cutin synthesis
from monomers.

within the cells must be transported to the extracellular
location where they are converted into cutin acids and
incorporated into polymer. This extracellular location of
the generation of potentially toxic materials such as epoxy
acids and the rapid incorporation of such compounds into
the polymer would indeed explain how the plant is protected
from such toxic materials generated by its own metabolism.

The esterified phenolic acids found in cutin, most
probably are incorporated into cutin by an acyl transferase
which transfers *p*-coumaroyl and feruloyl moieties to the
free hydroxyl groups of cutin from the corresponding CoA
esters which are produced by activating enzymes already
demonstrated in plant tissues.[36,37,59,83,114]

Regulation of Cutin Synthesis. As cutin is the
structural component of the cuticle, it might be expected
that the rate of synthesis of cutin would be related to
the rate of expansion of the tissue. In fact, incorporation
of exogenous [1-^{14}C]palmitic acid into cutin in *V. faba*
leaves was roughly proportional to the rate of their growth;[62]
fully expanded leaves incorporated the labeled acid into
cutin extremely slowly and even the small amount of ^{14}C
incorporated into the insoluble residue was in palmitic acid
and not in hydroxylated products. Furthermore, exogenous
16-hydroxy[G-^{3}H]palmitic acid was readily incorporated into
the dihydroxy acid of cutin only in young expanding leaves.[74]
These results suggest that hydroxylation reactions are under
fairly strict physiological control. Even esterification
of exogenous 10,16-dihydroxy[G-^{3}H]palmitic acid in *V. faba*
leaves showed such a control.[74] Similar physiological
control was also found in other tissues; in apple fruits as
well as in apple leaves, incorporation of exogenous labeled

precursors into cutin occurred only in expanding tissue, but not in fully expanded tissue.[76]

How the physiological state of the plant controls cutin synthesis is not known. The regulation would be expected to be in the control of the level of the enzymes involved. In fact, the enzymes of cutin biosynthesis, which have been examined thus far, show much higher levels of activity in the young expanding tissue than in the fully expanded tissue. For example, enzyme preparations from nearly mature spinach leaf showed a much lower level of 18-hydroxy-octadecenoyl-CoA epoxidase than similar preparations from young leaves.[24] Particulate preparations from the epidermis of young leaves of *V. faba* contained nearly ten times the acyl-CoA:cutin transacylase activity as that found in similar preparations from fully expanded leaves.[23] When the levels of the other enzymes involved in cutin synthesis become known, this pattern is likely to hold.

Developmental control in cutin biosynthesis appears to be reflected in the monomer composition of cutin. In embryonic shoots of *V. faba* about 70% of the disubstituted C_{16} acid contains a hydroxyl group at C-9, whereas in the mature leaf this isomer constitutes only about 10%.[64,65] The 9-hydroxyisomer of the cutin of the embryonic tissue has been characterized by the presence of an aldehyde function at the ω-carton. This association of 16-oxo-9-hydroxyhexadecanoic acid with the very young tissue is also reflected in the changes in monomer composition, observed in developing leaves (Table 7). Thus for example, cutin from the meristem contains 50% 9-hydroxyisomer, in which more than 80% is the 16-oxo derivative. These compositional changes, which were first deduced from deuterium labeling studies, were also clearly shown by biosynthetic studies. CrO_3 oxidation of the C_{16} alkane triols, obtained by $LiAlH_4$ treatment of cutin, derived from [1-^{14}C]palmitic acid in leaves of different ages, was used as a technique to examine the developmental changes in the biosynthesis of positional isomers. As illustrated in Table 8, labeled dicarboxylic acids obtained by such treatment quite clearly showed that the 9-hydroxy isomer is associated with very young leaves, and during development the 10-hydroxy isomer becomes clearly dominant. In addition to changes in the proportion of the two isomers, we have also measured the amounts of each isomer present during development (Fig. 17). In very small

Table 7.   Developmental changes in the composition of disubstituted hexadecanoic acid in *V. faba* cutin.

Leaf size (Half-width) (mm)	Positional isomer (%)			
	9-Hydroxy		10-Hydroxy	
	16-Oxo	16-OH	16-Oxo	16-OH
0	40.9	11.1	13.9	34.1
2-3	31.4	8.6	19.3	40.7
6-7	25.8	6.7	19.9	47.7
11-13	13.2	5.7	22.0	59.2
18	6.1	2.8	12.1	79.8

Data taken from reference 65.

Table 8.   Developmental changes in incorporation of $[1-^{14}C]$-
palmitic acid into the positional isomers of 9- or 10,16-
dihydroxypalmitic acid in *V. faba* cutin.

Leaf size (half-width) (mm)	Radioactivity in triol (cpm x 10^{-6})	Distribution of ^{14}C among the dicarboxylic acids derived by CrO_3 oxidation of the triols* (%)		
		C_8	C_9	C_{10}
1.5	0.78	29.1	49.5	21.4
4.5	1.09	29.8	54.8	15.4
8.5	1.68	17.0	40.5	42.3
12	2.70	10.8	50.2	39.0
15	3.40	10.8	45.5	43.7
18	0.96	7.8	44.6	47.7

Data taken from reference 65.

*The C_{16} triol degraded was derived by hydrogenolysis of
the dihydroxy C_{16} acid.

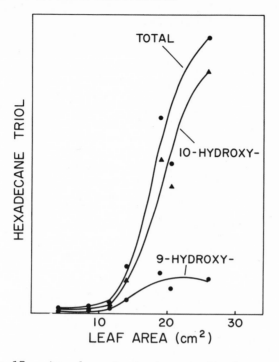

Figure 17. Age-dependent changes in the amount and the isomer composition of the hexadecane triol fraction from the hydrogenolysate of *Vicia faba* cutin.[68] The amount of the total triol was determined by gas chromatography and is represented by arbitrary units; isomer composition was estimated by mass spectrometry.[65]

leaves, which contain very little cutin, both the 9-hydroxy and the 10-hydroxy isomer are present in comparable quantities, and as the tissue expands, deposition of the 9-hydroxy isomer ceases while the synthesis of the 10-isomer becomes very rapid. As the 9-hydroxy isomer is quite specifically converted into the 16-oxo derivative, we tentatively conclude that the two isomers are synthesized by two different

hydroxylating enzymes with different positional specificity.
Recent analysis of cutin from various parts of *V. faba*
plant showed that the isomer composition of the dihydroxy
C_{16} acid of cutin is the same in the petiole and the stem
as it is in the leaf, whereas the underground part of the
stem contains substantially more of the 9-hydroxy isomer
than that found in other parts.[28] The only obvious difference
between the underground stem and the above-ground stem is
that the former is virtually dark-grown, while the latter
is light-grown. In fact, the above-ground stem of *V. faba*
plants, grown in the dark for two weeks, showed an isomer
composition similar to that of the under-ground stem, and
light appears to reverse this isomer ratio. Thus, it is
possible that the 10-hydroxylating enzyme in *V. faba* requires
light for its synthesis and/or activity. However, effects
of physiological and environmental changes on cutin synthesis
are virtually unexplored and warrants much further investigation.

   Suberin. Biosynthesis of Monomers. The phenolic
constituents of suberin would be derived from phenylalanine
in a manner similar to that observed in lignin synthesis
(discussed in Chapter 5) and the enzymes involved in such
transformations, such as phenylalanine ammonia lyase, and
the hydroxylases for the aromatic acids are known to be
induced prior to suberization during the wound-healing
process.[15,47,97,98,113,122] Therefore, I shall limit the
discussion in this chapter to the aliphatic components.
Since the major aliphatic monomers of potato suberin appears
to be derived from oleic acid, young potato tubers were
coated with an emulsion of [1-^{14}C]oleic acid and incubated.[68]
After extraction of the soluble lipids, the residue was
radioactive. This insoluble material, upon $LiAlH_4$ treatment
gave rise to labeled octadec-9-ene-1,18-diol, which would
be derived from ω-hydroxyoleic acid and/or the corresponding
dicarboxylic acid, the two major aliphatic components of
potato suberin. However uniform samples of intact potato
tubers are not conveniently obtained throughout the year.
Since the monomer composition of the aliphatic components
of suberin deposited at the wound periderm of potato tuber
tissue is identical to that of the natural skin (suberin),[72]
we tested whether such a tissue disk system is suitable for
biosynthetic studies on suberin. When [1-^{14}C]oleic acid
was incubated with suberizing potato tuber discs and the
soluble lipids were removed by thorough extraction with

solvents, the residual insoluble material was labeled.[27]
Upon hydrogenolysis, the bulk of this ^{14}C was released
into chloroform soluble products. Chromatographic analyses
of the labeled material and chemical degradation of the
isolated components showed that octadecenol and octadec-
9-ene-1,18-diol contained the major part of the label
incorporated into the insoluble material. Thus, oleic acid
was incorporated into ω-hydroxyoleic acid and/or the corresponding dicarboxylic acid of suberin. Suberizing potato
slices also incorporated [1-^{14}C]acetate into an insoluble
material. Hydrogenolysis followed by radio chromatographic
analyses of the products showed that ^{14}C was contained in
alkanols and alkane-α,ω-diols. In the former fraction a
substantial proportion of the label was contained in aliphatic chains longer than C_{20}, which are known to be common
constituents of suberin. Radio gas chromatography and
chemical degradation of the labeled diol fraction showed
that the major component was octadec-9-ene-1,18-diol, with
smaller quantities of saturated C_{16}, C_{18}, C_{20}, C_{22}, and C_{24}-
α,ω-diols. Soluble lipids derived from [1-^{14}C]acetate in
the aged tissue also contained labeled very long acids from
C_{20} to C_{28}, but no labeled ω-hydroxy acids or dicarboxylic
acids were detected. Thus it appears that the major
aliphatic monomers of suberin are incorporated into the
polymer as soon as they are synthesized just as previously
observed in the case of cutin monomers. In any case, aging
potato disks appear to be a suitable experimental material
with which to study the biochemistry of suberization.

   Enzymes Involved in Suberization. The chemical composition of aliphatic monomers of suberin suggests that the
key steps involved in their biosynthesis are chain elongation, ω-hydroxylation, and conversion of ω-hydroxy acids
to the corresponding dicarboxylic acids (Fig. 18). The
in-chain substituted minor components of suberin such as
the epoxy and di- and tri-hydroxy acids are most probably
produced in a manner similar to that involved in cutin
biosynthesis discussed in earlier sections.

   Although a chain elongating enzyme has not been isolated
from suberizing tissues, chain elongation has been demonstrated in cell-free extracts from other plant tissues.[69,84]
From the results obtained from such studies, it appears
highly probable that thioester of a fatty acid is elongated
using malonyl-CoA as the C_2 donor and NADPH as the preferred

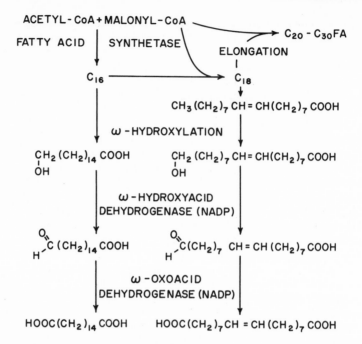

Figure 18. A probable pathway for the biosynthesis of the major aliphatic monomers of suberin.

reductant. It is probable that the first elongation step from C_{16} to C_{18} involves an acyl carrier protein derivative,[56] but the further elongation might use the CoA ester as the substrate.[16] However, the chemical nature of the acyl substrate has not been fully elucidated. The fatty alcohols found in suberin are probably produced by an acyl–CoA reductase similar to that described in other plant tissues,[57,58,61,63] but no such reductase has been yet found in suberizing plant tissues. Although ω–hydroxylase would appear to be a key enzyme involved in suberin biosynthesis, such an enzyme system has not been demonstrated in cell-free preparations from suberizing tissues. However, an ω–hydroxylating enzyme was demonstrated in the microsomal preparation from *V. faba* (see cutin biosynthesis section in this chapter), and probably a similar enzyme system is involved in suberin synthesis.

As dicarboxylic acids constitute a major component of
suberin, enzymes which catalyze oxidation of ω-hydroxy
acids to dicarboxylic acids should be present in suberizing
tissues.   In fact, cell-free extracts obtained from suberizing potato disks catalyzed conversion of 16-hydroxy[G-^{3}H]-
palmitic acid to the corresponding dicarboxylic acid with
NADP or NAD as the cofactor, with a slight preference for
NAD.[2]   This enzymatic activity, which was found mainly in
the 100,000g supernatant fraction, showed a pH optimum near
9.5 and an apparent Km of 5 x 10^{-5} M for 16-hydroxypalmitic
acid.[3]   Like most other dehydrogenases, this ω-hydroxy acid
dehydrogenase was inhibited by thiol-directed reagents.
Conversion of 16-oxo[G-^{3}H]palmitic acid to the corresponding
dicarboxylic acid was also catalyzed by this extract,
suggesting that such an oxo acid is an intermediate in the
conversion of the ω-hydroxy acid to the dicarboxylic acid.
In fact, the two steps involved in this conversion are
catalyzed by two enzymes separable by a gel filtration
procedure.[4,70]   The larger ω-oxo acid dehydrogenase highly
prefers NADP as the cofactor.   The smaller ω-hydroxy acid
dehydrogenase was similar in size to alcohol dehydrogenase
and, therefore, the two enzymes could not be resolved by
the gel filtration technique.   However, DEAE-cellulose ion
exchange chromatography could be used to resolve the ω-
hydroxy acid dehydrogenase from the alcohol dehydrogenase.
[4,66]   This ω-hydroxy acid dehydrogenase preferred NADP as
the cofactor, whereas the alcohol dehydrogenase preferred
NAD.

Regulation of Suberin Synthesis.   As the tissue ceases
growth, suberin synthesis ceases or occurs only at a reduced
rate.   Young growing potato tubers incorporated exogenous
precursors into suberin, while with mature tubers such
incorporation could not be demonstrated.[68]   The composition
of the aliphatic components of suberin does not appear to
change substantially with the age of the tissue, at least
in potato tubers, as the transparent suberin layer of very
young tubers, as well as the opaque brown layer of mature
tubers, contained 18-hydroxyoctadecenoic acid and the
corresponding dicarboxylic acids as the major components.[68]

The finding, that when a potato tuber is wounded, cells
at or near the wound which represent internal tissue and do
not synthesize suberin in the intact tuber, begin to
synthesize suberin,[72] illustrates another aspect of control

of suberin biosynthesis. The recent finding that suberization also occurs during the wound-healing of aerial parts
of plants such as fruits and leaves[26] shows that the wound-
induced synthesis of suberin is a general phenomenon in
plants. The time-course of deposition of suberin in the
periderm formed at the wound surface of potato tuber slices
shows that there is a lag period of 48 to 72 hr from the
time of wounding before suberin synthesis starts.[72] Measurement of the activity levels of one of the enzymes involved
in suberization, namely ω-hydroxy acid dehydrogenase, shows
that the activity of this enzyme is barely detectable in
the fresh tissue and activity appears at its maximal level
just prior to the period of rapid deposition of suberin
(Fig. 19). The rates of incorporation of exogenous
precursors into suberin also followed a similar pattern
in that after a lag period, rates increased rapidly and
after the period of rapid deposition of suberin monomers,
the rate of incorporation decreased.[4,27] This pattern
strongly suggests that the wounding process directly or
indirectly triggers the induction of the enzymes involved
in suberin synthesis. Preliminary experiments with cyclo-
heximide and actinomycin D strongly suggest that protein
synthesis is involved in this process and that transcrip-
tional and translational events involved in suberization
occur mainly between 72 and 96 hr after preparation of the
slices.[3] Since suberization occurs in aerial, as well as
underground, tissues, irrespective of the nature of the
natural protective polymer of the tissue, it appears
probable that some chemical, produced at or near the wound,
triggers a series of biochemical processes which result in
the induction of the suberin-synthesizing enzymes. Neither
the chemical nature of this suberization-inducing factor
nor its mechanism of action is known. However, recently it
was found that extensive washing of the potato slices
immediately or 24 hr after slicing severely inhibited suberization, strongly suggesting that some chemical, produced
at the wound, might be the suberization-inducing factor.[28]
This observation indicates that it might be possible to
isolate and characterize this factor. One additional
observation which might be helpful in elucidating the
mechanism of induction of suberization is that abscisic acid
stimulates suberization, indicating a possible hormonal
control of suberization.[28] Furthermore, analysis of the
material washed out of the freshly prepared potato disks
by high pressure liquid chromatography revealed that both

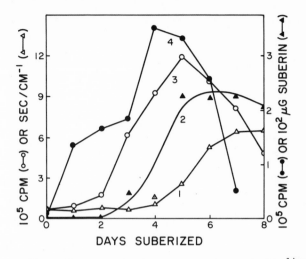

Figure 19. Changes in incorporation of [1-^{14}C]acetate into suberin (O—O); ω-hydroxy fatty acid dehydrogenase activity (●—●); level of the aliphatic monomers of suberin measured as octadecene-1,18-diol (▲—▲); and resistance of the tissue surface to diffusion of water vapor (△—△), during aging of potato slices.[3,27,72]

free and bound abscisic acid was washed out by water from the disks.[111] Exogenous abscisic acid did not fully restore the suberization rates in washed tissue. Even though it is not known whether the removal of abscisic acid is the sole reason for the severe inhibition of suberization resulting from the washing of the disks, the above results suggest that abscisic acid might play a key role in the regulation of suberin synthesis.

POSSIBLE ROLE OF LIPID
POLYMERS IN PATHOGENESIS

It has been pointed out that the development of cuticle might have been a crucial factor in the colonization of

land by plants.[18] However, the occurrence of cutin does
not appear to be limited to land plants; the insoluble
residue from a sea grass, *Zoestra marina*, which grows
submerged on coastal shorelines, contains covalently
attached dihydroxy C_{16} acids and the C_{18}-family of acids,
strongly suggesting that the cutin of this plant is similar
to that of the land plants (unpublished results). *Lemna
gibba*, an aquatic angiosperm, also has cutin similar to
that found in most other higher plants. The location of
the lipid polymers on the surface of plant tissues suggests
the obvious function of protection of the plant from
environmental factors unfavorable to the plant. These
include prevention of water loss as well as protection from
toxic chemicals and pathogenic microorganisms in the environment.[85] Since the cuticle is composed of the insoluble
polymeric material, as well as the soluble waxes, it is
difficult to assign the above functions to specific
cuticular components.

  Bacteria enter plant tissues through natural openings
and wounds, whereas fungi are known to penetrate the protective barrier, namely the cuticle.[85,115] The mode of
penetration has been the subject of controversy for many
decades, ever since Wiltshire proposed that a cuticle
dissolving substance was excreted by *Venturia* when it
infects apple and pear,[121] and it is still uncertain
whether fungal penetration is accomplished simply by the
physical thrust of the growing infection peg or by enzymatic
dissolution of the polymer. At the present time there
appears to be no concensus regarding the mechanism of
fungal penetration into plants. Scarcity of well conceived
experiments and eagerness to draw conclusions not warranted
from the data, have contributed to the confusion. Arguments
in favor of and opposed to the concept that fungal penetration into a plant is aided by cuticle hydrolyzing enzymes
excreted by the pathogen have been discussed in detail by
Martin,[85] and I shall not attempt to review the credibility
of these arguments. More recently, the importance of cutin
hydrolyzing enzymes in pathogenesis was emphasized in a
review devoted entirely to such enzymes.[115] Results of
recent electron microscopic studies on the penetration of
Botrytis cenerea into *V. faba* suggested an enzymatic
penetration[90] (Fig. 20).

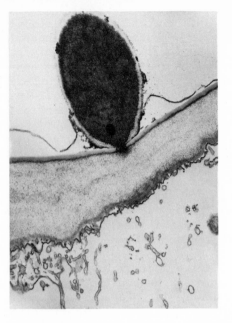

Figure 20. An electron micrograph showing the infection peg of *Botrytis cinerea* penetrating through the
cuticle of a *Vicia faba* leaf. Courtesy of Professor W. E.
McKeen of the University of Western Ontario, Canada.[90]

Enzymes Which Depolymerize Cutin. Phytopathogenic
fungi do penetrate the cuticle and if an enzymatic digestion is involved in this process, extracellular enzymes are
probably involved in this process. A pathogenic fungus
which lands on the plant surface finds itself in a situation
somewhat analogous to a culture medium containing cutin
as the sole source of carbon, although other organic
materials are also found on the plant surface. The finding
that a variety of microorganisms can grow on cutin as the
sole source of carbon[38,43,94,107] supports the contention
that phytopathogenic organisms produce extracellular enzymes
capable of hydrolyzing cutin. Heinen and coworkers have
published a series of papers[39-46] on cutin-degrading enzymes
called "cutinase complex," which reportedly contained cutin
esterase, and some ill-defined oxidases called "stearic
acid oxidase," "oleic acid oxidase," and "linoleic acid
oxidase." In addition, another enzyme called "carboxycutin
peroxidase" was supposed to participate in cutin degradation.

However, these interpretations are difficult to understand in the light of modern concepts of biochemistry and the present knowledge of cutin structure except for the obvious conclusion that some cutin hydrolyzing enzyme must have been produced by the fungi studied by these investigators. The techniques used to study cutin hydrolyzing enzymes were often not reliable, as the production of acid from crude cutin preparations observed could be due to hydrolytic enzymes other than cutinase. The products of hydrolysis were not carefully examined. An enzyme preparation was classified as an exotype cutin esterase because it released only small amounts of dihydroxyeicosanoic acid, which was called the main component of tomato cutin.[106,107] In fact, by far the main component of tomato cutin is dihydroxy-hexadecanoic acid and it is difficult to understand how one arrives at the above conclusions.

Recently, cutin hydrolyzing enzymes were studied with radioactive cutin as the substrate.[94] Such techniques gave unequivocal evidence for production of cutinase by phyto-pathogenic fungi. From the extracellular fluid of *Fusarium solani* f. *pisi* grown on cutin as the sole source of carbon, two isozymes of cutinase were isolated using standard techniques of protein fractionation (Table 9). These two enzyme preparations were homogeneous, as judged by electro-phoresis and sedimentation equilibrium centrifugation.[95,96] The properties of the two enzymes were quite similar. Both cutinases gave similar molecular weight values near 22,000 by Sephadex G-100 gel filtration, sedimentation equilibrium centrifugation, amino acid composition, and sodium dodecyl sulfate polyacrylamide electrophoresis. The amino acid composition of cutinase I was very similar to that of cutinase II. Neither enzyme has free SH groups and both contain one disulphide bridge.[82] Both cutinases also showed similar catalytic properties; with tritiated apple cutin as the substrate, they showed similar substrate concentration dependence, protein concentration dependence, time-course profiles, and pH dependence profiles with an optimum near 10.0. Both enzymes catalyzed hydrolysis of methyl hexadecanoate, cyclohexyl hexadecanoate, and to a much lesser extent hexadecyl hexadecanoate. Neither enzyme catalyzed hydrolysis of 9-hexadecanoyloxy-heptadecane, triglycerides, cholesteryl hexadecanoate, hexadecyl cinnam-ate, and α- or β-glucosides of p-nitrophenol. Both enzymes catalyzed hydrolysis of indoxyl acetate and short-chained

Table 9.   Purification of cutinase and non-specific esterase* from *F. solani pisi*.

	Total activity (%)		Specific activity	
	Cutinase	PNP Hydrolase	Cutinase (10^4 dpm/μg)	PNP Hydrolase (μmol/mg)
Extracellular fluid	100	100	1.2	11
0 to 50% $(NH_4)_2SO_4$ ppt	67	40	2.0	12
First Sephadex G-100	61	26	4.1	17
QAE-Sephadex	36	5	7.6	123
Second Sephadex G-100		5		340
SE-Sephadex Isozyme I Isozyme II	30 13		7.8	

*The non-specific esterase was measured using *p*-nitrophenyl palmitate (PNP) as the substrate.

Data taken from reference 95.

esters of *p*-nitrophenol, but not long-chained (C_{16} and C_{18}) esters. Both enzymes gave the highest V with *p*-nitrophenyl acetate. Neither cutinase I nor cutinase II showed any metal ion requirement and neither enzyme was inhibited by thiol-directed reagents such as *p*-chloromercuribenzoate, N-ethyl maleimide, and iodoacetamide. Both cutinase I and cutinase II contained an "active serine" as judged by the severe inhibition of both enzymes by diisopropylfluorophosphate, and paraoxon. Treatment of the two enzymes with [^{3}H]diisopropylfluorophosphate resulted in covalent attachment of the radioactive diisopropylphosphoryl moiety to the enzymes. Both enzymes contained one "active" serine per molecule, and reaction of this serine with the inhibitor completely inactivated the enzyme in both cases.

The two isozymes of cutinase from *F. solani pisi* show discernable differences in molecular structure and catalytic properties. Cutinase I, upon treatment with [^{3}H]diisopropylfluorophosphate followed by sodium dodecyl sulfate electrophoresis, revealed a single labeled peptide at the 22,000 molecular weight region, whereas cutinase II, upon similar treatment, showed, in addition to the major labeled peptide at 22,000 molecular weight, two smaller peptides with molecular weights of 10,600 and 9,800 (Fig. 21). Only the peptide at 10,600 contained ^{3}H, suggesting that a small portion of cutinase II contained a proteolytic clip and that inspite of the clip the serine at the active site was reactive. A further difference between the two cutinases was demonstrated by the observation that cutinase I catalyzed hydrolysis of model substrates indicated above about three times as rapidly as did cutinase II. This difference was also reflected in the sensitivity of the two enzymes to "active serine"-directed reagents; for example, cutinase I was significantly more sensitive to inhibition by diisopropylfluorophosphate and paraoxon. The structural difference between the two enzymes could also be demonstrated by immunological techniques. Although cutinase II cross-reacted with rabbit anticutinase I, spurs revealed that the two enzymes were not immunologically identical (Fig. 22).

Extracellular fluid from a variety of other phytopathogenic fungi grown on cutin also revealed cutin depolymerizing activity. Cutinases which are similar to those isolated from *F. solani pisi* have been isolated in homogeneous condition from some of these pathogens. For example, a single cutinase was isolated from the extracellular fluid of a wheat pathogen, *Fusarium roseum culmorum*, using Sephadex G-100, QAE-Sephadex, and SP-Sephadex chromatography.[109] The molecular weight of this cutinase, which was electrophoretically homogeneous, was 24,300, as estimated by SDS polyacrylamide gel electrophoresis. The catalytic properties of this enzyme, such as time course of hydrolysis of cutin, protein concentration dependence, substrate concentration dependence, pH dependence profile, specificity for the hydrolysis of model substrates (such as the esters of *p*-nitrophenol), and the involvement of active serine in catalysis were similar to those observed with the enzymes from *F. solani pisi*. Furthermore, the amino acid composition of the *F. roseum* enzyme was quite similar to that of

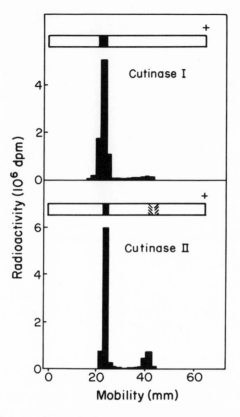

Figure 21.   Sodium dodecyl sulfate polyacrylamide disc gel electrophoresis of [^{3}H]Dip-F-treated cutinases I and II.[96]

the *F. solani pisi* enzymes. However, the enzyme from *F. roseum* did not form a precipitant line with the rabbit anti-cutinase I from *F. solani pisi*, clearly demonstrating that the two enzymes are not identical.[109] On the other hand, cutinase from *F. roseum* was equally sensitive to inhibition by rabbit anticutinase I, as were the two cutinases from *F. solani pisi*. Even though cutinases produced by the other pathogens have not been studied as extensively as those discussed above, an electrophoretically homogeneous enzyme has been obtained from *Fusarium roseum sambucinium*, and this enzyme was also found to be quite similar to the other cutinases.[82] The *F. roseum* enzymes also contained

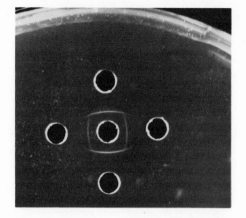

Figure 22. Ouchterlony double-diffusion plate illus-
trating the immunological relationships between cutinase I
and II. Center well, immunoglobulin fraction from the
serum of rabbits immunized with cutinase I isolated from
F. solani pisi; top and bottom wells, cutinase I from *F.
solani pisi*, 100 µg/ml; left and right wells, cutinase II
from *F. solani pisi*, 150 µg/ml.[111]

proteolytic clips and the proportion of the enzyme containing
the clip was somewhat variable, particularly in the enzyme
preparation obtained from *F. roseum culmorum*.

In the case of all of the phytopathogens thus far
examined, the procedure used to purify the cutinases also
yielded a nonspecific esterase which was electrophoretically
homogeneous.[82,95,96,109] The molecular weight of this
enzyme was about 52,000 and SDS electrophoresis indicated
that this enzyme also contained some proteolytic clips.
This esterase catalyzed the hydrolysis of all of the model
substrates which were hydrolyzed by the cutinases as
discussed above. In addition, unlike the cutinases, this
enzyme also hydrolyzed long-chain (C_{16}) esters of *p*-nitro-
phenol, but it did not catalyze the hydrolysis of cutin.
The nonspecific esterase also contains an "active" serine
which is involved in catalysis, as shown by the same
techniques as those used for the studies on the cutinases
discussed above. Although it could be readily demonstrated
that the production of this esterase was induced by growth

on cutin, this enzyme could not be assigned any specific
function in cutin degradation because it did not catalyze
the hydrolysis of cutin or the hydrolysis of oligomers
generated from cutin by cutinase.[95,96] The cutinases
generated the oligomers and hydrolyzed them further to give
the monomers. Some recent studies raise the possibility
that the nonspecific esterase might be a "procutinase" in
which a large amount of carbohydrates prevents the active
site of the enzyme from access to the polymer substrate,
cutin, but not small substrates such as esters. This
tentative conclusion is based mainly on the following
observations: (a) Ouchterlony double diffusion analysis
showed that the nonspecific esterase from *F. solani pisi*
cross-reacted with rabbit anticutinase I and immunoelectro-
phoresis showed that this cross-reactivity was not due to
contamination of the nonspecific esterase by cutinase.[111]
(b) The amino acid composition of the esterase was fairly
similar to that of cutinase I. (c) The nonspecific esterase
which has twice the molecular weight of cutinase contains
about 50% carbohydrate, and therefore it appears that the
nonspecific esterase might be cutinase with an equal amount
of carbohydrate attached to it.[82]   (d) In both cutinase
and the nonspecific esterase, the linkages between the
carbohydrates and the protein are similar; carbohydrates
are attached glycosidically to serine, threonine, β-hydroxyphenylalanine, and β-hydroxytyrosine.[82]

Novel Structural Features of Cutinase. Attempts to
determine the primary structure of cutinase I from *F. solani
pisi* lead to the finding that this enzyme is a glycoprotein
containing about five moles of monosaccharides per mole of
the enzyme.[81] Treatment with alkali resulted in an increase
in absorbance at 241 nm, indicating generation of dehydroalanine and/or dehydrothreonine as expected from β-elimination of the carbohydrates glycosidically linked to the
hydroxyl group of serine and/or threonine. Alkaline NaB^3H_4
treatment of cutinase gave rise to radioactive protein and
carbohydrates. Hydrolysis of the protein followed by amino
acid analysis revealed that alanine and α-aminobutyric acid
were radioactive, as expected from glycosidic linkages at
serine and threonine. Unexpectedly, phenylalanine and
tyrosine were also found to be labeled (Fig. 23). These
results suggested that carbohydrates were attached glycosidically also to β-hydroxyphenylalanine and β-hydroxytyrosine in this enzyme. In accordance with this hypothesis,

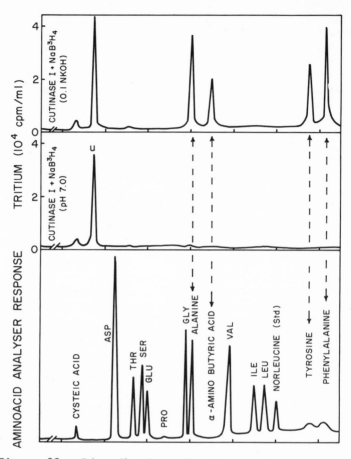

Figure 23. Distribution of radioactivity among the amino acids in the hydrolysate of the protein isolated by Bio-gel P-2 gel filtration of NaB^3H_4-treated cutinase I from *F. solani pisi*. The bottom tracing shows an amino acid analyzer pattern obtained from a hydrolysate of untreated cutinase.[81]

the labeled phenylalanine was found to be a 1:1 mixture of D̲ and L̲ isomers, and the radioactivity was found to be equally distributed between the α- and β-positions. Similar results were also obtained with the other labeled amino acids. Thus, this enzyme contains two novel linkages which have not been found heretofore in any other glycoprotein.

Another novel structural feature of this enzyme is that the N-terminal is blocked by what appears to be glucuronic acid which is attached to the amino group by an amide linkage.[82] This conclusion is based on the following observations: (a) Treatment of cutinase I with alkaline NaB³H released the glycosidically linked carbohydrates from the protein, but a labeled carbohydrate was still attached to the protein. Upon acid hydrolysis, the labeled protein gave rise to labeled gulonic acid in addition to the labeled amino acids discussed above. (b) Treatment of cutinase I under neutral conditions with NaB³H₄ resulted in incorporation of ³H into the protein which upon acid hydrolysis gave rise to only one labeled component, which was identified as gulonic acid. Therefore, the reducing group of glucuronic acid must have been free while the acid was attached to the protein. (c) The N-terminal amino group of the protein did not react with dansyl chloride, suggesting that it was blocked. (d) Direct evidence that glucuronic acid was attached to the N-terminal amino group was obtained with cutinase treated with NaB³H₄ (under neutral conditions). This labeled protein upon digestion by pronase gave a labeled ninhydrin negative compound which upon hydrolysis gave glycine. From this experimental evidence it was tentatively concluded that the amino group of N-terminal glycine was attached to glucuronic acid by an amide linkage and the model presented in Figure 24 is proposed for cutinase structure.

The isolation and characterization of cutinase from the pathogen allowed us to use an immunological technique to test whether the pathogen produces cutinase during the infection process. During the infection period of pea stem by *F. solani pisi*, ferretin-conjugated anticutinase I was applied and sections of the tissue were examined by electron microscopy (Fig. 25). Disruption of the cuticular layer around the region of infection very strongly suggested that the infecting fungus did, in fact, use cutinase in gaining entry into the plant.[105] Unconjugated ferretin did not show any such finding, showing that the interaction was dependent on the presence of anticutinase I. This demonstration of the production of cutinase by an infecting fungus should lead to further tests with other pathogens and hosts so that the role of cutinase in phytopathogenesis can be elucidated.

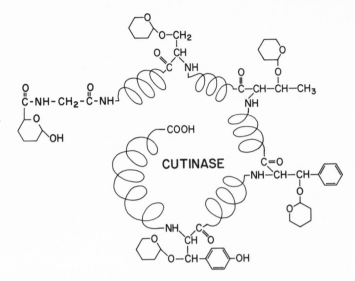

Figure 24. A schematic representation of the glycoprotein nature of cutinase I isolated from *F. solani pisi*.
The pyranoid rings represent a carbohydrate.[81,82]

One interesting aspect concerning the defensive role
of the lipid polymers is the possibility that during
hydrolysis of the polymers by microorganisms, covalently
attached chemicals toxic to microorganisms might be
released and thus prevent microbial penetration and/or
growth. The recent finding that phenolic materials are
covalently attached to cutin and suberin[84] and the observation that at least certain pathogenic fungi such as *F. solani
pisi* produce hydrolytic enzymes which release not only
aliphatic monomers, but also phenolic materials[82] suggest
that this aspect might provide clues concerning natural
defense mechanisms of plants. However, limited information
is available concerning natural antimicrobial agents which
are covalently attached to the protective polymers cutin

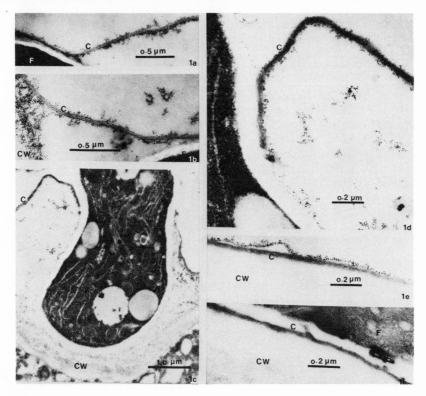

Figure 25.   (a & b) Electron micrographs of the infected
pea epicotyl treated with ferritin-conjugated anticutinase-
I 15 hr after inoculating with conidial suspension of *F.
solani pisi*.   The fungal mycelia shown here are growing
between the cuticle and the cell wall of the epidermis.
Disruption of lamellar cuticular structure is obvious and
the spherical pieces of the lamella separated from the
cuticle can be seen in (b).   Ferritin granules are found in
close association with the cuticle.   (c) Electron micrograph
of a fungus penetrating the cuticle of pea epicotyl 12 hr
after inoculation with a conidial suspension of *F. solani
pisi*.   Treatment with ferritin conjugated anticutinase
resulted in intense localization of ferritin granules at and
in the vicinity of the site of penetration.   (d) The top
left portion of c at higher magnification, showing the
ferritin granules attached to the cuticle.   (e & f) Electron
micrographs of infected pea epicotyl treated with ferritin-
conjugated anticutinase-I (e), or treated with ferritin-
conjugated immunoglobulin from a nonimmunized rabbit (f).
(Continued on next page.)

Ferritin granules are attached to the cuticular layer in e,
but hardly any ferritin can be seen in the control, f.   In
all cases C, cuticle; CW, cell wall; F, *Fusarium*.

and suberin. We have demonstrated that pentachlorophenol
covalently attached to cutin can severely inhibit the growth
of *F. solani pisi*, which readily grows with this polymer
as the sole carbon source.[101] A recent finding, that cutin
and suberin contain covalently attached phenolic compounds,[99]
raises the possibility that such phenolic components might
play the defensive role indicated above.   Furthermore,
enzymes capable of hydrolyzing phenolic esters are known
to be produced by microorganisms,[1,5,6,55] although the
possible role of such enzymes in host–pathogen interactions
has not been investigated.

    Deposition of suberin in the wound periderm presents
another aspect of the defensive role which lipid polymers
might play in the life of the plant. A defensive role for
suberin is also suggested by the observation that suberization is a fundamental process involved in the wound–healing
of plants, irrespective of the nature of the natural protective polymer of the tissue. Time-course of development of
suberin in potato tissue correlated with the development of
resistance of the cut surface to water loss,[72] and in all
cases we have thus far examined, changes in suberin content
was always correlated with the diffusion resistance. Thus
for example, the enhanced rate of deposition of suberin
resulting from absicic acid treatment and the inhibition
of synthesis of suberin resulting from thorough washing of
the tissue were correlated to the changes in diffusion
resistance. Therefore, one function of the aliphatic
components of suberin appears to be prevention of water loss.

CONCLUSION

    Although the major aliphatic monomers of cutin and
suberin from several species of plants have been identified,
very little is known about the intermolecular structure of
these polymers. Also unknown is the nature of the phenolic
constituents of suberin. The models proposed in this
chapter are based on very little information and they are
intended only as working models which need to be tested.

Biochemical pathways involved in the synthesis of the major
monomers of cutin and suberin have been fairly well established, but the enzymes involved have not been studied in
any great detail. Regulation of synthesis of these polymers
is not understood. Even though cutin depolymerizing enzymes
have been isolated from several phytopathogens, the role
of these polymers in pathogenesis is far from clear. Thus
we have only begun to probe this area just enough to show
that the stage is set for the exploration of the various
aspects of structure, biosynthesis, and function of these
unique phytopolymers—a largely neglected but important
area of plant biochemistry.

ACKNOWLEDGMENTS

        I thank Ms. Linda Brown for assistance in preparing
this manuscript and skilled technical assistance in much
of the work from this laboratory discussed in this chapter.
The work on lipid polymers in this laboratory was supported
in part by grants GB-43076X and BMS 74-09351 from the
National Science Foundation.

REFERENCES

1. Adachi, O., M. Watanabe, and H. Yamada. 1968. *Agr.*
       Biol. Chem. 32:1079.
2. Agrawal, V. P. and P. E. Kolattukudy. June 1974.
       29th Annual Northwest Reg. Meeting of the Am. Chem.
       Soc., Abs. 202.
3. Agrawal, V. P. and P. E. Kolattukudy. 1977. *Plant*
       Physiol. in press.
4. Agrawal, V. P. and P. E. Kolattukudy, unpublished
       results.
5. Aoki, K., R. Shinke, and H. Nishira. 1976. *Agr.*
       Biol. Chem. 40:79.
6. Aoki, K., R. Shinke, and H. Nishira. 1976. *Agr.*
       Biol. Chem. 40:297.
7. Bredemeijer, G. and W. Heinen. 1968. *Acta Bot. Neerl.*
       17:15.
8. Brieskorn, C. H. and P. H. Binnemann. 1974. *Z.*
       Lebensm. Unters. Forsch. 154:213.
9. Brieskorn, C. H. and P. H. Binnemann. 1975. *Phyto-*
       chemistry 14:1363.

10. Brieskorn, C. H. and J. Böss. 1964. *Fette Seifen Anstrichmittel 66*:925.

11. Brieskorn, C. H. and L. Kabelitz. 1971. *Phytochemistry 10*:3195.

12. Brieskorn, C. H. and W. Schneider. 1961. *Z. Lebensm. Unters Forsch. 115*:513.

13. Caldicott, A. B. and G. Eglinton. 1976. *Phytochemistry 15*:1139.

14. Caldicott, A. B., B. R. T. Simoneit, and G. Eglinton. 1975. *Phytochemistry 14*:2223.

15. Camm, E. L. and G. H. N. Towers. 1973. *Phytochemistry 12*:1575.

16. Cassagne, C., R. Lessire, and P. Blanchardie. 1976. Symposium on "Lipids and Lipidpolymers in Higher Plants: Karlsruhe, 1976." Botanical Institute (Plant Physiology), University of Karlsruhe. Abstracts pp. 13-14.

17. Chevreul, M. 1815. *An. Chim. Phys. 96*:141.

18. Crafts, A. S. 1961. "The Chemistry and Mode of Action of Herbicides," p. 28. Interscience, New York.

19. Crisp, C. E. 1965. "The Biopolymer Cutin." Ph.D. thesis, University of California, Davis.

20. Croteau, R. and P. E. Kolattukudy. 1973. *Biochem. Biophys. Res. Commun. 52*:863.

21. Croteau, R. and P. E. Kolattukudy. 1974. *Arch. Biochem. Biophys. 162*:458.

22. Croteau, R. and P. E. Kolattukudy. 1974. *Arch. Biochem. Biophys. 162*:471.

23. Croteau, R. and P. E. Kolattukudy. 1974. *Biochemistry 13*:3193.

24. Croteau, R. and P. E. Kolattukudy. 1975. *Arch. Biochem. Biophys. 170*:61.

25. Croteau, R. and P. E. Kolattukudy. 1975. *Arch. Biochem. Biophys. 170*:73.

26. Dean, B. B. and P. E. Kolattukudy. 1976. *Plant Physiol. 58*:411.

27. Dean, B. B. and P. E. Kolattukudy. 1977. *Plant Physiol.*, in press.

28. Dean, B. B. and P. E. Kolattukudy, unpublished results.

29. Deas, A. H. B., E. A. Baker, and P. J. Holloway. 1974. *Phytochemistry 13*:1901.

30. Deas, A. H. B. and P. J. Holloway. 1976. Symposium on "Lipids and Lipidpolymers in Higher Plants: Karlsruhe, 1976." Botanical Institute (Plant Physiology), University of Karlsruhe. Abstract pp. 103-104.

31. Eglinton, G. and D. H. Hunneman. 1968. *Phytochemistry* 7:313.
32. Eglinton, G., D. H. Hunneman, and A. McCormick. 1968. *Organic Mass Spectrometry* 1:593.
33. Fremy, E. and V. Urbain. 1882. *An. Sci. Nat. (Bot.)* 13:360.
34. Fremy, E. and V. Urbain. 1885. *C. r. hebd. Seanc. Acad. Sci., Paris* 100:19.
35. Freudenberg, K. 1968. *In* "Constitution and Biosynthesis of Lignin" (K. Freudenberg and A. C. Neish, eds.), pp. 47–122. Springer-Verlag, Inc., New York.
36. Gross, G. G., R. L. Mansell, and M. H. Zenk. 1975. *Biochem. Physiol. Pflanzen* 168:41.
37. Gross, G. G. and M. H. Zenk. 1974. *Eur. J. Biochem.* 42:453.
38. Hankin, L. and P. E. Kolattukudy. 1971. *Plant Soil* 34:525.
39. Heinen, W. 1960. *Acta Bot. Neerl.* 9:167.
40. Heinen, W. 1962. *Arch. Mikrobiol.* 41:268.
41. Heinen, W. 1963. *Acta Bot. Neerl.* 12:51.
42. Heinen, W. 1963. *Enzymologia* 25:281.
43. Heinen, W. and H. De Vries. 1966. *Arch. Mikrobiol.* 54:331.
44. Heinen, W. and H. F. Linskens. 1960. *Naturwissenschaften* 47:18.
45. Heinen, W. and I. van den Brand. 1961. *Acta Bot. Neerl.* 10:171.
46. Heinen, W. and I. van den Brand. 1963. *Z. Naturforsch.* 186:67.
47. Hill, A. C. and M. J. C. Rhodes. 1975. *Phytochemistry* 14:2387.
48. Holloway, P. J. 1972. *Chem. Phys. Lipids* 9:158.
49. Holloway, P. J. 1972. *Chem. Phys. Lipids* 9:171.
50. Holloway, P. J. 1973. *Phytochemistry* 12:2913.
51. Holloway, P. J. and A. H. B. Deas. 1971. *Phytochemistry* 10:2781.
52. Holloway, P. J. and A. H. B. Deas. 1973. *Phytochemistry* 12:1721.
53. Huelin, F. E. 1959. *Aust. J. Biol. Sci.* 12:175.
54. Heulin, F. E. and R. A. Gallop. 1951. *Aust. J. Scient. Res.* Ser. B 4:526.
55. Iibuchi, S., Y. Minoda, and K. Yamada. 1968. *J. Agr. Biol. Chem.* 32:803.
56. Jaworski, J. G., E. E. Goldschmidt, and P. K. Stumpf. 1974. *Arch. Biochem. Biophys.* 163:769.

57. Khan, A. A. and P. E. Kolattukudy. 1973. *Arch. Biochem. Biophys. 158*:411.
58. Khan, A. A. and P. E. Kolattukudy. 1975. *Arch. Biochem. Biophys. 170*:400.
59. Knobloch, K. H. and K. Hahlbrock. 1975. *Planta Medica* suppl. 1975:102.
60. Kolattukudy, P. E. 1970. *Biochem. Biophys. Res. Commun. 41*:299.
61. Kolattukudy, P. E. 1970. *Biochemistry 9*:1095.
62. Kolattukudy, P. E. 1970. *Plant Physiol. 46*:759.
63. Kolattukudy, P. E. 1971. *Arch. Biochem. Biophys. 142*:701.
64. Kolattukudy, P. E. 1972. *Biochem. Biophys. Res. Commun. 49*:1040.
65. Kolattukudy, P. E. 1974. *Biochemistry 13*:1354.
66. Kolattukudy, P. E. *In* "Lipids and Lipid Polymers in Higher Plants" (M. Tevini, ed.), in press. Springer-Verlag, Berlin.
67. Kolattukudy, P. E. and V. P. Agrawal. 1974. *Lipids 9*:682.
68. Kolattukudy, P. E. and L. Brown, unpublished results.
69. Kolattukudy, P. E. and J. S. Buckner. 1972. *Biochem. Biophys. Res. Commun. 46*:801.
70. Kolattukudy, P. E., R. Croteau, and J. S. Buckner. 1976. *In* "Chemistry and Biochemistry of Natural Waxes" (P. E. Kolattukudy, ed.), pp. 290-334. Elsevier-North Holland Press, Amsterdam.
71. Kolattukudy, P. E., R. Croteau, and T. J. Walton. 1975. *Plant Physiol. 55*:875.
72. Kolattukudy, P. E. and B. B. Dean. 1974. *Plant Physiol. 54*:116.
73. Kolattukudy, P. E., K. Kronman, and A. J. Poulose. 1975. *Plant Physiol. 55*:567.
74. Kolattukudy, P. E. and T. J. Walton. 1972. *Biochemistry 11*:1897.
75. Kolattukudy, P. E., T. J. Walton, and R. Kushwaha. 1971. *Biochem. Biophys. Res. Commun. 42*:739.
76. Kolattukudy, P. E., T. J. Walton, and R. P. S. Kushwaha. 1973. *Biochemistry 12*:4488.
77. Lai, Y. Z. and K. V. Sarkanen. 1971. *In* "Lignins: Occurrence, Formation, Structure and Reactions" (K. V. Sarkanen and C. H. Ludwig, eds.) pp. 165-240. John Wiley and Sons, Inc., New York.
78. Lee, B. 1925. *Ann. Bot. 39*:755.
79. Lee, B. and J. H. Priestley. 1924. *Ann. Bot. 38*:525.

80. Legg, V. H. and R. V. Wheeler. 1925. *J. Chem. Soc.* 1412.
81. Lin, T.-S. and P. E. Kolattukudy. 1976. *Biochem. Biophys. Res. Commun. 72*:243.
82. Lin, T.-S. and P. E. Kolattukudy, unpublished results.
83. Lindl, T., F. Kreuzaler, and K. Hahlbrock. 1973. *Biochem. Biophys. Acta 302*:457.
84. Macey, M. J. K. and P. K. Stumpf. 1968. *Plant Physiol. 43*:1637.
85. Martin, J. T. 1964. *Annu. Rev. Phytopathol.* 2:81.
86. Martin, J. T. and B. E. Juniper. 1970. "The Cuticles of Plants." St. Martin's Press, Inc., New York.
87. Matic, M. 1956. *Biochem. J. 63*:168.
88. Mazliak, P. 1968. *In* Progress in Phytochemistry, Vol. 1 (L. Reinhold and Y. Liwschitz, eds.) pp. 49-111. Interscience, New York.
89. Mazliak, P. and J. Pommier-Miard. 1963. *Fruits 18*:177.
90. McKeen, W. E. 1974. *Phytopathol. 64*:461.
91. Oesch, F. 1973. *Xenobiotica 3*:305.
92. Orgell, W. H. 1955. *Plant Physiol. 30*:78.
93. Priestley, J. H. 1943. *Bot. Rev. 9*:593.
94. Purdy, R. E. and P. E. Kolattukudy. 1973. *Arch. Biochem. Biophys. 159*:61.
95. Purdy, R. E. and P. E. Kolattukudy. 1975. *Biochemistry 14*:2824.
96. Purdy, R. E. and P. E. Kolattukudy. 1975. *Biochemistry 14*:2832.
97. Rhodes, M. J. C. and L. S. C. Wooltorton. 1975. *Phytochemistry 14*:1233.
98. Rich, P. R. 1975. *Biochem. Soc. Trans 3*:980.
99. Riley, R. and P. E. Kolattukudy. 1975. *Plant Physiol. 56*:650.
100. Riley, R. and P. E. Kolattukudy, manuscript in preparation.
101. Riley, R. and P. E. Kolattukudy, unpublished results.
102. Rodriquez-Miguens, B. and I. Ribas-Marques. 1972. *An. R. Soc. Esp. Fis. Quim. 68*:303.
103. Sarkanen, K. V. and H. L. Hergert. 1971. *In* "Lignins: Occurrence, Formation, Structure and Reactions" (K. V. Sarkanen and C. H. Ludwig, eds.), pp. 43-94. John Wiley and Sons, Inc., New York.
104. Schneider, W. 1960. "Beiträge zum Chemischen Aufau der Apfelschale." Ph.D. thesis, Julius-Maximillian University, Würzburg.
105. Shaykh, M. M., C. L. Soliday, and P. E. Kolattukudy, unpublished results.

106. Shishyama, J., F. Araki, and S. Akai. 1970. *Plant Cell Physiol.* *11*:323.

107. Shishyama, J., F. Araki, and S. Akai. 1970. *Plant Cell Physiol.* *11*:937.

108. Sitte. P. 1955. *Mikroskopie 10*:178.

109. Soliday, C. L. and P. E. Kolattukudy. 1976. *Arch. Biochem. Biophys.* *176*:334.

110. Soliday, C. L. and P. E. Kolattukudy, manuscript in preparation.

111. Soliday, C. L. and P. E. Kolattukudy, unpublished results.

112. Swan, E. P. 1966. *Pulp Paper Mag. Can.* *67*:T456.

113. Tanaka, Y., M. Kojima, and I. Uritani. 1974. *Plant Cell Physiol.* *15*:843.

114. Vance, C. P., A. M. D. Nambudiri, C.-K. Wat, and G. H. N. Towers. 1975. *Phytochemistry 14*:967.

115. van den Ende, G. and H. F. Linskens. 1974. *Annu. Rev. Phytopathol.* *12*:247.

116. von Höhnel, F. 1877. *Sber. Akad. Wiss, Wien 76*:507.

117. Walton, T. J. and P. E. Kolattukudy. 1972. *Biochem. Biophys. Res. Commun.* *46*:16.

118. Walton, T. J. and P. E. Kolattukudy. 1972. *Biochemistry 11*:1885.

119. Wardrop, A. B. 1971. *In* "Lignins: Occurrence, Formation, Structure and Reactions" (K. V. Sarkanen and C. H. Ludwig, eds.), pp. 19–41. John Wiley and Sons, Inc., New York.

120. Wattendorff, J. 1974. *A. Pflanzenphysiol.* *72*:119.

121. Wiltshire, S. P. 1915. *An. Appl. Biol.* *1*:335.

122. Zucker, M. 1972. *Annu. Rev. Plant Physiol.* *23*:133.

Chapter Seven

SECONDARY CHANGES IN WOOD

W. E. HILLIS

Forest Products Laboratory, Division of
Building Research, CSIRO, Highett, Victoria
3190, Australia

INTRODUCTION

During the ontogenetic development of trees several
important anatomical changes take place when the tree
passes from juvenile to adult to overmature stage of development. For example, in both angiosperms and gymnosperms
the average length of the xylem elements, at any stem
height, increases from the pith outward for a number of
years and then become more or less constant. In angiosperms, the number of vessels decreases and their individual
size increases relatively steadily until the adult form
of the tree is attained. Also for a number of years the
percentage, in a growth ring, of the small-diameter latewood
cells with thick walls increases with the increasing age
of the tree. In addition the wood formed by an overmature
Pseudotsuga menziesii tree (300 years old) is noticeably
different from the wood formed when the same tree was
50 years old. Wood formed when the tree is overmature has

narrower xylem rings than younger wood and its rings are
made up essentially of thin-walled cells with negligible
latewood and a higher lignin and lower α-cellulose content.[109]

Many of the properties of a particular piece of
wood are established shortly after the cambial cells have
differentiated but as mentioned above these properties can
be different at various stages of the life span of the
tree. The basic chemistry, fine structure and thickness
of the cell wall remain unchanged after differentiation
as well as the proportions and characteristics of different
elements in the wood. These latter can have a considerable
effect on wood quality, particularly in the case of angiosperms with their very wide range of size and number of
vessels, rays, parenchyma, etc. Together with their spatial
arrangements these elements determine the suitability of
a particular wood for solid or fibrous end-use.

Important secondary changes can take place in the
structural material after biosynthesis. These changes
can modify wood quality significantly. The environment or
forest practices or genetic influences, or the interaction
of all three, operating at the time of cambial activity
can indirectly cause these changes in wood tissues. For
example, abnormal growth stress in the periphery of hardwoods can result in broken fibers and low quality wood
in the interiors of trees.[151,261] However, in the discussion which follows attention will be directed to those
situations involving the non-structural components and not
primarily to the structural materials.

After wood has been formed in the living tree,
aging begins to occur. Its effects can be slight in the
outer pale-colored annulus of the trunk. For most tree
species this sapwood is similar having a very pale or
almost colorless appearance when freshly cut. Other
important changes associated with the aging of sapwood
parenchyma commonly take place abruptly over a few rows of
cells, sometimes resulting in tissues with markedly
different properties from the surrounding tissues. These
changes are initiated by internal situations, and their
nature is determined genetically and modified by both
internal and external environments. Mechanical and biological injuries can result in secondary wood changes which
superficially resemble the above changes. Some injuries

can result in the formation of exudates or extracellular secondary components.

Secondary changes can provide the tree with chemical or physical defenses which delay degradation and can affect wood quality. The influence of many of the changes is well established for a number of species, but there are indications that plantations of fast-grown trees are producing woods with unfamiliar secondary wood changes. Moreover, the increasing use of wood-production forests for various community needs could lead to a lowering of wood quality because of secondary wood changes resulting from injury, transference of disease, etc. Furthermore, in the next few years there will be an increasing use of hardwoods, particularly of those from tropical regions, for commercial purposes. However, the 2,000 to 3,000 different species potentially available from tropical forests differ in properties from the common commercial species, not only because of their differing anatomy, but also because of several types of secondary changes, many of which have not previously been encountered. In recent years attention has been given to the accumulation of non-structural materials because of their influence on wood quality. However, changing situations with regard to global resources, energy, pollution, and social costs may cause more attention to be given to these materials as future sources of valuable chemicals.

The study of the nature and causes of the various secondary changes found in wood is a complex one and conclusions drawn from one species and one set of circumstances may not apply elsewhere. Precise knowledge of the causes will assist the efficient production and conversion of one of the world's most important resources into products of high quality.

DESCRIPTION OF SECONDARY
WOOD CHANGES

Transformation of Sapwood into Heartwood. Sapwood is "the portion of the wood that, in the living tree, contains living cells and reserve material (e.g. starch)."[184] Fibers and tracheids die as soon as the lignification phase is completed although there is an exception with *Tamarix aphylla*.[91] The transverse and longitudinal parenchyma

(which vary between 4 to 50% by volume of the wood) remain
viable for many years. These cells serve as an avenue for
translocation of water and minerals, maintain metabolic
processes and store foods. After a certain time the
parenchyma may die and heartwood is formed. Usually the
transformation occurs after a few years but living cells
have been reported in sapwood of considerable age, such as
115 year old *Acer saccharum*,[105] and only sapwood has been
found in large trees of *Alstonia scholaris*.[49]

The width of sapwood in a mature tree can vary considerably, occupying the whole tree or only the outermost
growth rings surrounding an approximately conical shaped,
central core of dead heartwood cells. This variation in
width occurs in different trees of the same genera or
species and even in different positions in the same tree.
Moreover, the width in a cross section of a stem of a tree
is not always uniform and the number of sapwood rings can
be greater at the stump than higher in the tree.

The characteristics of sapwood are not uniform across
the annulus. The movement of sap is much less in inner
sapwood than in outermost portions. The amounts of primary
metabolites, which may be either starch and sugars or fats
or both, stored in the sapwood are low (usually less than
7% starch by weight) and the amounts vary between species,
across the sapwood and during periods of tree growth. These
changes reflect the dynamic situation of competing metabolic
sinks which change continually to meet the needs of a
particular zone and indeed of the whole tree, for seed
production, foliage, etc. There is little available information on this dynamic relationship of metabolites. It has
been reported that, as aging progresses, the nuclei in
parenchymal cells gradually degenerate and disintegrate,
and the cells progressively lose their organelles, vitality
and ability to utilize oxygen and lose nitrogen-containing
compounds, starch, sugars, and biotin.[39] However there is
evidence of increased metabolism at the heartwood periphery
(see below).

The sapwood of the trunk, branches and roots of many
uninjured trees changes abruptly in appearance and in
function after a certain stage to form heartwood. This
interior core of "heartwood" has been defined as "The inner
layers of wood which, in the growing tree, have ceased to
contain living cells and in which the reserve materials

(e.g. starch) have been removed or converted into heartwood substances. It is generally darker in color than sapwood, though not always clearly differentiated."[184] The color of heartwood can range between light yellow, pink, red, brown, and dark brown and once its formation begins the process is spatially continuous. Among those trees which form heartwood regularly, slight visual differences can be observed, for example, between *Acacia, Pinus* and *Eucalyptus* species. These differences include the regularity of the heartwood boundary, evenness of color, presence of a transition zone, etc. It would be premature to classify heartwoods into a number of classes until more physiological and biochemical characteristics can be defined.

The proportion, and even the existence, of heartwood in a mature tree varies with the family, genus and sometimes species.[65,142] Often the age of the sapwood cells transformed into heartwood appears to be characteristic of the species.[136] Within a species, under normal circumstances, the amount and rate of heartwood formation, which shows appreciable genetic control,[258,259] is influenced by tree age,[131] growth rate,[149,150] environment, and silvicultural practice[314] (see section on features of heartwood formation).

Heartwood formation commences in many species about 5 feet above ground[66,333] and the proportion of heartwood in some species remains greatest at this level. The periphery of the approximately conical-shaped central core of heartwood often undulates vertically and horizontally and can cut across parts of several growth rings as abrupt tongues.

Less moisture is found in heartwood than in sapwood. The decrease can be abrupt and considerable, from about 150% or more to about 50% or less as in *Picea glauca* and *Pinus* spp.[58,66,203,278] There are reports[138,140,270,352] of heartwood of some hardwood species (such as *Carya, Ulmus, Fraxinus,* and *Populus* spp.) containing more moisture than the sapwood and this could be due to wetwood formation (see subsection on wetwood). However, a detailed examination of three *Eucalyptus* spp. has shown that a significant increase in the moisture content of heartwood took place at the boundary with sapwood.[55a] In some woods, such as *Fagus sylvatica*, a steady loss of water takes place across the stem and at the sapwood-heartwood boundary 60% of water is present.[365] The gas in the heartwood (and sapwood) of a number of trees contains a larger proportion of carbon dioxide and a smaller amount of oxygen than that found in air.[42,66,190,236]

In those coniferous species containing suitable pit structures, aspiration takes place on heartwood formation. Bordered pits apparently aspirate when they are situated between a tracheid containing water and another tracheid containing a gas. An important study has been made of the variation in aspiration of the bordered pits in the early wood of different zones of sapwood and heartwood of *Pinus radiata*.[111] Dominant trees were found to have about 15 growth rings of sapwood and suppressed trees about 10 growth rings and both about two rings of transition zone. The percentage of pits aspirated gradually increased across the sapwood from a low figure to 40 to 50% at the boundary with the transition zone. In this zone a marked increase to more than 90% aspiration occurred and this increased slightly through the heartwood. It was calculated[111] that when more than 50% of the bordered pits of *P. radiata* are aspirated, liquid water transport through the tracheids in the sapwood and transition zone does not occur.

Balloon-like structures called tyloses grow out of the horizontal ray parenchyma into the lumen of vessels when heartwood is formed in certain angiosperms. As a result of differences in osmotic pressure the inner surfaces of the ray cell extend through the pit aperture to initiate tyloses formation. This occurs in species having pit aperture diameters greater than 10 μm. Consequently it is an important feature of heartwood formation in many angiosperms. Tyloses can form abruptly in the transition zone,[13] often proliferate greatly and vary considerably in structure, and may have thin walls or thick lignified secondary walls. Tyloses also form in the sapwood of these species in response to wounding, invasion by fungal pathogens or viral infection.[48,94,103] In eucalypts they form before extractives appear.[248]

The extractives which are responsible for the distinctive color of heartwood are present in the cell wall of the tissues but mainly in the lumen of the radial parencyma and also in the longitudinal parenchyma, vessels and sometimes the fibers and specialized cells. Knowledge of the precise

location of extractives in tissues would provide understanding of various features of wood quality and disease
resistance.

The amount of extractives shows an overall increase
from pith to heartwood periphery followed by an abrupt
decrease in amount in sapwood. Large amounts of polyphenolic extractives can be present in some species (e.g. over
30% in some eucalypt heartwoods). Large amounts (up to
29%) of an arabinogalactan have been found in the heartwood
of *Larix* sp.[62,140]

Extractives in the lumen have a considerable effect
on the chemical properties but little on the physical
properties of wood apart from density. Most attention in
the past has been given to lumen-contained extractives
which sometimes fill the lumen completely. In some instances
extractives form a coating on the cell wall and over the
pit apertures, interfering with the permeability and pene-
trability of the wood[207,341] and promoting collapse.[229]

The transformation of sapwood to heartwood is accompanied by necrosis of the xylem parenchyma and rupture of
the protoplasm. In a number of species, extractives enter
the cell wall of the heartwood tissues and the effects of
this are aparent in the physical or chemical properties.[92,145,329,338] Extractives influence shrinkage and recovery
of *Eucalyptus delegatensis* wood[260] and *Pseudotsuga menziesii*
heartwood during drying,[293] and other physical properties.[231,232,329,338] The extractives in the cell wall are probably
largely responsible for the differences in dimensional
stability and durability between sapwood and heartwood[142]
but there is little information on the amount present. As
much as 75% of the total heartwood extractives has been
estimated to be in the cell wall of *Sequoia sempervirens*[329]
but this figure may be too high.[11] Secondary components
which enter the wall on the death of the cell cannot be
removed completely by neutral solvents and some appear to
combine with components of the wall. There are very few
data on possible changes in the amount and nature of hemicelluloses in the cell wall when heartwood is formed.[319]

The hydrated S2 layer of the tracheid cell walls in
Pinus resinosa contain 25% free space[22] and the water in
the capillaries of the wet walls of other woods has a
volume of more than 40% (on a dry wood basis) with about

half the water being in the free state.[323] A number of
workers have produced evidence that the free space in wet
cell walls is in capillaries having cross sections from
160 to 600 nm.[142,323] This is large enough to allow molecules of significant size to enter; by way of example, the
widest part of a glucose molecule is 63 nm and that of a
flavonoid monomer about 150 nm. It has been suggested that
some species which do not always contain colored heartwood,
such as ash or beech, contain polymerized extractives that
are too large to penetrate the cell wall so that shrinkage
of these heartwoods on drying is much greater than in oak,
which regularly forms colored heartwood.[27]

   Transition Zone and Intermediate Wood. In 1894,
Hartig described a thin pale zone which appeared between
sapwood and heartwood in oak and *Fagus sylvatica*.[117] An
illustration of the narrow colorless, dry-wood transition
zone found between sapwood and heartwood of *Taxus baccata*
in late winter was reported by Craib in 1923[65] and attention has been drawn to the possibility of a zone of intensified metabolism at the heartwood periphery.[49] Despite
these earlier reports, the transition zone in woody species
has received relatively little attention, although its
existence in a wide range of species has been reported.[136,]
[140] The zone can be quite distinct and even when the
periphery of the heartwood in a cross section is erratic and
changes direction sharply over several growth rings, the
width of the transition zone (about 2 to 5 mm) remains about
the same. Failure to detect the transition zone in some
species may be due to its seasonal occurrence (it has
been found from mid-summer to winter in oak and *Fagus
sylvatica*[117]), or to the location of the hearwood in the
tree. It is noteworthy that a zone with very similar
characteristics exists in *P. radiata* around the injury
resulting from the *Sirex noctilio-Amylostereum areolatum*
complex.[304] The "ripewood" zone which surrounds some
heartwoods is possibly the same as a transition zone[66] and
in Japanese larch it forms when cambial activity decreases.[166]

   The colorless appearance of the transition zone is
probably due to its lower moisture content which has been
reported to be less than that in the adjacent sapwood in
Abies and *Ulmus* species, *Fagus sylvatica*, *Dacrydium cupres-
sinum*, and *Cryptomeria japonica*[66,352,353,354,359,365] and
sometimes even less than that in heartwood in *Taxus
baccata*.[65] The amount of extractives present in the

transition zone is low in pine,[111],[145] eucalypt and probably
in other species, but increased amounts of nicotinic acid
amide, biotin, pyridoxine and in some cases protein nitrogen
have been reported.[362]

Some trees have an inner layer of the sapwood that is
intermediate in color and some other characteristics
between sapwood and heartwood. This "intermediate wood"[28],
[99] can be wide (10 cm) in trees such as *Sloanea woollsii*
and *Diospyros pentamera*.[49] The relationship of transition
zone to intermediate zone has to be determined.

Discolored Wood. Heartwood is usually distinguished
from sapwood by its different color (although the difference
may be slight) and its lower moisture content. However
there are several types of color differences in wood, some
of which appear regularly in stems of particular species,
but the colored zones are irregular in shape and do not
show uniformity in color intensity across a stem. Attempts
have been made to devise a nomenclature for these colorations. The heartwood which regularly forms in certain
trees (such as *Quercus* and *Pinus* spp.) has been defined as
obligatory colored or regular heartwood. Irregularly
formed heartwood (as in *Fraxinus excelsior*) has been termed
facultatively colored heartwood and in this type the extractives do not penetrate the cell wall. Some trees have
light colored heartwood with a low moisture content and
small amounts of extractives ("ripewood" as in *Abies alba*)
and this wood is known also as light heartwood. Sapwood
trees have been called trees with retarded formation of
heartwood.[25],[26],[27],[28]

Some areas of colored wood do not arise from the
inherent characteristics of the species and the age-envir-
onment interaction. They arise primarily from external
influences such as injury, although the growth habits of
some trees can predispose them to the influence of external
factors. The color differences in "discolored wood" are
mainly the results of microbiological activity with or
without the involvement of an unusually large amount of
water as in "wetwood."[310] "Discolored" wood is a poor
description of the wood as it gives little idea of the
changes that have occurred and the extent to which the
tissues do or do not accumulate water. Indeed it is likely
that after more study of the discolorations encountered in
wood a wetwood situation will be recognized more frequently.

In some cases the quality of the wood is lower where there
is coloration, or the changes represent an intermediate
stage to more deleterious effects. On the other hand,
colorations of particular types considerably enhance the
value of furniture timbers. Once the causes of colorations
are accurately understood more selective silvicultural
techniques or other forms of treatment can more easily be
developed to control such aberrations.

A number of events occur from the time of wounding
to the formation of discolored wood, to the total decomposition of the tissues. These events are continuous over
time but there are three main stages in the sequence which
can stop at any stage but can only proceed by passing
through the previous stage. The situation was recently
reviewed.[310]

Stage I includes all those processes associated with
the host response to wounding and involve both the tree
and the environment. Discoloration may occur in the xylem
as a result of chemical processes, including the formation
of different components and oxidation resulting from exposure
to air. If trees are vigorous and the wounds are not
severe, no further change takes place. More discoloration
is associated with branch wounds than with any other type
of wound and it is usually in the center of the tree.
However if the vigor of the tree is low or if the wounds
are sufficiently severe then the processes may go on'to
Stage II.

Stage II includes those events that occur when microorganisms overcome chemical protection barriers and invade
the xylem. These pioneer invaders are usually, but not
always, bacteria and non-hymenomycetous fungi. Along with
the discoloration of the tissues, the pH and moisture
content increase and inorganic materials accumulate. The
discoloration of the wood can be intensified as a result
of interactions between invading microorganisms and living
xylem cells. A host response to the invasion occurs
usually, although woundwood forms in *Pinus sylvestris* from
the beginning of cambial activity to the termination of
winter.[219] Where the vigor of the tree is low, the wounds
severe and the microorganisms aggressive, then decay
follows,[249,326] as in Stage III.

In the State III, decay microorganisms, especially
hymenomycetes, invade and degrade cell-wall substances.
At this stage all xylem cells are dead and after the pioneer
microorganisms invade the wood many others follow and compete
for the remaining portions of the tissues.

Light and dark bands or streaks are frequently observed
in the sapwood and heartwood of a number of species, particularly tropical ones, but the cause has received little
study. No microorganisms were found in the light bands in
the heartwood of *Thuja plicata* but the non-decay fungi
found in the dark bands were probably responsible for the
discoloration and reduction in the durability of the wood.
[16,81,93,286] Streaks or patches of light and dark wood
can be due to injury and death of the cambial cells. In
such cases, the sapwood between the cambium and heartwood
periphery is not transformed to heartwood although the
sapwood on all sides of the injured region is transformed.
The paler colored "included sapwood" is little changed
from normal sapwood and is susceptible to decay, etc.

A variation in the color of heartwood known as "target
pattern" has been observed in *Thuja plicata*,[243] *Pseudotsuga
menziesii*,[195] *Eucalyptus marginata*, etc. The pale colored
rings are deficient in the extractives normally found in
heartwood but the cause of the decrease is not known,
although it has been suggested that it is the result of
either defoliation or the demands of heavy seed formation.
Occasionally a distinct outer heartwood zone is evident.
In the case of *Eucalyptus marginata* the zone was found to
contain the same amount of alkali-soluble extractives as
the inner zone, but more water solubles; apparently there
was a sharp increase in the degree of polymerization of
some of the components in the inner heartwood.[133]

Wetwood. In freshly cut stems of some species,
wetwood (when present) is usually recognized by a darker
or wetter appearance, although appearance is not a reliable
indication of moisture content in *Populus* spp.[238] Wetwood
contains more water than the inner sapwood zone, and often
has a "fermentation" color. Sometimes a gas, largely carbon
dioxide or methane, can build up a considerable pressure
within the tissues. It usually has a higher pH and density
than the surrounding tissues. Wetwood occurs independently
of normal heartwood and can exist as small or large zones
or streaks anywhere in the sapwood or heartwood in a large

number of softwoods or hardwoods.[18,90,118,300,352,353] In
Abies alba, heartwood forms before wetwood.[19] Wetwood
possesses a weak respiration activity[18] which is absent
from heartwood. Because the mode of initiation of wetwood
and its appearance and characteristics differ within and
between softwood and hardwood trees there is no reliable
definition of it.

When dried, wetwood is darker than its surrounding
tissues and is sometimes associated with brightly colored
streaks. Only a slight change in most of the properties
of unaffected wood has been noticed, although in aspen
there is a significant lowering of mechanical properties.[124]
Mineral stains or deposits are frequently greater in wetwood areas, and often the deposits are of calcium carbonate.
There is often an association of wetwood with shakes.
Zones of wetwood can obviously affect drying schedules (in
particular those using high temperatures) of timbers, such
as *Tsuga* spp. and *Abies* spp., and this adverse situation
is accentuated by the lower permeability of the zones.[18,199,300] Excessive shrinkage can occur with wetwood on
kiln drying.[238] The concentration of phenolic components
in the wetwood of young *Abies alba* is 3 to 5 times, and
in older trees 10 to 20 times, higher than in the sapwood.[18]
In *Tsuga heterophylla* wetwood a fivefold increase in extractives content in comparison with normal heartwood has been
reported, with the extractives having the same composition
in both tissues.[300] However, in another case only a
slight increase was reported.[208]

Various aspects of wetwood were extensively reviewed
in 1961,[118] and have received further attention subsequently.[18,300,352,353,354] Wetwood is considered to be
caused frequently by the action of bacteria which enter
the stem of the tree through wounds or particularly through
branch stubs.[43,118] It appears from other evidence that
bacteria are not the direct cause of wetwood and they
become established after the wetwood conditions occur,
which in *Populus* spp. can exist anywhere in the stem without an association with injuries of any type[18,199,216]
although infection may occur through the roots.[19] It has
been demonstrated that water diffuses through branch stubs
in *Abies* spp. into the interior of the tree. Obviously
those trees which characteristically have persistent
branch stubs without concomitant formation of resin or
other blocking components will be prone to wetwood and to

the effects of bacterial infections particularly in hot humid climates with high rainfall.

In some situations, bacteria have no discernible effect on the cell wall[18][348] although in the wetwood of *Populus* spp. and *Abies* spp. they degrade pit membranes.[217,289] The increase in pH has been ascribed to ammonia arising from breakdown of the proteins in the cell wall or lumen of the parenchyma,[118] but recently it was suggested that bacteria may contribute.[18] In *Araucaria* species, bright red or pink streaks or zones are commonly found and in *A. hunsteinii* these colored streaks are often associated with zones of colorless wetwood and on extraction yield degradation compounds of lignin.[146] The mechanical properties of these colored areas are unaffected.

The possibility that minor amounts of lignins and hemicelluloses in the cell wall may serve as substrates in the formation of extractives in wetwood deserves further attention. With our present state of knowledge it is not possible to state whether the extractives are formed before, during or after wetwood formation, by microbiological activity, or what is the nature of the substrate.

Shakes. A shake is a naturally-occurring radial or tangential split or crack frequently found in the stem of standing trees of several softwood and some hardwood species. It is characterized by failure along the middle lamella with loose fibers and ray ends projecting from the surface of the shake. A number of interacting factors appear to be responsible for the formation of shakes[228,237,300] and associations have been noticed between shake development and microorganisms[339] and wetwood.[165]

The cavities of heart-shakes in many species are empty, but there are some others, particularly in the angiosperms growing in tropical regions, which are partly or completely filled with inorganic or organic materials, of a powdery, crystalline, or viscid nature.[66,165] Calcium salts are of common occurrence and sometimes large masses of calcium carbonate ("stone") are found as in *Chlorophora excelsa*; calcium oxalate and malate are also found.[189] Aluminum succinate has been found in the shakes of *Cardwellia sublimis*.[343]

Resin-enriched Areas. Frequently in the interior of
old pine trees, resin-enriched areas or patches are
observed.[35] The source of the resin is probably the
horizontal resin canals of the wood which are continuous
into sapwood where the epithelial cells lining the canals
are fully functional. Resin canals in sapwood can exert
pressures of several atmospheres so that resin is forced
into those areas which have an increased permeability or
have been ruptured.[112] Presumably, resin in the canals is
forced through the ruptured rays cells into latewood
tracheids to give the resin-soaked appearance. The mechanism
of this enrichment is unknown but sapwood canal resin of
pine can be forced into dead heartwood after its formation.
Knots of *Araucaria angustifolia* also can become filled
with resinous materials.[6] Resin enrichment can result
from injection of suspensions of pitch canker fungus or
chemicals; one of the most effective of the latter is the
herbicide "Paraquat" (1,1'-dimethyl-4,4'-bipyridinium ion).[283]

Damaged Wood Resulting from Attack by Insects or
Microorganisms. The hypersensitive reactions which can
occur in tissues surrounding wounds by living organisms
can be quite extensive and are considered in Chapter 10
(see also[23]). However in the context of other aspects
reviewed in this chapter, it is noteworthy that a transition
zone has been observed surrounding regions in *Picea abies*
and *Pinus taeda* that have been affected by *Fomes annosus*[302],
[303] and in *P. radiata* affected by the *Sirex-Amylostereum*
complex.[305]

Discrete Deposits and Inorganic Elements. Some secondary changes involve the formation of discrete deposits by
a small group of cells or by specialized cells. Usually
these deposits are found in the heartwood. A specialized
function is shown by the epithelial cells surrounding the
resin canals of conifers in the formation of materials
different from those of the parenchyma.[234] Other specialized cells include the tanniniferous tubes in Myristicaceae,
latex tubes in *Alstonia* sp. and the Euphorbiaceae, oil
cells in *Cryptocarya* spp. and also resin plugs in the
parenchyma of Araucariaceae and Podocarpaceae.[66] Some
timbers are readily recognized by the deposits in some of
their vessels such as the yellow flecks in *Intsia bijuga*
and white flecks in *Castanospermum australe*. White flecks,
known as floccosoids, in the wood of *Tsuga heterophylla*
are due to the occlusion of clusters of tracheids with

lignans.[208] Crystals of ellagic acid are formed in
specially developed parenchyma cells of the Myristaceae.[50,
95] Whereas starch grains and polymerized polyphenols
frequently occur together in the same cell, crytals of
calcium salts and polyphenols appear to be mutually
exclusive as are crystals and silica.[301]

Numerous types of inorganic deposits are often present
as discrete particles in the heartwood. The most common
inorganic constituent in the form of crystals is calcium
oxalate, which occurs in specially developed crystalliferous
cells in both vertical and ray parenchyma. Sometimes the
crystals are comparatively large. Calcium carbonate is
present as deposits in a number of species.[51,52,301] Globules of silica can be found in the parenchyma of a range
of species in both sapwood and heartwood. Occasionally
flecks or thin streaks of silica are found in the vessels
of *Tectona grandis* and a number of tropical species,
while aluminum succinate has been found in the vessels of
Cardwellia sublimis.[72]

The main mineral constituents of wood are salts of
calcium, potassium, and magnesium, but many other elements
are present in minor amounts. The acid radicals are
carbonates, phosphates, silicates, sulfates, and oxalates
and probably acidic groups of components of the cell wall.
The inorganic material occurs scattered throughout the
cell wall, or as accumulations in the form of crystals
as mentioned above, or as large amorphous lumps.

Usually the amount of ash is higher in sapwood than
in heartwood, but the amount of certain elements is selectively different. During the transformation of sapwood of
Robinia pseudoacacia and *Maclura pomifera* to heartwood,
the ash fell about 30%, but the amount of phosphorus
dropped about 95%, while the calcium levels fell 24% and
41% respectively; the other elements remained constant.[115]
In eucalypts, the phosphorus content in heartwood dropped
by 75 to 98% and potassium by 50 to 98.5%.[13] Magnesium
and manganese tended to be concentrated in the heartwood
of pine, unlike calcium which steadily decreased with the
succeeding growth rings.

An increase in inorganic components in discolored
tissues of sapwood has been reported by many workers.
These accumulations are commonly calcium carbonate, but

other components can be present. A significant aspect of
discolored wood is that the amount of inorganic material
is higher than in the surrounding tissue. Thus, discolored
wood of sugar maple had 6 times more ash than normal sapwood;
however, there was a 9-fold increase in calcium but a 56%
reduction in potassium.[85] There can also be an increase
in manganese with this species.[311] The differences in
these three aspects in shagbark hickory are much greater.[84]
In the discolored wood of *Robinia pseudoacacia* a 136%
increase in ash as compared with normal sapwood has been
observed.[115] Potassium increased 61%, calcium 100%, and
magnesium 168%, but phosphorus decreased 35%; a similar
pattern was observed with *Maclura pomifera*.

   Exudates Resulting from Injury. In many softwoods and
hardwoods, an injury to the developing cambial cells gives
rise to the formation of traumatic intercellular canals.
In softwoods these canals result from a cleavage or separation of cells in a manner similar to that which occurs in
the formation of the more normal vertical canals. Epithelial
cells surrounding the cavity may become thickened and
pitted. The longitudinal traumatic canals in softwoods
are generally arranged tangentially in cross section. They
differ from the more normal canals in that they are usually
restricted to the early part of the growth ring. The
cavities often become filled with resinous material which
exudes when the cavity is ruptured.

   The existence of resin pockets in *Pinus radiata* has
been attributed to water stress rather than to strong winds.
The split which initiates their formation takes place in
the cambial zone; the shape is then determined by the resin
pressure of the tree and the volume of resin available.[63]
On the other hand, star-shaped, centrally-placed shakes
filled with resin have been closely associated with strong
winds. As much as 15 gallons of resin have been found in
pockets in *Pseudotsuga menziesii*.[88]

   The canals or "gum" ducts produced by hardwoods are
also generally the result of damage to the cambium brought
about by abrasion, insects, fire, etc. They may be distinguished from the more normal type of intercellular canal
because they are usually in tangential arrangement and the
cavities are filled with gum-like deposits. In addition,
these traumatic canals may often be considerably larger
than the vessels and are typically quite wide in the

tangential direction. The exudates from *Prunus*, *Acacia*, *Grevillea* spp., etc., are composed of carbohydrates and are correctly known as gums.

In the case of the genus *Eucalyptus*, numerous traumatic canals are formed in many species (usually as a result of different types of damage to the cambium) and they are referred to as *gum veins*, or more correctly, *kino veins*.[136] They are one of the major causes of degrade in eucalypt timbers.[186] These veins are an anatomically distinct system of tangentially anastomosing lacunae containing polyphenols very largely of the flavonoid type. The transition from normal wood to a vein is very abrupt, taking place over less than four rows of cells. Those veins studied most are found in the Myrtaceous genera *Angophora* and *Eucalyptus* but kino is also obtained from *Pterocarpus marsupium* and *Butea frondosa*. During initiation of kino veins, the cambium produces layers of anomalous parenchyma which are generally isodiametrically or tangentially elongated rather than the usually axially elongated xylem elements. These parenchyma become organized into lacunae and bridges. The lacunae are wedge-shaped with the base at the interface with the previously formed wood and the apex towards the cambium. The bridges progressively return to normal xylem and are widest in those positions closest to the cambium. The vertical length (up to 5 m or more) of the vein is much greater than the tangential width and when the point of initiation can be detected it is close towards the bottom of the vein. On most occasions the vein begins to secrete kino (or the cells rupture and allow kino to escape) within three weeks of injury and continues to do so for only a few months. However, if the cambium has been extensively stimulated kino may be secreted for several years. The parenchyma bridges are broken and pockets are formed which can contain several liters of kino. The outer surfaces of active kino pockets have a number of hemispherical protruberances and eventually the cell walls of the kino vein or pocket become thickened and suberized.[68,77,187,313]

The propensity of eucalypts to form kino varies throughout the genus. In the bloodwood group (e.g. *E. calophylla*) kino is frequently found in appreciable quantity; in the ash group (*E. regnans*, *E. delegatensis*, etc.) it is found less frequently whereas in *E. microcorys*, *E. maculata*, etc. it is uncommon.[136] It is not known whether the

tendency to produce kino is controlled by the nature of
the bark protecting the cambium, by the physiological state
of the cambium or by environmental factors. Kino veins are
rarely found naturally in eucalypts younger than 3 years.
There are observations that some races of *E. obliqua* exude
kino more freely than others, and also that kino veins are
more frequently found when the natural growth conditions
are less than optimum or when the conditions provide more
stress than the natural habitat.[68] An examination of
several individuals of *E. regnans* on two sites indicated
that variation in the degree of development of kino veins
is influenced more by environmental than by genetic factors.[76]
A detailed study of the causes of kino formation in selected
regrowth *E. regnans* from one locality showed that more
than 15% of the veins were associated with dead knots,
17% with insect damage and more than 39% had no obvious
association with either cause.[235] It is also known that
fires are a cause of kino formation (e.g.[186]).

Veins and pockets are also found in members of the
Dipterocarpaceae; the contents in these cases are usually
solidified dammar, copal or oleoresin. Large latex canals
are formed in certain timbers of the Apocynaceae and in
sawn timber appear as lens-shaped openings on the tangential
surface.[221]

FEATURES OF HEARTWOOD FORMATION

Factors Affecting the Extent of Formation. Because
the presence of heartwood affects the quality and value of
timber, extensive examinations have been made of a few
important species.

Heartwood usually commences to form at about breast
height and, in the mature tree, its area (relative to disc
area) at this level is greater than at the stump in *Pinus*
spp.,[215,242,269,271] *Pseudotsuga menziesii*,[314,344] *Thuja
plicata*[345] and other species.[66] The relative enlargement
of heartwood at this level does not always occur in *Pseudotsuga menziesii*,[314] *P. taeda*[242] and *Quercus* species.[333]
In one study of *Pinus radiata* the average of the heartwood
area at zero and 2.4 m heights provided a more reliable
basis for calculations than either area separately.[149,150]
Heatwood tapers from the level of initiation towards both
the crown and the butt, but the extent of taper towards the

apex differs with different trees. With *Pseudotsuga menziesii* sapwood thickness decreases from butt to apex and is
relatively thin near the base of the crown.[314] In 39-year
old *Pinus radiata* trees there were 29 growth rings of sapwood at ground level and 16 rings at the 20 m level.[14]

The age of the tree when sapwood tissues change to
heartwood varies widely. In *Cryptomeria japonica* it is
6 to 8 years,[179] in *Robinia* spp. 2 to 3 years, in *Nothofagus
cunninghamii* about 50 years, and in *Alstonia scholaris*
over 100 years. For a large proportion of the *Eucalyptus*
spp. transformation to heartwood takes place when the sapwood width is 1.5 to 2.5 cm. Frequently, 5 to 7 growth
rings are present in this width although more rings may be
present and up to 29 rings have been detected in *E. dalrymp-
leana* growing at a high altitude. On the other hand, fast-
grown *E. camaldulensis* in Israel have 5 to 7 growth rings
in a sapwood width of 9 cm. The number of sapwood rings
can vary widely within a genus or with the age of the tree.
With *Pinus elliottii* there are usually 10 growth rings of
sapwood; with *P. radiata* there are frequently about 14 rings
when grown in Australia and New Zealand,[111] but with *P.
ponderosa* the number may vary between 36 and 200. In *P.
sylvestris* grown in southern, central and northern Sweden
the numbers of sapwood rings were 25, 40 and 70 respectively.[108] The number of rings of sapwood in *Juglans nigra*
of a wide variety of ages, size and growth rates was between
3 to 23 years and the width between 4 to 8 mm. In *Prunus
serotina*, the corresponding figures were 3 to 34 and 4 to
44. There was a wide variation between and within sites,
and sapwood was narrower with fewer growth rings in the
higher positions than near the butt.[253]

It is generally considered that there is a close
association between the age of sapwood tissues and the
formation of heartwood in a species. Two regressions
have been derived for the relationship between age and
amount of heartwood expressed as a percentage of the disc
diameter for *Thujopsis dolabrata* and *Chamaecyparis pisi-
fera*.[185] In *Pinus sylvestris, Picea abies, Pseudotsuga
menziesii,*[66] *P. taeda,*[242] and *Liquidambar styraciflua*[177]
the heartwood percentage increases directly with the age
of the tree. It has been concluded that the increase in
heartwood proportion with age of *P. sylvestris* is due to
diminution of the "vitality" of the tree.[211] The number
of sapwood rings (Y) in *Prunus serotina* and *Juglans nigra*

is correlated with tree age (X), the relationship for the
first species being $\log_{10} Y = 0.1414 + 0.5962 \log_{10} X$ ($r^2 =$
0.68) and for *J. nigra* $\log_{10} Y = 0.3023 + 0.5254 \log_{10} X$
($r^2 = 0.64$).[258] A marked decrease in the number of sapwood
rings occurs in *Nothofagus cunninghamii* with an increase
in size of the tree. On the other hand, 20-year old *Cryptomeria japonica* trees have about 9 years' sapwood whereas
50-year old trees have 18 to 22 years' sapwood. The maximum rate of heartwood diameter increase in *C. japonica* is
attained at a greater age on poorer sites than on better
sites.[179] With *Thuja plicata* a strong correlation was
found between sapwood thickness and radius of the section
within the tree with the maximum width (about 25 mm)
occurring at section radii of 26 cm.

The variations in the amount of sapwood in some species
indicate the involvement of several factors. One estimate
of the gross heritability of heartwood content in a small
number of clones of   *Pinus radiata*   indicated a low
genetic control[258] and a later examination of clonal
material of three different ages showed a significant
heritability of the proportion of the area in heartwood.[259]
Nevertheless wide variation in the extent of heartwood
formation in *P. radiata* has been observed. In one stand
of 30-year old trees, the transformation age of sapwood
was between the usual 14 years and 28 years,[41] with up to
33 years of sapwood occurring in suppressed trees.[145] It
has been proposed that heartwood is formed to regulate the
amount of sapwood in the tree[13] and a causal relationship
between the ratio of wall material to space and the formation of heartwood has been suggested.[41]

Highly significant trends or regressions between the
crown class of the tree, stem diameter and the diameter or
volume of heartwood or width of sapwood have been found
with *Pseudotsuga menziesii*,[314] *Pinus glabra*,[222] *Cryptomeria
japonica*,[179] *Pinus taeda*,[200,269] and *P. radiata*.[149,150]
Crown class is a relative assessment for a particular
stand and it is influenced by growth rate, bole size and
bark thickness.[179,314] A highly significant correlation
has been found between heartwood diameter (d_h) and stem
diameter under bark (d) for *Cryptomeria japonica* in the
regression $d_h = -5.6607 + 0.9018d$.[178] A strong correlation
has been found between the total surface area of sapwood
and tree diameter of *Juglans nigra*.[254] In one stand of
Pinus radiata the relationship between the width of sapwood

(y) and diameter of stem (x) was found to be y = 0.694 x
-1.67.[41] In a study of *P. radiata* trees from two different
areas it was fount that

$$\text{heartwood diameter} = 0.1055 \, D_b + 0.03114A \times D_5 - 2.83$$

where A is age, D_5 is diameter after 5 years' growth and
D_b is diameter at breast height outside bark,[149,150] i.e.
those trees with the most rapid growth in the first 5 years
had the greatest diameter of heartwood. In a study of
Pseudotsuga menziesii, the radial growth in the last
10 years was the most significant growth variable and the
relationship between sapwood thickness (T) and the radial
growth for the last 10 years in inches (G) and diameter
outside bark (D) was found to be

$$T = 0.21 + 0.070D + 0.73G.[314]$$

With *Thuja plicata* it was found that the faster the growth
rate the wider the sapwood;[345] with *Juglans nigra* and *Prunus
serotina* there is a strong and consistent relationship
between growth rate and sapwood thickness.[253] For a given
age *Pinus taeda* trees having the fastest growth rate during
the last 10 years have a relatively smaller proportion of
heartwood and this relationship is maintained irrespective
of the nature of the site.[242] A relationship has also
been shown between rate of growth and sapwood width in
spruce, silver fir and sessile oak.[39] A decrease in site
quality results in a decrease in heartwood volume of *P.
sylvestris* of the same age[113] and an increase in the sapwood
transition age of *Cryptomeria japonica*,[179] but it has no
effect on *Pseudotsuga menziesii*.[344]

Studies of the effects of the level of humidity and
availability of water have yielded contradictory results.[66]
High humidity and availability of water appear to slow down
the formation of heartwood in *Pinus sylvestris*.[271] On the
other hand, factors conducive to heartwood formation in
P. radiata are those which are less affected by periodic
drought conditions, a well distributed and adequate rainfall, and absence of drying winds or conditions of low
humidity.[111]

Pruning of *Cryptomeria japonica* increases the heartwood ratio and volume[179] in agreement with work on *Populus*
spp.[290] and *Pseudotsuga menziesii*.[314] Turpentining

appreciably increases the volume of heartwood in longleaf pine.[70]

An examination has been made of the extent of spatial and of temporal control of heartwood formation in *Juglans nigra* and *Prunus serotina*.[253] Temporal control refers to the death of the xylem parenchyma cells solely in response to the age of the cells, and evidence indicates this form of control has the strongest influence in *J. nigra*. In spatial control the death of the xylem parenchyma is solely in response to an increase in the distance between them and the outer portions of the tree stem and cell age has no effect. Data on the influence of growth rate on sapwood width and number of rings in the sapwood of *Prunus serotina* indicate that both temporal and spatial types of control are important in that species.

Factors Affecting the Amount of Extractives Present. Evidence will be presented in the section on differences in non-structural components to support the view that heartwood extractives are formed in the sapwood adjacent to the heartwood periphery. Consequently possible factors limiting the amount present are the lumen volume in the wood (and to a lesser extent the capillary space in the cell wall), the amount of translocated carbohydrate available at this region (and to a lesser extent the stored starch and fatty acid esters) and the factors controlling the activity of the parenchyma at the time the photosynthates are available.

In this section attention is given to the influence of anatomical features and growth conditions. No close relationship could be found[135] between the amount of extractives and differing lumen volumes in alternating bands of normal and tension wood of an *Acacia* sp. with different lumen volumes. (Although extractives can be found in the lumen of vessels and fibers as well as in the parenchyma, they are formed predominantly in the latter.) Tension wood of *Angophora costata* had a lower volume of axial parenchyma than normal wood of the same tree and also a lower polyphenol content.[154] The total alcohol-soluble materials in *Pseudotsuga menziesii* showed a positive relationship with the volume of ray parenchyma.[125] However in a detailed study of the number of parenchyma cells and their relative volumes per unit weight of wood in *Juglans nigra* and *Quercus rubra*, no causal relationship could be

found with the quantity of phenolic heartwood extractives
in either species.[251]

Cross-sectional discs containing tension wood have
been studied to gain an indication of the influence of the
availability of carbohydrate at the heartwood periphery on
the amount of extractives formed. The extra amount of
translocated carbohydrate required to form the inner gelatinous layer in the tension wood fibers during cambial
growth would result in a smaller amount of carbohydrate
available for storage or formation of heartwood extractives.
This viewpoint was supported by a number of analyses,[154]
and it is in agreement with a number of observations that
the quantity of heartwood extractives is negatively correlated with rate of diameter growth.[280,335] However it was
concluded from a study of living stumps of *Pseudotsuga
menziesii*[125] that formation of heartwood extractives is
insensitive to major changes in the rate of radial growth
and availability of carbohydrates. A recent detailed study
of the between-tree variation in the amount of total heartwood extractives in *Quercus rubra* and *Juglans nigra* failed
to show any relationship with growth rate.[251]

There are a number of observations of variations of
heartwood color in trees of one species growing in different
localities, including a study on *Liriodendron tulipifera*
and *Prunus serotina*.[324] A detailed examination of *Juglans
nigra* showed that the relationship of heartwood color to
soil properties was greater than it was either to tree age
or to diameter-growth rate (rings per inch).[255] It has
been frequently observed that the amount of extractives in
heartwood increases with the distance from the pith or the
age of the tree when the extractives were formed.[136] In
many cases, while the extractives show this overall increase
they do not increase steadily in amount, but rather increase
at different rates during various periods so that streaks
or bands of differing color intensity are apparent. An
examination of 480 *Pinus echinata* trees of different ages
growing under a variety of conditions showed that the age
of the tree influences extractives content more than any
other variable, and that, while the environmental conditions
contribute to the variation in content, no one environmental
variable is a significant factor.[274]

The oleoresin yield and composition of turpentine from
slash pine are under strong genetic control. Increases as

high as 31% in oleoresin yield and 11% in rate of growth
have been obtained from progeny of tested clonal seed
orchards.[82,97,262,315,316,364] On the other hand, inheritance of yield of oleoresin in other species appears to be
under weak genetic control.[337]

EFFECT OF CHANGES ON WOOD QUALITY

     Variations in the anatomical and fine structural
features of wood have been reviewed on several occasions.[189,
201,267,334] Much less attention has been given to secondary
changes which can occur in the normal characteristics of
the tree or log. The considerable differences in wood
quality due to secondary changes that can occur in different
parts of one log, or between different trees of the same
species, can have significant influence on utilization.
Secondary changes described in an earlier section may have
increasing importance in the future with significant amounts
of wood coming from unfamiliar species growing in tropical
regions, from low quality forests, from familiar species
harvested from multiple purpose forests where damage can
be at a higher level than in wood production forests, and
from intensively cultivated, rapidly grown production
forests. Changes in morphology arising from cambial injury
affect solid wood uses and there are various wood grading
rules to quantify the extent of the veins, pockets, and
shakes, etc., that can be allowed in construction timbers.
In some cases the presence of these irregularities is
desired for decorative purposes.

     In addition to the effect on appearance caused by
anatomical features, the variations in color from sample to
sample and within a sample are important to the decorative
wood industry. The color of wood is due to accumulation of
extractives in cell lumen or cell walls. Color changes
resulting from a particular cause can be markedly different
in different species such as the changes caused by light
irradiation which have been attributed to heartwood extractives in several species.[231,232,357] Some chemical components of wood have a pleasant aroma for humans but are
repellant to other forms of life. Examples are found in
colored zones of species such as *Camphora*, *Santalum*,
Callitris, *Juniperus*, *Cedrela*, *Chamaecyparis* spp., *Dacrydium
franklinii*, *Acacia acuminata*, etc. Not all odors are
pleasant (e.g. *Ocotea bullata*) and several may taint food.

Some colored woods contain components conveying a greasy feel, such as triterpenes and sterols in *E. microcorys* ("tallow wood")[29,64] and in *Eugenia gustavioides*.[351] Extractives have been related to frictional behavior[239] and wear of cutting tools.[55,158,198,240]

A number of species contain in heartwood hydrophobic components which may have some preservative influence but can affect adhesion in the manufacture of plywood, particleboard and laminated wood and can affect the stability of surface films. Other components if present in sufficient quantity can also affect bond strength by affecting adhesion of the glue to the wood surface, or they can affect the strength of the adhesive by diluting it at the interface or by reacting with it to affect setting.[5,53,142,188,265,272,277,350] Some extractives can improve the stability of surface films by restricting the degree of penetration or by absorbing ultraviolet light. Other extractives, either in the wood structure or in veins, pockets, etc., can result in a rapid breakdown of such films.[16,96,104,212,213,294,297]

Sugars, "tannins" and phenolic extractives can affect the hardening of cement in the preparation of wood-cement boards.[295,296,358] Extractives in some woods can affect the rate and extent of polymerization of the plastic component of wood-plastic composites.[75,263,355,356]

It is not possible to generalize on the effect of extractives on shrinkage although sapwood generally shrinks less than the heartwood of the same species.[201] Extractives appear to be more hygroscopic than the cell wall but they also have a bulking effect at low temperatures. The difficulties encountered in relating content of extractives to variations in fiber saturation point could be due to differing amounts in the cell wall and differing chemical natures.[54,60,293,331,338] The density or specific gravity of wood increases with content of extractives in several cases[194,328] but not in *Pinus pinaster*.[336] The results of studies on *Sequoia sempervirens*, *Thuja plicata* and *Robinia pseudoacacia* indicated that increasing amounts of extractives are usually associated with increased modulus of rupture, compression parallel to the grain and shock resistance,[218] but later detailed studies showed that, in redwood, modulus of rupture was independent of the amount of extractives

present whereas modulus of elasticity decreased with
increasing amounts of extractives.[11]

The presence of extractives in wood reduces its permeability and affects fluid flow and the resultant treata-
bility with preservatives, penetration of pulping chemicals,
and drying.[12,47,196,207] A positive correlation has been
found between the amount of extractives and thickness
shrinkage in *Sequoia sempirvirens*[89] and collapse in *Thuja
plicata*.[229] Extractives may act as plasticizers during
drying,[245,330] particularly at temperatures above 55 C
where the amount of extractives has been related directly
to shrinkage and collapse in redwood.[69]

The amount and type of extractives found in woundwood
and heartwood are a significant factor in the confinement
of the spread of wood destroying organisms in the living
tree and in wood placed in destructive conditions.[299] The
presence of extractives can also have a significant effect
on pulp and paper making.[136,142] In some cases they provide
a valuable byproduct in the form of "tall oil" from conifers;
in other cases, e.g. *Nauclea* spp., they prevent the production of a satisfactory pulp. The extractives in a number
of tropical species give different types of pitch on
pulping and cause considerable color reversion.[327] Extractives can influence manufacture at practically every stage;
some accelerate the blunting and corrosion of tools, and
corrode pulp digesters; some cause considerable color
changes, reduce the yield of pulp and lower its quality,
increase chemical consumption, inhibit pulping reactions,
and affect the recovery of pulping chemicals.[15,136,140,141,273]

NATURE OF DIFFERENCES IN
NON-STRUCTURAL COMPONENTS

Chemotaxonomy of Heartwood Extractives. Heartwood and
other tissues which have undergone secondary changes contain
a wide variety of extractives or extraneous substances, such
as polyphenols (including the "tannins"), resins, fats, oils,
salts of organic acids, etc. Among the most common of
these extractives are the polyphenols, which are present
in all heartwoods, and many sapwoods, from trace to large
amounts. They are largely present in polymerized forms
particularly when the components contain vicinal-trihydroxy

moieties.   There is little precise information on the
nature of the polymerized polyphenols.

Extractives are biosynthesized from intermediates
resulting from the breakdown of sugar in primary metabolism.
One important intermediate is pyruvate, which leads to
acetyl coenzyme A, a major precursor, from which the tri-
carboxylic acid cycle is entered as well as two other
pathways which lead to formation of two important classes
of extractives.   One path leads to malonyl-CoA and in turn
to fatty acids, and the other to acetoacetyl-CoA and isoprenoid substances (e.g. terpenoids, steriods, etc.).
Some woods contain extractives that are largely formed
from pyruvate.   Another major intermediate is shikimic
acid and the shikimic acid-prephenic acid pathway leads to
formation of the C_6, C_6-C_1, C_6-C_2, C_6-C_3 phenolic compounds.
Several groups of heartwood extractives (e.g. flavonoids,
stilbenes, isoflavonoids) are produced from a combination
of the acetate and shikimic acid pathways.[136,140]

The biosynthetic pathways of extractives are under
genetic control and the normal heartwood extractives of
many species have distinctive compositions.   Samples of
the same species collected from widely separated areas have
the same qualitative composition indicating that environment
has no pronounced effect.   It is important in any experimental study of heartwood formation that the extractives
be analyzed.   A number of investigations have considered
that, because of similarity in color, the microscopic
appearance of the cellular contents or the solubility in
unselective solvents, wound-induced discolorations were
extensions of normal heartwood into sapwood.   There can be
important differences in composition between the normal
heartwood extractives formed from internal stimuli associated
with aging and the extractives in discolored sapwood induced
by wounding.

The composition of the mixture of components in extractives of different heartwoods has been used for chemotaxo-
nomic purposes.   The polyphenols are a most useful class of
compounds for this purpose.   Among the genera studied are
Pinus (and other conifers[87]), *Prunus*,[119] *Acacia*,[57] *Nothofagus*,[159] and *Eucalyptus*.[143]   The polyphenols found in
heartwoods of each of these genera are distinctive.   Some
families have distinctive components; for example the
Cupressaceae are the sole source of tropolones such as

β-thujaplicin (I).[103] In most cases the extractives of
heartwood are "conservative" and show only small changes
in composition throughout series of groups of related
species. Nevertheless, two *Nothofagus* species whose wood
anatomy was practically identical contained extractives
with markedly different constitution.[159]

Occasional varieties of three species containing
significantly different wood extractives from the normal
are known and are found in genera such as *Cinnamomum*,
Pinus, *Pterocarpus*,[139] *Ocotoea*,[107] and *Acacia*.[57]

The major portion (80% or more) of the polyphenolic
extractives of heartwood contains polymeric materials
which cannot be resolved chromatographically at the present
time. However those components which can be resolved
characterize normal heartwood in trees of the same species,
although the composition of extractives does not remain
unchanged in the living tree. The decay resistance of the
interior of a tree decreases with time and this could be
due in part to the lower amount of polyphenols formed
when the tree was in the juvenile form. However, some
work indicates that as a tree ages certain heartwood polyphenols are converted by enzyme systems or by other agents
to quinones and other forms with reduced toxicity.[8,9,287]
This proposal is supported by the recent discovery of two
phenol-oxidizing enzymes in the heartwood of *Pinus radiata*.[307]

Distinctive or Specific Cellular Components. A number
of distinctive variations exist in the chemical and chemo-
morphological changes which take place in the formation of
non-structural components. The evidence indicates a hier-
archy of secondary components, an order in formation, a
varying sensitivity to changes in the direction of biosynthesis, and a specialized function of some parenchyma.
More knowledge on the chemomorphology (the precise location

I β- Thujaplicin

of extractives in tissues) of these variations will provide
greater understanding of aspects of disease resistance and
of wood quality.

The function of some parenchyma is entirely the formation of crystals of calcium oxalate or of calcium carbonate.
Other parenchyma can have more than one function. This is
shown in *Syncarpia glomulifera* where silica globules are
formed in young sapwood and later the same parenchyma are
filled with polyphenolic extractives at the heartwood
periphery.[72]

Parenchyma can form complex polyphenols with considerable selectivity. Some of the vessels of several samples
of *Intsia bijuga* are completely filled with pure robinetin
whereas adjacent vessels may be empty (Fig. 1) or filled
with a mixture of polyphenols such as robinetin (II),
dihydromyricetin (III), myricetin (IV), naringenin (V),
3,5,4'-tri- and 3,5,3',4'-tetrahydroxy-stilbenes (VI and
VII), and polymerized leucocyanidin (VIII).[161] The vessels
in another *Intsia* species contained a mixture of flavonols
including mainly fisetin (IX), with robinetin (II), 3,5,4'-
trihydroxystilbene (VI) and 3,7,3',5'-tetrahydroxy flavone
(X).[180,181] It is noteworthy that the biosynthesis of
vessel contents in the latter *Intsia* species is not as
selective as in *I. bijuga* and is along slightly different
pathways. The major polyphenols in *Castanospermum australe*
lack the commonly occurring C5- oxygen function and are
7-hydroxy-4'-methoxyisoflavone (XI), 7-hydroxy-6,4'-dimethoxy-
isoflavone (XII) and bayin--the 8-C-β-D-glucopyranosyl-7,4'-
dihydroxyflavone (XIII).[79,80,153] It is remarkable that
some vessels of *C. australe* can be completely filled with
pure bayin which belongs to a different class of flavonoids
from the other major polyphenols.[164] The contents of
different tracheids in the small clusters of cells which
make up floccosoids in *Tsuga heterophylla* can differ.
Some contain very largely conidendrin (XIV), others largely
hydroxymatairesinol (XV) and others an unidentified lignin.
[208,268a] The white specks on the surface of the veneers
of *Haplormosia monophylla* appear to be due to sakuranetin
(XVI). The heartwood contains 6,8-di-C-glucosylapigenin
(XVII), which is the major component of the bark exudate,
and other flavonoids and stilbenes.[122,123] Yellow-brown
crystals of ellagic acid (XVIII) can form in heartwood ray
cells of eucalypts when the cell adjoins a vessel.[50] The
parenchyma which form tyloses in *Eucalyptus botryoides*
appear to occupy a particular position in the rays.[248]

Figure 1. A longitudinal section through vessel
elements of wood from *Intsia bijuga*. One vessel completely
filled with pure robinetin (light region in center of
photograph) lies adjacent to an empty vessel (dark region
to the left of the filled vessel).

     In some cases visual observations indicate a sequential
formation of extractives at the boundary of heartwood or
damaged regions. Ellagitannins and gallotannins are the
major resolvable components of the extractives of the sapwood of *Eucalyptus polyanthemos*. Large amounts of 3,4,3'-
tris-O-methylellagic acid 4'-glucoside (XIX) form in the
outer transition zone and produce the characteristic fluorescence under ultraviolet light seen in this region. The
fluorescence of XIX in inner layers of the transition zone
and heartwood is masked by the large amounts of leucocyani-
dins formed in these regions.[162] It is noteworthy that
XIX is a distinctive component of the clear-cut group of
eucalypts to which *E. polyanthemos* belongs.[162]

     Composition of Shake Deposits and Damaged Woods. Heart
shakes, frost checks, etc., can contain pure compounds, or
a mixture of a small number of components, which may not
be present in significant amounts in heartwood extractives.
The lignans gmelinol (XX) and iso-olivil (XXI) have been

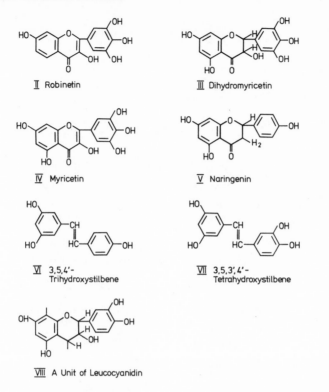

II Robinetin

III Dihydromyricetin

IV Myricetin

V Naringenin

VI 3,5,4'-
Trihydroxystilbene

VII 3,5,3',4'-
Tetrahydroxystilbene

VIII A Unit of Leucocyanidin

obtained respectively from a white deposit in the shakes of
Gmelina leichhardtii, and from the resinous deposits in the
shakes of *Olea cunninghamii*.[30] Particles occurring close
together on the surface of a frost check in *Tsuga heterophylla* were found to consist of lignans, either almost pure
matairesinol (XXII), or largely hydroxy-matairesinol (XV)
or mostly conindendrin (XIV).[208] Crystalline XXII has been
found in ring shakes of *T. mertensiana*.[16a] XXII, XIV and
quercetin (XXIII) have been isolated from the heartshake
of *Podocarpus spicatus*.[31,32] Almost pure crystals of podocarpic acid (XXIV), an aromatic diterpene acid, have been
found in the heartshakes of three *Dacrydium* species which
possessed colored heartwoods. Except for trace amounts in
a thin band of the tissues surrounding the shakes, this
acid was absent from the rest of the wood.[165] The yellow
powdery deposit in the shake of an *Afzelia* species contained
very largely kaempferol-3-rhamnoside (XXV) which was not
present in the heartwood.[197] The heartshake of *Cardwellia*

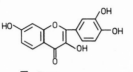

IX   Fisetin

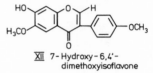

X   3,7,3′,5′-
Tetrahydroxyflavone

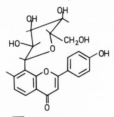

XI   7-Hydroxy-4′-
methoxyisoflavone

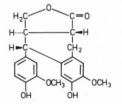

XII   7-Hydroxy-6,4′-
dimethoxyisoflavone

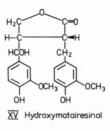

XIII   Bayin

XIV   Conidendrin

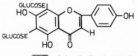

XVI   Sakuranetin

XV   Hydroxymatairesinol

XVII   6,8-di-C-glucosylapigenin

sublimis contains deposits of aluminum succinate.[72] The
yellow deposit in the shake of *Intsia bijuga* from New Cale-
donia is pure II but in the shake of an *Intsia* species
from Borneo it is a mixture of II, VI, IX, X, and XXII.[182]
Crystalline camphor is found within heartshakes of *Dryo-
balanops aromatica* and *D. lanceolata*.[56,71] Large crystals
of methyl thujate have been found in the heartwood of
Thuja plicata.[209a]

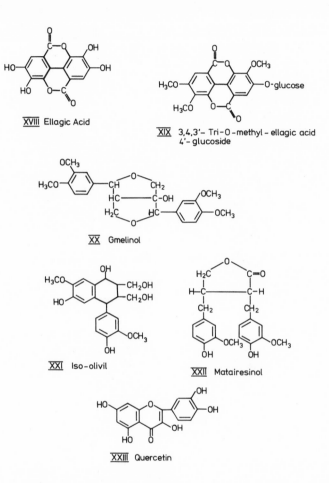

XVIII Ellagic Acid

XIX 3,4,3'- Tri-O-methyl-ellagic acid 4'-glucoside

XX Gmelinol

XXI Iso-olivil

XXII Matairesinol

XXIII Quercetin

Like heartwood, discolored sapwood does not contain living cells but the two types of wood can be significantly different in appearance in *Quercus alba, Acer saccharinum, Juglans nigra, Maclura pomifera,* and *Robinia pseudoacacia* and in the broad chemical characteristics of their components.[114],[115] There have been detailed examinations of sapwood infected by different fungi and a marked change in the direction of the biosynthesis of extractives has been found. Sapwood of *Prunus domestica* affected by the fungus

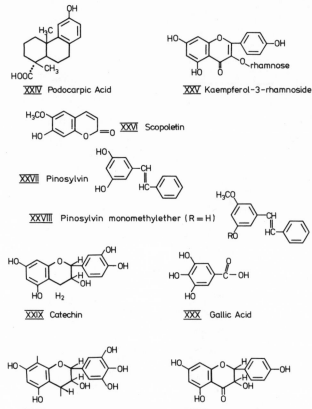

XXIV Podocarpic Acid

XXV Kaempferol-3-rhamnoside

XXVI Scopoletin

XXVII Pinosylvin

XXVIII Pinosylvin monomethylether (R = H)

XXIX Catechin

XXX Gallic Acid

XXXI A Unit of Leucodelphinidin

XXXII Dihydrokaempferol

Stereum purpureum contains considerable amounts of the coumarin scopoletin (XXVI) which is present in only very small amounts in sapwood[160] and absent from normal heartwood of other *Prunus* species.[119] When *Trametes versicolor* attacked *Prunus jamasakura*, large amounts of an abnormal constituent, the lignan iso-olivil (XXI), were formed.[120] *Pinus radiata* sapwood affected by *Amylostereum areolatum* lacks the flavonoids found in heartwood but contains two fungicidal pinosylvins (XXVII and XXVIII R=H) in a ratio different from that in heartwood.[156] The fungus *Peridermium pini* induces the formation of pinosylvin dimethyl ether (XXVIII R=CH₃), not found elsewhere in *Pinus sylvestris*, as well as large amounts of resin.[347] On the other hand

induced heartwood produced in *Pinus* spp. following the
application of sulfuric acid does not contain the phenolic
constituents of normal heartwood.[7]

It is evident that some parenchyma have very specific
biosynthetic functions and these must be under genetic
control as is the composition of normal heartwood extractives. The above examples show that the extractives of
affected wood resulting from different forms of damage can
be chemically different from normal heartwood. Consequently,
the processes leading up to the death of the living cells
can have a great influence on the composition of the
extractives and are apparently also under genetic control.

Exudates. After certain trees are injured, exudates
of various types frequently appear on the outside of the
tree where they dry to soft or brittle solids. Liquids
(of varying consistencies) are also found in veins or
pockets within the wood. Frequently these exudates are
referred to as "gums" which in some cases gives a misleading
indication of their composition. There are certain similarities in the mode of formation of the different types of
exudate. They appear within 3 weeks after injury to the
cambium in the spring season. They frequently result when
the trees are grown under adverse conditions of cold, heat
(fires) and lack of moisture, or as a result of insect or
other mechanical damage, infection, etc.[136]

True gum exudates are composed largely of polysaccharides
and when allowed to dry, form clear glassy masses, which
vary in color from brown to pale yellow. They are more or
less soluble in water or swell to a jelly, but are insoluble
in alcohol.

Resins are pale-colored materials more or less soluble
in various organic solvents but insoluble in water. Their
composition varies considerably and the components are
usually of a terpenoid nature.[136] "Pocket resin" of
Douglas fir can have a different composition from normal
resin.[88,110]

When cambium of eucalypts is injured or affected,
normal wood formation ceases and there is an abrupt reorgani-
zation of the tissues. There is also a marked change in
the direction of polyphenol biosynthesis resulting in the
formation of kino, a polyphenolic extracellular fluid.

Those cells immediately surrounding the kino vein are
filled with leucoanthocyanins which are weakly present
elsewhere.[134]

    The main constituent of kino is commonly polymerized
leucoanthocyanin although there is at least one kino (from
E. moluccana syn. *E. hemiphloia*) containing polymerized
catechins (XXIX) as the major component.[148] Fresh kino
from a number of species does not contain monomeric components, apart from trace amounts of gallic (XXX) and ellagic
(XVIII) acids, and appears to be pure polymerized leucodelphinidin (XXXI)[137] giving ruby red-colored solutions in
alcohol. One group of kinos contains small amounts of
alcohol-insoluble material considered to be of a carbohydrate nature. Another group contains small amounts of
dihydrokaempferol (XXXII) and the lignan eudesmin (XXXIII).
A small number of kinos can be recognized by the significant
amount of monomeric flavonoids present[136,163] and in one
kino appreciable amounts of flavonoid C-glycosides.[148]

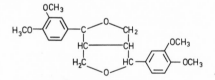

XXXIII Eudesmin

XXXIV Vanillin

XXXV Benzoic Acid

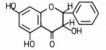

XXXVI Pinobanksin

XXXVII Pinocembrin

ORIGIN OF EXTRACTIVES

   Theories of Heartwood Formation. Frequently heartwood
forms in a tree after a certain stage of growth and consequently most theories consider its formation to be a simple
aging process. This view was, until recently, supported
by the results of cytological and biophysical examinations.
Among the criteria used in these studies have been altera-
tion in shape and size of the nuclei, the reaction with
vital stains and the number of organelles present. It has
been concluded that with increasing distance from the
cambium, the ray parenchyma cells undergo irreversible
changes which result in the degradation of the protoplasm,
disorganization of the cells' organizing system and altera-
tion of metabolism in the transition zone if nuclei are
still present.[28,99,101,102,129,175,257,264,317] However,
the cell volume and surface area of nuclei and nucleoli in
ray parenchyma of trees such as walnut, oak, maple, and
hornbeam may be larger in the zone adjacent to the heartwood boundary than in middle sapwood.[175] The vitality of
the sapwood cells as measured by electrical resistance,
conductivity and plasmolysis has been found to decrease
across sapwood and reach zero-value at the heartwood
boundary.[246,247] It has been proposed[99] that aerobic conditions in outer sapwood change with the gradual degradation
of nuclei to semi-anaerobic conditions with the decline in
sugar content in inner sapwood. In the inner transition
zone the phenols change through oxidative-polymerization
reactions. There was little agreement in these studies
with the postulation of a zone of intense metabolism at the
periphery of heartwood.[49] The possibility of seasonal
differences in the metabolic activity of the inner sapwood
cells received little attention, as the concept of steady
degeneration of the sapwood parenchyma over a period of
years made such seasonal differences seem unlikely.

   Other theories propose the translocation of secondary
metabolites from the cambial tissues and parenchyma to the
heartwood periphery as a type of internal excretion of
biochemical "waste" products.[27,86,320,321] It is considered
that these "excretory substances" are translocated in non-
toxic concentrations along the rays towards the interior
where they accumulate in amounts which rise to lethal levels
in trees which contain heartwood but apparently not in
trees or species which do not form heartwood. When the
innermost parenchyma die and further translocation is

impeded, the continued accumulation of toxic components at
the heartwood periphery causes death of the parenchyma and
gradual outward movement of the heartwood. There is partial
support for these ideas in that flavonoid glycosides which
occur in small quantities in sapwood are rarely present
in heartwood although the aglycones are present in larger
quantities. Theories based on translocation of secondary
metabolites do not recognize the differences (which are
sometimes considerable) in composition of the extractives
in the cambial regions, sapwood and heartwood.[140,147]

Some workers (e.g. [233]) have associated heartwood
formation with non-pathogenic fungi; hyphae have been found
in the central tissues up to the heartwood boundary in
some species.[49,365] In an analogous situation, the accumulation of pinosylvin (XXVII) and its monomethyl ether
(XXVIII) in *Pinus taeda* has been induced by the injection
of a toxic sesquiterpene isolated from culture filtrates
of the fungus *Fomes annosus*.[17] However, neither fungal
hyphae nor other evidence of fungal initiation have been
detected in a number of heartwoods.

As an alternative theory, it has been suggested that
air enters tree trunks from broken branch stubs and other
wounds and accumulates in the interior. The water content
of the wood is lowered and this in turn may affect the
metabolism of the living parenchyma cells so that heartwood
is formed.[116,174,193,276,365]

On the basis that heartwood transformation is a development process rather than a deterioration of cell function
with age, it has also been suggested[13] that a heartwood-
inducing substance moves centripetally along the rays from
the cambium. This substance reaches the threshold level
for heartwood initiation in cells adjacent to the impermeable
heartwood core. At present the mechanisms of heartwood
formation have not been established. However, the association of moisture stress with heartwood formation in some
species is becoming more evident.

Site of Formation and Precursors of Extractives. Evidence of several types supports the conclusion that extractives or secondary metabolites are formed *in situ*. One
type of evidence revolves around the fact that the extractives in heartwood, woundwood or discolored wood, sapwood,
and cambial tissues, can belong to different classes of

compounds. Studies of the biosynthesis of various secondary
compounds show that end-products of metabolism cannot be
transformed into other types without degradation and
rebuilding through primary metabolites. A number of energy-
demanding stages would be required for these processes
with the result that secondary metabolites would be less
effective precursors of heartwood extractives than sugars
or fats.

The phenolic extractives of sapwood of *Pinus radiata*
are present in very small amounts and consist of vanillin
(XXXIV), substituted benzoic acids (XXXV) and traces of
pinosylvin monomethyl ether (XXVIII). On the other hand
heartwood contains appreciable amounts of pinosylvin
(XXVII) and its methyl ether (XXVIII) and the flavonoids,
pinobanksin (XXXVI) and pinocembrin (XXXVII).[156] The areas
of living sapwood of *P. radiata* affected by the attack of
Sirex noctilio and its symbiotic fungus, *Amylostereum
aereolatum*, contain a larger amount of resinous and phenolic
extractives than normal heartwood. As noted above the
two pinosylvins, with their reputed fungitoxic properties,
are the only phenolics present although their proportions
differ from those in heartwood.[156] The rhytidome and
phloem of *P. radiata* contain appreciable amounts of leucoanthocyanins or proanthocyanidins (VIII) yielding cyanidin
in contrast to the heartwood.[126,132,226]

The major portion of the monomeric polyphenols in the
cambium of *Eucalyptus astringens, E. sieberi, E. regnans,*
etc. is composed of ellagitannins (polyphenolic acids
esterified with glucose) although catechins (XXIX) and leucoanthocyanins (XXXI) are also present.[145,147] The wood of
juvenile *E. globulus* contains very low amounts of polyphenols.[325]

Sapwood of *E. astringens* contains catechins and leucoanthocyanins in a much greater proportion to the ellagitannins than in cambium. In heartwood, appreciable amounts
of stilbenes are also present[147] and the bark contains
large amounts of leucodelphinidin (XXXI).[132] Sapwood of
Calophospermum mopane contains a low concentration of
flavonoids of the phloroglucinol-catechol type. The heartwood also contains small amounts of these flavonoids but
much larger amounts of resorcinol-catechol type.[78]

As mentioned above, compounds of considerable molecular complexity have been found almost pure or in a concentrated form in single cells or small groups of cells. Adjacent or neighboring cells may be empty or contain different compounds. Materials of an aromatic nature were found to be absent from the ray parenchyma of *Tsuga heterophylla* even when lignans were present in large amounts in tracheids on the other side of connecting half-bordered pit pairs.[208]

It is evident from these few examples that fundamentally different metabolic processes can operate in different tissues and even in neighboring cells in the tissues. The mechanism for this situation can be most readily explained on the basis that the extractives in heartwood or injured wood are formed *in situ* from primary metabolites. These metabolites would be carbohydrate and possibly, to a limited extent, stored fatty acids and their esters. Some carbohydrate is stored in the parenchyma of sapwood but most would be translocated from the cambium; consequently the amount of extractives in wood after secondary changes have taken place will depend on the photosynthates reaching that region at the time of change.

The influence on photosynthesis in forest trees of light, temperature, soil moisture, soil fertility, and disease has been reviewed.[136,209] Photosynthates are translocated through the sieve elements of the phloem to other parts of the tree where a portion is respired to provide energy and to the active cambial region for conversion into wood elements, or (in the case of injury) into secondary metabolites. The remainder of the photosynthate is stored as starch or fatty substances in sapwood or is transformed into secondary metabolites. The nature of the stored primary metabolites may be predominantly fatty materials (as in *Pinus, Pseudotsuga, Aesculus,* etc.), predominantly starch (as in *Quercus, Eucalyptus, Fraxinus,* etc.) or both (as in *Salix, Prunus,* etc.).[136,209]

The amount of stored materials varies according to species, season, and the demands of the tree but can reach more than 7% starch or 2.5% fatty substances.[136,209,363] Although there are reports of small amounts of free sugars in heartwood,[136] starch disappears at the heartwood boundary. The amount of starch varies across the sapwood sometimes being greater in middle sapwood than on either side.[73,74,154,167,168,206] It is interesting to note that, in view of the implication of the involvement of water stress in secondary changes, hydrolysis of starch has been associated with water deficit.[136,220,291] It should be noted, however, that starch content can decrease for some distance from the heartwood periphery and before a marked fall in moisture content occurs. Starch in *Robinia pseudoacacia* sapwood changes from an amylose type in outer sapwood to an amylopectin type in inner sapwood.[168] The starch content of inner sapwood of *Angophora costata* trees is insufficient to account for the quantity of polyphenols present in heartwood so that it is probable that the major part of the heartwood polyphenols is formed from translocated carbohydrates which have not been utilized in growth.[154] Consistent with this proposal is the common observation that rapid growth and efficient utilization of available carbohydrates are often associated with heartwoods which are lighter in color than normal and contain low amounts of polyphenols.[140] There are observations of a general nature that trees containing low amounts of reserve primary metabolites are susceptible to disease as for example in the case of *Citrus sinensis*.[257]

The presence of readily translocatable mono- and disaccharides in significant amounts has been reported on several occasions.[74,136,140,167,168] The distribution of sugar in *Angophora costata* in autumn was different from that at other seasons and was more than double at the heartwood periphery than at the cambium.[154] This observation may be significant in view of recent findings that the greatest metabolic activity exists in the autumn season at the heartwood periphery of a related myrtaceous species.[145] Soluble carbohydrates can be translocated quickly and even when labeled glucose was applied to the cambial region of a eucalypt in the spring season, radioactivity was translocated to the heartwood periphery and accumulated within a period of 19 days.[152] Twelve days after labeled sucrose was administered to the cambial region of *Prunus yedoensis*, the labeled flavonoid in the transition zone had a higher activity than the same compound in inner or outer sapwood.[121] Accordingly not only are sugars readily translocated to the transition zone but they are also quickly transformed to polyphenols.

Data relating to the radial concentration gradients of lipid material are less extensive than those for starch and sugar and to an extent are conflicting.[206] The composition of lipids in the interior of heartwood-forming trunks of hardwood species appears to be qualitatively different from those which exhibit no heartwood formation.[126,130,172] Fats are important reserve materials as they contain more than twice the amount of energy per unit weight that is found in carbohydrates and can be converted to sugars in wood of *Tilia cordata*.[170]

   Mechanism of Formation. Extensive anatomical study has revealed the predominant role played by ray cells in heartwood formation with a probable intensified degree of biochemical activity at the periphery, whereas the axial parenchyma appear to be largely a storage tissue.[48] As noted above, some cytological data indicate that the sapwood cells adjacent to heartwood are incapable of enhanced activity, but the increase in volume and surface area of nuclei and nucleoli in the parenchyma cells adjacent to the heartwood boundary observed by others may indicate an increased vitality.[175] Some workers[106,129] have been unable to detect any increase in the respiratory activity of inner sapwood of several species, while other studies on *Quercus pedunculata*, *Robinia pseudoacacia*, *Pinus radiata*, etc., have revealed an increase in respiratory activity of inner sapwood relative to middle or outer sapwood.[171,175,250,308,359,360] The respiratory quotient of the transition zone in *Robinia pseudoacacia* indicates that carbohydrates are the main substrate.[171] The most marked increase in respiration on storage occurred with the rings adjacent to the colorless transition zone[359,360] although bacterial contamination may have contributed to the respiration levels in some of these studies because of the long storage times. Further indication of increased activity at the heartwood boundary has been obtained from data on the accessibility of hydroxyl groups of cellulose.[55] Accessibility across disks of Douglas fir decreased from outer to middle sapwood followed by a rapid rise to a maximum at the sapwood-heartwood boundary and then a rapid drop in heartwood. This situation in the cell walls at the heartwood boundary should facilitate permeation by extractives when the tonoplast ruptures and facilitate reaction with polyphenols.

   Cytological examinations to determine the organelles responsible for particular syntheses in the ray parenchyma cells situated in sapwood close to the heartwood boundary,[20,44,59,268,342,346] have resulted in few conclusions.

Other factors can be proposed as determinants of the
formation of heartwood extractives. These include activation, formation and inhibition of enzymes, hormone supply,
selective synthesis, aeration, and carbon dioxide concentration and water movement. In a detailed study of the formation of heartwood in *Juglans nigra* and *Quercus rubra*[251] it
was found that all data, including within- and between-
species comparisons, physiological conditions in the transition zone, or near the heartwood periphery, had a much
greater influence on the rate and duration of aromatic
biosynthesis than the number or volume of active parenchyma
units.

It has been reported that the DNA content increases
towards the heartwood periphery[127] as do the amounts of
nicotinic acid amide, biotin, pyridoxine,[361,362] and in
some cases protein-N.[298,362] A rapid decrease in ATP
content takes place in the centripetal direction of the
heartwood-forming *Robinia pseudoacacia* whereas the content
remains the same or increases in *Betula verrucosa* or *Tilia
cordata* which do not form heartwood.[171] The activity of
some enzymes has been found to increase in inner sapwood
or at the heartwood periphery. These enzymes, which are
necessary for the biosynthesis of extractives, include
malic and glucose-6-phosphate dehydrogenases (which also
had higher activities in the dormant season),[129,308,362]
an amylase,[167,171,202,227] an aldolase,[167] enzymes controlling starch hydrolysis,[202 227] peroxidases[73,145,210,227,342]
(one of which was concentrated in the torus),[210] and
phenol oxidases.[362] There was a concomitant decrease in
the activities of starch synthetase and phosphorylase.[167]
An attempt to block the utilization in *Rhus* sapwood of the
acetate units of the citric acid cycle, by using certain
enzyme inhibitors, resulted in the formation of heartwood
flavonoids.[155] Histological tests indicated that in the
transition zone of *Pinus radiata*[308] and *Cryptomeria japonica*[129]
the pentose phosphate shunt activity increased more than did
the citric acid cycle. In contrast the activity of phenylalanine ammonialyase decreased progressively inward and was
absent in the heartwood.[128] This enzyme showed its highest
activity in the cambial region of *Cryptomeria* and *Chamaecyparis* spp. where it was primarily related to the production
of a lignin precursor. The lack of evidence for increased
activity at the heartwood periphery could be due to the
stage of the growing season at which the samples were taken.
Alternatively it is possible that phenylalanine ammonia-

lyase is not an obligatory enzyme for the biosynthesis of polyphenols.[157]

Little attention has been given to the factors which could increase the activity of a number of enzymes in the transition zone. The role of auxin in the formation of heartwood extractives has been considered previously[140,362] but so far only the involvement of the plant growth substance ethylene has been demonstrated. Ethylene is produced by the transition zone of *Pinus radiata* in larger amounts than by adjacent sapwood.[306] The peak of production takes place in the dormant season of late winter and in this regard agrees with another study which indicated that heartwood formation in larch takes place mainly when cambial growth ceases.[166] The transition zone of *Eucalyptus tereticornis* produces more ethylene than sapwood[144] and the transition zone surrounding *Sirex-Amylostereum* lesions in *P. radiata* produces appreciable amounts of ethylene.[305] Ethylene production near the heartwood boundary reaches a maximum in the early dormant season in *Juglans nigra* sapwood and during the late growing season in *Prunus serotina* sapwood.[252,254] Many studies have shown that ethylene acts as a regulatory hormone in a variety of physiological changes occurring at many stages in the ontogeny of plants.[2,275] Very small amounts (1 to 5 ppm and less) of ethylene effectively trigger a wide range of changes dependent upon the tissue involved. Ethylene increases, with a short lag period, RNA and protein synthesis[225] and the activity or *de novo* synthesis of a number of enzymes. These include phenylalanine ammonia-lyase,[45,178,281] polyphenoloxidase,[318] α-amylase,[191] cellulase,[3] and--particularly--peroxidase,[183,282] as well as other enzymes involved in phenol synthesis.[309] Ethylene increases the rate of respiration,[176,279,281] and in the transition zone of *P. radiata* an increase in respiration (relative to surrounding sapwood)[308] took place over a period similar to the increase in ethylene production.[306]

Evidence is accumulating of an association between the presence of ethylene and the formation of materials similar to or identical with those in heartwood. The capacity of the dry transition zone surrounding heartwood and *Sirex* lesions to produce ethylene clearly precedes phenol synthesis, but not resin accumulation.[305,306,308] The amount of ethylene present has been related to the concentration of a polyphenol formed in carrot tissues[46] and dominant *P. radiata* produced more ethylene in response to mechanical

injury and much more, at a faster rate, in response to
Sirex injury, than did suppressed trees.[305] Heartwood
polyphenols have been formed in fresh blocks of *P. radiata*
sapwood stored in a container continuously ventilated with
humidified air containing ethylene (5 ppm).[306] As these
polyphenols are fungitoxic, the results are in accordance
with previous reports which suggested a positive relationship between ethylene production and disease resistance.[46,
318] Other evidence of the effect of ethylene has been
obtained from the administration of the ethylene-releasing
chemical chloroethyl-phosphonic acid (Ethephon) to the
sapwood of a *Rhus* sp. which resulted in the enhanced
formation of polyphenols similar in composition to those
found in heartwood.[144] Ethephon administered to the cambial
region of several *Eucalyptus* species resulted in copious
formation of the polyphenolic exudate, kino,[144,256] whereas
injection into the sapwood of a *Prunus* sp. resulted in the
copious exudation of a carbohydrate gum and the formation
of small amounts of polyphenols in woody tissues.[144] Resin
enrichment also resulted from Ethephon injection into
sapwood of *Pinus radiata* trees.[256] Extensive trials have
shown that the administration of ethylene-releasing chemicals
to the trunk of trees of *Hevea braziliensis* results in
considerably enhanced yields of rubber.[4,266] Synthetic
auxins such as naphthalene acetic acid (NAA), 2,4-di- and
2,4,5-tri-chlorophenoxyacetic acid stimulate and prolong
ethylene production in higher plants[100,173,275,292] or
stimulate the formation of polyphenols[21,46,67,144,230] or
rubber.[4,266] Ethylene has also been associated with the
formation of tyloses in a eucalypt.[144] Other work indicates
however that the balance between auxin and ethylene is more
important than the absolute amounts of either.[36,224]

    The physiological situation in the transition zone or
at the heartwood boundary is complex and requires further
study. There is a need for oxygen[223] and at low oxygen
concentration the sensitivity of the tissue to ethylene is
decreased. In addition, carbon dioxide is competitive at
the receptor site of ethylene, particularly at low oxygen
levels.[38,224] Accordingly it is noteworthy that a very
high CO_2 concentration, little oxygen and no ethylene have
been found in the gases of the inner sapwood of *Acacia
mearnsii* collected early in the winter season.[42] It has
been suggested[40] that synthesis of compounds from CO_2 could
take place but no evidence of this could be found in the
inner parts of sapwood or in heartwood[169] or in fresh

blocks of radiata pine sapwood ventilated with humidified
air containing 27.5% carbon dioxide.[306] Injury to the outer
sapwood of *Pinus resinosa*, desiccation and application of
solid carbon dioxide (at -78 C) resulted in the formation
of larger amounts of heartwood polyphenols[192] than did
bore-holes in the sapwood of *P. densiflora.*[129] Because
freezing is known to enhance the production of ethylene[61]
the effect may have been a response to ethylene rather than
to carbon dioxide.

     The Initiation of Ethylene Production. Ethylene can
be produced as a result of various stresses, particularly
dehydration.[1,83,241,244] Some stress conditions have been
associated with heartwood formation and although factors
other than the presence of ethylene may be responsible for
the secondary effects, at least situations suitable for
ethylene formation exist in the tree.

     Dehydration has been associated frequently with heart-
wood formation[204] and the loss of water from the relatively
dry transition zone before polyphenols are formed supports
this viewpoint. Moreover trees growing in semi-arid regions
frequently have narrow sapwoods with few growth rings while
those in tropical regions frequently have wide sapwoods or
lack heartwood. However experimental proof of the effect
of dehydration is difficult. Contradictory conclusions
have been drawn concerning the influence of moisture stress
on different trees of the same species.[111,136] Whereas
partial dehydration of short lengths of beech-stem resulted
in the formation of tyloses and extractives[365] a similar
trial using *Rhus* sapwood has been unsuccessful.[145] Studies
on the pruned branches of the bifurcated stems of different
species has also failed to yield conclusive evidence of
increased heartwood formation in the stressed branch.
Living stumps, which had grown under severe physiological
stress, usually showed the normal time-period change of
sapwood to heartwood but in some cases heartwood did not
form.[125] However it is noteworthy that with blocks of fresh
P. radiata sapwood used for *in vitro* polyphenol synthesis
in a stream of ethylene, the polyphenols were formed predominantly in the partially dried zone near the surface.[305]
In addition, desiccation and the influence of solid carbon
dioxide has resulted in the formation of heartwood polyphenols in the sapwood of *P. resinosa* during the dormant
season.[192] Following the administration of Paraquat to
pine trees, ethylene production and respiration rate

increased and reached maximum values just before a dramatic
increase in oleoresin production in the affected wood[349]
which occurred after drying of the region. The resin-
enriched areas had moisture contents below that of the
heartwood and were surrounded by a dry zone.[33,312] More
definitive studies of the influence of moisture stress on
the extent of heartwood formation are required to determine
the part played by this factor.

The dry zone surrounding the lesions caused by *Sirex
noctilio* and *Amylostereum areolatum* in outer sapwood of
P. radiata releases large amounts of ethylene, particularly
in relation to the very small amount of living parenchyma
present. The clearly differentiated transition zone lacks
both polyphenols and fungal hyphae. The major role played
by the host in this host-pathogen relationship is shown by
the observation that dominant pines produced ten times the
amount of ethylene produced by suppressed pines, as was
also the case with mechanical injury. Suppressed trees
did not produce polyphenols with either injury; apparently
the ethylene did not reach the threshold level.[305]

The association between fungal hyphae and heartwood or
discolored wood has been noticed occasionally.[49] The hyphae
may cause the death of the parenchyma cells, and it has
been found that the amount of ethylene evolved in other
tissues is proportional to the degree of fungal invasion or
host damage.[10] The greatest evolution of ethylene takes
place at the edge of the injured region in potatoes where
it is generated by the dying cells, or the cells adjacent
to those dying.[10] Evidence that some microorganisms themselves liberate ethylene is accumulating[10,98,180] although
Amylostereum areolatum is one that does not.[305] In these
situations the role of ethylene remains in doubt and its
production may be an inseparable result of protoplasmic
disorganization.[10]

THE ROLE OF ETHYLENE IN THE
FORMATION OF CELL WALL AND
POLYPHENOLS

When trees grow away from the vertical situation and
are under mechanical stress an internal redistribution of
growth regulators takes place. In the case of tension
wood formation in angiosperms, auxin levels are lower than

normal.[205] The walls of the fibers growing in the cambia
of these trees are thicker than normal and the thickness,
as well as the amount of an inner "gelatinous" layer of
cellulose, can vary.

Correlations have been found recently in several tree
species between levels of stress, a 30 to 300% increase in
internal ethylene levels, increased diameter and shorter
height growth.[34,214,285] Ethylene results in more secondary
xylem production in the trunks of *Pinus radiata* and *Liquidambar styraciflua*.[284] Current studies on seedlings of
Eucalyptus gomphocephala show that the tension wood half
of seedlings grown in a horizontal position contain higher
internal levels of ethylene than the opposite half. These
horizontally-grown stems also produce higher amounts of
ethylene than seedlings grown vertically.[256] Studies on
the initiation of kino formation with ethylene suggest that
the occurrence of kino veins would be more frequent in
tension wood areas of eucalypts, but there is no evidence
for this supposition. These apparently contradictory
observations draw attention to the complexity of the situation, i.e. the association of low cambial auxin levels
with tension wood, versus the association of tension wood
formation with high levels of ethylene. Because ethylene
is formed in areas of high auxin level[37] it is evident that
the associations are not simple and more work is required
to determine the inter-relationship and interaction of
ethylene at low concentrations with other plant hormones
in the control of variations in wood quality.

CONCLUSIONS

The darker colored woods or secondary changes in the
middle of the stems of woody plants can arise in different
ways. A number of internally controlled processes influenced
by environmental situations result in the formation of heartwood. On the other hand, factors external to the plant,
such as mechanical injury and microorganisms, result in
the formation of discolored wood, wetwood, etc. In some
cases these latter changes can modify preformed heartwood.

Different types of heartwood appear to exist although
satisfactory definition of them is not yet possible. The
death of the sapwood parenchyma cells in some trees appears
to be largely controlled by the age of the cells. In other

cases, the distance from the cambium appears to have more influence on the death of the cell. In addition the growth rate at particular stages can influence the amount of heart-wood formed.

The extractives in colored or discolored wood are formed *in situ* from carbohydrate or lipid substrates and the amount formed is significantly influenced by the physio-logical conditions impinging on the parenchyma at the boundary of the region. In some species the transition zone adjacent to the boundary has an enhanced metabolic activity, usually in the dormant season. Ethylene has been associated with this activity and evidence from other systems indicates that in some species ethylene may play a key role. Its formation could be triggered by water stress.

Increasing attention to experimentally controlled studies will elucidate the mechanisms responsible for the movement of water and inorganic salts from the heartwood boundary and of carbohydrates centripetally to it during the period when heartwood is formed. These studies will determine those periods during the growth of the trees which are important in affecting the volume of heartwood formed, the effect of growth on this volume and amount of extractives present in heartwood. They will determine the importance of a dry transition zone in this transformation in different species, the balance between ethylene, other hormones and carbon dioxide in the control of parenchymal activity, and the various stages of different metabolic pathways to secondary metabolites.

Precise understanding of the mechanisms controlling secondary changes in the wood of living trees will facili-tate the biological control of wood quality and significant improvements in the technology of wood utilization. This understanding will also facilitate the development of improved production of chemicals by biological means. Such chemicals include those found in tree exudates of various types and in barks, those formed by artificial induction in sapwood and those produced in tissue cultures of paren-chyma cells with specialized functions. These new approaches offer a more complete utilization of forest resources.

ACKNOWLEDGMENTS

I am grateful to Dr. J. S. Fitzgerald and Dr. N. D. Nelson for reading the manuscript and making suggestions.

REFERENCES

1. Abeles, F. B.   1972.   *Ann. Rev. Plant Physiol. 23*:259.
2. Abeles, F. B.   1972.   Ethylene in Plant Biology, Academic Press, New York.
3. Abeles, F. G. and G. R. Leather.   1971.   *Planta 97*:87.
4. Abraham, P. D., P. R. Wycherley, and S. W. Pakianathan. 1968.   *J. Rubber Res. Inst. Malay 20*:291.
5. Akaike, Y., T. Nakagami, and T. Yokota.   1974.   *J. Japan. Wood Res. Soc. 20*:224.
6. Anderegg, R. J. and J. W. Rowe.   1974.   *Holzforschung 28*:171.
7. Anderson, A. B.   1946.   *Ind. Eng. Chem. 38*:450.
8. Anderson, A. B., T. C. Scheffer, and C. G. Duncan. 1963.   *Holzforschung 17*:1.
9. Anderson, A. B., T. C. Scheffer, and C. G. Duncan. 1963.   *Chem. Ind. (London)* pp. 1289–90.
10. Archer, S. A. and E. C. Hislop.   1975.   *Proc. Assoc. Appl. Biol. 81*:121.
11. Arganbright, D. A.   1969.   *Wood & Fiber 2*:367.
12. Bailey, P. J. and R. D. Preston.   1969.   *Holzforschung 23*:113.
13. Bamber, R. K.   1976.   *Wood Science & Technology 10*:1.
14. Bamber, R. K.   (personal communication).
15. Barton, G. M.   1973.   *TAPPI 56*:115.
16. Barton, G. M. and B. F. MacDonald.   1971.   *Canad. Forestry Service Publ. No. 1023.*
16a. Barton, G. M. and J. A. F. Gardner.   1962.   *J. Org. Chem. 27*:322.
17. Bassett, C., R. T. Sherwood, J. A. Kepler, and P. B. Hamilton.   1967.   *Phytopathology 57*:1046.
18. Bauch, J., W. Hüll, and R. Endeward.   1975.   *Holzforschung 29*:198.
19. Bauch, J. and G. Tiedemann.   1976.   IUFRO XVI Congress, Norway.
20. Baur, P. S. and C. H. Wilkinshaw.   1974.   *Can. J. Bot. 52*:615.
21. Berlin, J. and W. Barz.   1971.   *Planta 98*:300.
22. Berlyn, G. P.   1969.   *Am. J. Bot. 56*:498.
23. Berryman, A. A.   1972.   *Bio Science 22*:598.

24. Birch, A. J., G. K. Hughes, and E. Smith. 1954.
    Aust. J. Chem. *7*:83.
25. Bosshard, H. H. 1965. *Holzforschung 19*:65.
26. Bosshard, H. H. 1966. *Internat. Ass. Wood Anat. News
    Bull. 1*:11.
27. Bosshard, H. H. 1968. *Wood Sci. Technol. 2*:1.
28. Bosshard, H. H. 1974. Holzkinde. Zur Biologie,
    Physik und Chemie des Hllzes Vol. 2, pp. 148-185.
    Birkhauser Verlag Basel-Stuttgart.
29. Bottaril, F., A. Marsili, and I. Morelli. 1972.
    Phytochemistry 11:2120.
30. Briggs, L. H. and A. G. Frieberg. 1937. *J. Chem.
    Soc.*   271.
31. Briggs, L. H. and B. F. Cain. 1959. *Tetrahedron
    6*:143.
32. Briggs, L. H., R. C. Cambie, and J. L. Hoare. 1959.
    Tetrahedron 7:262.
33. Brown, C. L. and L. E. Nix. 1975. *Forest Sci. 21*:359.
34. Brown, K. M. and A. C. Leopold. 1973. *Can. J. For.
    Res. 3*:143.
35. Buckland, N. J., O. T. Dalley, and C. J. Mathieson.
    1953. *Appita 7*:165.
36. Burg, S. P. 1968. *Plant Physiol. 43*:1503.
37. Burg, S. P. and E. A. Burg. 1966. *Proc. Natl. Acad.
    Sci. 55*:262.
38. Burg, S. P. and E. A. Burg. 1967. *Plant Physiol.
    42*:144.
39. Busgen, M. and E. Munch. 1929. Transl. by T. Thomson,
    Chapmall Hall, London p. 123.
40. Carrodus, B. B. 1970. *New Phytologist 70*:939.
41. Carrodus, B. B. 1972. *New Phytologist 71*:713.
42. Carrodus, B. B. and A. C. K. Triffett. 1975. *New
    Phytologist 74*:243.
43. Carter, J. C. 1945. *Bull. Ill. Nat. Hist. Survey
    23*:407.
44. Chafe, S. C. and D. J. Durzan. 1973. *Planta 113*:251.
45. Chalutz, E. 1973. *Plant Physiol. 51*:1033.
46. Chalutz, E., J. E. De Vay, and E. C. Maxie. 1969.
    Plant Physiol. 44:235.
47. Charuk, E. V. and A. F. Razumova. 1974. *Holtztech-
    nologie 15*:3.
48. Chattaway, M. M. 1949. *Aust. J. Biol. Sci. 2B*:227.
49. Chattaway, M. M. 1952. *Aust. For. 16*:25.
50. Chattaway, M. M. 1953. *Aust. J. Biol. 1*:27.
51. Chattaway, M. M. 1955. *Tropical Woods 102*:55.
52. Chattaway, M. M. 1956. *Tropical Woods 104*:100.

53. Chen, C-M. 1970. *Forest Prod. J. 20*:36.
54. Choong, E. T. 1969. *Wood & Fiber 1*:124.
55. Chow, S. Z. 1972. *TAPPI 55*:539.
55a. Christensen, F. J. Personal communication.
56. Chu, Fei-Tan. 1975. Unpublished data.
57. Clark-Lewis, J. W. and I. Dainis. 1967. *Aust. J. Chem. 20*:2191.
58. Clark, J. and R. D. Gibbs. 1957. *Can. J. Bot. 35*:219.
59. Constabel, F. 1968. *Planta 79*:58.
60. Cooper, G. A. 1974. *Wood Science 6*:380.
61. Cooper, W. C., G. K. Rasmussen, and E. S. Waldon. 1969. *J. Rio Grande Valley Hort. Soc. 23*:29.
62. Côté, W. A., A. C. Day, B. E. Simpson, and T. E. Timell. 1966. *Holzforschung 20*:178.
63. Cowan, D. J. 1973. *New Zealand J. For. 13*:233.
64. Cox, J. S. C., F. E. King, and T. J. King. 1956. *J. Chem. Soc.*   1384.
65. Craib, W. G. 1923. *Royal Bot. Gardens Edinburgh Notes 14*:1.
66. Dadswell, H. E. and W. E. Hillis. 1962. *In* Wood Extractives (W. E. Hillis, ed.), pp. 3-55, Academic Press, New York.
67. Davies, M. E. 1972. *Planta 104*:66.
68. Day, W. R. 1959. *Empire For. Rev. 38*:35.
69. Demaree, L. A. and R. W. Erickson. 1976. *Wood Science 8*:227.
70. Demmon, E. L. 1936. *J. Forestry 34*:775.
71. Desh, H. E. 1941. *Malayan Forest Records No. 14.*
72. De Silva, D. and W. E. Hillis. Unpublished data.
73. Dietrichs, H. H. 1964. *Bundesforsch. Anst. fur Forst u. Holzwirtschaft. Mitt. 58*:1.
74. Dietrichs, H. H. 1964. *Holzforschung 18*:14.
75. Dietrichs, H. H. and B. M. Hausen. 1971. *Holzforschung 25*:183.
76. Doran, J. C. 1975. *Aust. For. Res. 7*:21.
77. Dowden, H. and R. C. Foster. Personal communication.
78. Drewes, S. E. and D. G. Roux. 1966. *J. Chem. Soc.* :1644.
79. Eade, R. A., W. E. Hillis, D. H. S. Horn, and J. J. H. Simes. 1965. *Aust. J. Chem. 18*:715.
80. Eade, R. A., I. Salasoo, and J. J. H. Simes. 1966. *Aust. J. Chem. 19*:1717.
81. Eades, R. A. and J. B. Alexander. 1934. *Can. Dep. Int. Forest Serv. Circ. 41*:15 pp.
82. Einspahr, D. W., R. E. Goddard, and H. S. Gardner. 1964. *Silvae Genetica 13*:103.

83. El-Beltagy, A. S. and M. A. Hall. 1974. *New Phytologist 73*:47.

84. Ellis, E. L. 1965. *In* Cellular Structure of Woody Plants (W. A. Côté, ed.), pp. 181, Syracuse Univ. Press, N.Y.

85. Ellis, E. L. 1967. Personal communication.

86. Erdtman, H. 1955. *Experimentia Suppl. 2*:156.

87. Erdtman, H. 1963. *In* Chemical Plant Taxonomy (T. Swain, ed.), pp. 89-125, Academic Press, London.

88. Erdtman, H., B. Kimland, T. Norin, and P. J. L. Daniels. 1968. *Acta. Chem. Scan. 22*:938.

89. Erickson, R. W. 1968. *Forest Prod. J. 18*:49.

90. Etheridge, D. E. and L. A. Morin. 1962. *Can. J. Bot. 40*:1335.

91. Fahn, A. and N. Arnon. 1963. *New Phytologist 62*:99.

92. Fengel, D. 1970. *Wood Sci. Technol. 4*:176.

93. Findlay, W. P. K. and C. B. Pettifor. 1941. *Empire Forest. J. 20*:64.

94. Foster, R. C. 1967. *Aust. J. Bot. 15*:25.

95. Foster, R. C. Personal communication.

96. Fracheboud, M., J. W. Rowe, R. W. Scott, S. M. Fanega, A. J. Buhl, and J. K. Toda. 1968. *Forest Prod. J 18*:37.

97. Franklin, E. C., M. A. Taras, and D. A. Volkman. 1970. *TAPPI 53*:2302.

98. Freebairn, H. T. and I. W. Buddenhagen. 1964. *Nature 202*:313.

99. Frey-Wyssling, A. and H. H. Bosshard. 1959. *Holzforschung 13*:129.

100. Fuchs, Y. and M. Lieberman. 1968. *Plant Physiol. 43*:2029.

101. Fukazawa, K. and T. Higuchi. 1965. *J. Japan. Wood Res. Soc. 11*:196.

102. Fukazawa, K. and T. Higuchi. 1966. *J. Japan Wood Res. Soc. 12*:221.

103. Gardner, J. A. F. 1962. *In* Wood Extractives (W. E. Hillis, ed.), pp. 317-330, Academic Press, New York.

104. Gardner, J. A. F. 1965. *Official Digest J. Paint Technol. Eng.* p. 698.

105. Good, H. M., P. M. Murray, and H. M. Dale. 1955. *Can. J. Bot. 33*:31.

106. Goodwin, R. H. and D. R. Goddard. 1940. *Am. J. Bot. 27*:234.

107. Gottlieb, O. R. and M. T. Magalhaes. 1960. *Perfumery Essent. Oil Record 51*:18.

108. Hagglund, E., S. Ljungren, H. Nihlen, and C. Sandelin. 1935. *Svensk Papperstidn 38*:454.

109. Hale, J. D. and L. P. Clermont. 1963. *J. Polymer Sci. Part C, 2*:253.

110. Hancock, W. V. and E. P. Swan. 1965. *Phytochemistry 4*:791.

111. Harris, J. M. 1954. *New Zealand Forest Serv. Forest Res. Inst., Tech. Paper No. 1, New Phytol. 53*:517.

112. Harris, J. M. 1961. *New Zealand For. Res. Note No. 28.*

113. Harsh, W. 1912. *CSIRO Translation No. 9955* pp. 51-57.

114. Hart, J. H. 1965. *Mich. Agr. Exp. Sta. Quart. Bull. 48*:101.

115. Hart, J. H. 1968. *Forest Sci. 24*:334.

116. Hartig, J. H. 1888. *Allg. Forst-u. Jagdztg. 64*:52.

117. Hartig, R. 1894. *Forst-Naturw. Zeitschrift 3*:255.

118. Hartley, C., R. W. Davidson, and B. S. Crandell. 1961. *U.S. Dept. Agr. Forest Serv. Forest Prod. Lab. Rept. 2215*:34 pp.

119. Hasegawa, M. 1958. *J. Japan Forestry Soc. 40*:111.

120. Hasegawa, M. and T. Shirato. 1959. *J. Japan Forestry Soc. 41*:1.

121. Hasegawa, M. and M. Shiroya. 1966. *Botan. Mag. Tokyo 79*:595.

122. Hayashi, Y., K. Sakurai, T. Takahashi, and K. Kitao. 1974. *Makuzai Gakkaishi 20*:591.

123. Hayashi, Y., K. Kitao, and A. Sato. 1968. *Wood Research No. 44*:68.

124. Haygreen, J. G. and S. S. Wang. 1966. *For. Prod. J. 16*:118.

125. Hemingway, R. W. and W. E. Hillis. 1970. *Wood Sci. Technol. 4*:246.

126. Hemingway, R. W. and W. E. Hillis. 1971. *Appita 24*:439.

127. Higuchi, T., K. Fukazawa, and S. Nakashima. 1964. *J. Japan Wood Res. Soc. 10*:235.

128. Higuchi, T. and K. Fukazawa. 1966. *J. Japan Wood Res. Soc. 12*:135.

129. Higuchi, T., K. Fukazawa, and M. Shimada. 1967. *Hokkaido Univ. Coll. Exp. Forest. Res. Bull. 25*:167.

130. Higuchi, T., Y. Onda, and Y. Fujimoto. 1969. *Wood Res. Inst., Kyoto Univ., Wood Res. Bull. 48*:15.

131. Higuchi, T., M. Shimada, and K. Watanabe. 1967. *J. Japan Wood Res. Soc. 13*:269.

132. Hillis, W. E. 1954. *J. Soc. Leather Trades Chem. 38*:91.

133. Hillis, W. E. 1956. *Aust. J. Biol. Sci. 9*:263.

134. Hillis, W. E. 1958. *Nature 182*:1371.
135. Hillis, W. E. 1960. *Holzforschung 14*:105.
136. Hillis, W. E. 1962. *In* Wood Extractives (W. E.
     Hillis, ed.), pp. 59–131, Academic Press, N.Y.
137. Hillis, W. E. 1964. *Biochem. J. 92*:516.
138. Hillis, W. E. 1965. *IUFRO Meeting* 1965 Section 41,
     Australia.
139. Hillis, W. E. 1966. *Phytochemistry 5*:541.
140. Hillis, W. E. 1968. *Wood Sci. Technol. 2*:241.
141. Hillis, W. E. 1969. *Appita 23*:89.
142. Hillis, W. E. 1971. *Wood Sci. Technol. 5*:272.
143. Hillis, W. E. 1972. *Appita 26*:113.
144. Hillis, W. E. 1975. *Phytochemistry 14*:2559.
145. Hillis, W. E. 1976. Unpublished.
146. Hillis, W. E. and A. Carle. 1959. *Nature 182*:1594.
147. Hillis, W. E. and A. Carle. 1962. *Biochem. J. 82*:435.
148. Hillis, W. E. and A. Carle. 1963. *Aust. J. Chem.
     16*:147.
149. Hillis, W. E. and N. Ditchburne. 1974. *Can. J. For. Res.
     4*:524.
150. Hillis, W. E. and N. Ditchburne. 1975. *Can. J. For. Res.
     5*:743.
151. Hillis, W. E., A. D. Hardie, and J. Ilic. 1973. *Proc.
     IUFRO Division V Meeting,* South Africa, p. 487.
152. Hillis, W. E. and M. Hasegawa. 1963. *Phytochemistry
     2*:195.
153. Hillis, W. E. and D. H. S. Horn. 1965. *Aust. J.
     Chem. 18*:531.
154. Hillis, W. E., F. R. Humphreys, R. K. Bamber, and
     A. Carle. 1962. *Holzforschung 16*:114.
155. Hillis, W. E. and T. Inoue. 1966. *Phytochemistry
     5*:483.
156. Hillis, W. E. and T. Inoue. 1968. *Phytochemistry
     7*:13.
157. Hillis, W. E. and N. Ishikura. 1970. *Phytochemistry
     9*:1517.
158. Hillis, W. E. and W. M. McKenzie. 1964. *Forest Prod.
     J. 24*:310.
159. Hillis, W. E. and H. R. Orman. 1962. *J. Linn. Soc.
     (Bot.) 58*:175.
160. Hillis, W. E. and T. Swain. 1959. *J. Sci. Food Agr.
     10*:533.
161. Hillis, W. E. and Y. Yazaki. 1973a. *Phytochemistry
     12*:2401.
162. Hillis, W. E. and Y. Yazaki. 1973b. *Phytochemistry
     12*:2969.

163. Hillis, W. E. and Y. Yazaki. 1974. *Phytochemistry*
      13:495.
164. Hillis, W. E. and Y. Yazaki. 1976. Unpublished data.
165. Hillis, W. E., Y. Yazaki, and J. Bauch. 1976. *Wood*
      Sci. Technol. 10:79.
166. Hirai, S. 1951. *Japan Forestry Soc. Trans 59*:231.
167. Höll, W. 1972. *Holzforschung 26*:41.
168. Höll, W. 1973. *Phytochemistry 12*:975.
169. Höll, W. 1974. *Can. J. Bot. 52*:727.
170. Höll, W. 1976. *IUFRO XVI Congress*, Norway.
171. Höll, W. and K. Lendzian. 1973. *Phytochemistry 12*:
      975.
172. Höll, W. and G. Poschenrieder. 1975. *Holzforschung*
      29:118.
173. Holm, R. E. and F. B. Abeles. 1967. *Plant Physiol.*
      42:30.
174. Huber, B. 1956. *In* Encyclopedia of Plant Physiology
      (W. Ruhland, ed.), p. 541, Berlin: Springer-Verlag.
175. Hugentobler, U. H. 1965. *Vierteljahrsschrift der*
      Naturforschenden Gesellschaft in Zurich 110:321.
176. Hulme, A. C., M. J. C. Rhodes, and L. S. Wooltorton.
      1971. *Phytochemistry 10*:1315.
177. Hunter, A. G. and J. F. Goggans. 1968. *TAPPI 51*:76.
178. Hyodo, H. and S. F. Yang. 1971. *Plant Physiol. 47*:
      765.
179. Ihara, N. 1972. *Bull. Kyushu Univ. Forests 46*:123.
180. Ilag, L. and R. N. Curtis. 1968. *Science 159*:1357.
181. Imamura, H., H. Fushiki, S. Ishihara, and H. Ohashi.
      1972. *Res. Bull. Fac. Agr. Gifu Univ. 33*:39.
182. Imamura, H., K. Nomura, Y. Hibino, and H. Ohashi.
      1974. *Res. Bull. Fac. Agr. Gifu Univ. 36*:93.
183. Imaseki, H. 1970. *Plant Physiol. 46*:172.
184. International Association of Wood Anatomists. 1957.
      Tropical Woods 105:1.
185. Ito, M. 1953. *Sci. Report Fac. Liberal Arts Educ.*
      Gifu Univ. 1:63.
186. Jacobs, M. R. 1955. *Comm. Forestry Bureau Bull. No.*
      20 Canberra, Aust.
187. Jacquoit, C. and N. Hervet. 1954. *Rev. Path. Veg.*
      Ent. Agric. Fr. 33:199.
188. Jain, N. C., R. C. Gupta, and B. R. S. Chauhan. 1974.
      Holzforschung und Holzverwertung 26:129–130.
189. Jane, F. W. 1970. *In* The Structure of Wood, Adam
      & Charles Black, London.
190. Jensen, K. F. 1967. *U.S.D.A., Forest Res. Note NE-74*.
191. Jones, R. L. 1968. *Plant Physiol. 43*:442.

192. Jorgensen, E. 1961. *Can. J. Bot. 39*:1765.
193. Jorgensen, E. 1962. *Forest. Chron. 38*:292.
194. Keith, C. T. 1969. *Forestry Chronicle 45*:338.
195. Kennedy, R. W. and J. W. Wilson. 1956. *For. Prod. J. 6*:230.
196. Kharuk, E. V. and G. S. Kovrigin. 1973. *For. Abst. 34*:219.
197. King, F. E., J. W. Clark-Lewis, and W. F. Forbes. 1955. *J. Chem. Soc.* 2948.
198. Kirbach, L. E. and S. Chow. 1976. *For. Prod. J. 26(3)*:44.
199. Knutson, D. M. 1973. *Can. J. Bot. 51*:498.
200. Koch, P. 1972. Utilization of the southern pine, U.S.D.A. Forest Service. *Agric. Handbook 420(1)*:87.
201. Kollmann, F. F. P. and W. A. Côté. 1968. Principles of Wood Science & Technology, Springer Verlag, N.Y.
202. Kondo, T. 1964. *J. Japan Wood Res. Soc. 10*:43.
203. Koran, Z. and W. A. Côté, Jr. 1965. *In* Cellular Ultrastructure of Woody Plants (W. A. Côté, Jr., ed.), pp. 319-333. Syracuse University Press, Syracuse, N.Y.
204. Kozlowski, T. T. 1964. Water Metabolism in Plants. Harper, New York.
205. Kozlowski, T. T. 1971. Growth and Development of Trees Vol. I & II. Academic Press, New York.
206. Kozlowski, T. T. and T. Keller. 1966. *Bot. Review 32*:293.
207. Krahmer, R. L. and W. A. Côté. 1963. *TAPPI 46*:42.
208. Krahmer, R. L., R. W. Hemingway, and W. E. Hillis. 1970. *Wood Sci. Technol. 4*:122.
209. Kramer, P. J. and T. T. Kozlowski. 1960. Physiology of Trees. McGraw-Hill, New York.
209a. Kurth, E. F. 1950. *J. Am. Chem. Soc. 72*:5778-9.
210. Lairand, D. B. 1963. *Drevarsky Vyskum Pt. 1*:1.
211. Lappi-Sëppäla, M. 1952. *Commun. Inst. Forestalis Fenniae Helsinki 40(25)*:26.
212. Lee, C. L., Y. Hirose, and T. Nakatsuka. 1975. *J. Japan Wood Res. Soc. 21*:107.
213. Lee, C. L., Y. Hirose, and T. Nakatsuka. 1975. *J. Japan Wood Res. Soc. 21*:249.
214. Leopold, A. C., K. M. Brown, and F. H. Emerson. 1972. *Hort. Science 7*:715.
215. Liese, J. 1936. *Forst 12*:37.
216. Liese, W. and G. Karnop. 1968. *Holz Roh- und Werkstoff 26*:202.

217. Lin, R. T., E. P. Lancaster, and R. L. Krahmer. 1973.
      Wood & Fiber 4:278.
218. Luxford, R. F. 1931. *J. Agr. Res. 42*:804.
219. Lyr, H. 1967. *Arch. Forstwesen 16*:51.
220. Magness, J. R., L. O. Regeimbal, and E. S. Degman.
      1933. *Proc. Amer. Soc. Hort. Sci. 29*:246.
221. Mantell, C. L., C. W. Kopf, J. L. Curtis, and E. M.
      Rogers. 1942. *In* The Technology of Natural Resins,
      Wiley, New York, p. 506.
222. Manwiller, F. G. 1972. *Southern For. Expt. Sta.
      Alexandra La. Rept FS-SO-3201-1.1.*
223. Mapson, L. W. 1970. *Endeavour 29*:29.
224. Mapson, L. W. and A. C. Hulme. 1970. *In* Progress in
      Phytochemistry 2:343-84 (L. Reinhold, Y. Linschitz,
      eds.).
225. Marei, N. and R. Romani. 1971. *Plant Physiol. 48*:806.
226. Markham, K. R. and L. J. Porter. 1973. *New Zealand
      J. Sci. 16*:751.
227. Matsukuma, N., H. Kawano, Y. Shibata, and T. Kondo.
      1965. *J. Japan Wood Res. Soc. 11*:227.
228. Meyer, R. W. and L. Leney. 1968. *Forest Prod. J.
      15*:51.
229. Meyer, R. W. and G. M. Barton. 1971. *Forest Prod. J.
      21*:58.
230. Miller, C. O. 1969. *Planta 87*:26.
231. Morgan, J. W. W. and R. J. Orsler. 1967. *Phytochem-
      istry 6*:1007.
232. Morgan, J. W. W. and R. J. Orsler. 1968. *Holzfor-
      schung 22*:11.
233. Münch, E. 1910. *Naturw. Ztschr. Forst-U. Landwirtsch.
      8*:533.
234. Mutton, D. B. 1962. *In* Wood Extractives (W. E.
      Hillis, ed.), pp. 337-63. Academic Press, New York.
      513 pp.
235. McCombe, B. Personal communication.
236. MacDougal, D. T., J. B. Overton, and M. S. Gilbert.
      1929. The Hydrostatic-Pneumatic System of Certain
      Trees. Carnegie Inst., Washington.
237. McGinnes, E. A., C. I-J. Chang, and K. Y-T. Wu. 1971.
      J. Polymer Sci. 36:153.
238. Mackay, J. F. G. 1975. *Wood and Fiber 6*:319.
239. McKenzie, W. M. and H. Karpovich. 1968. *Wood Sci.
      Technol. 2*:139.
240. McKenzie, W. M. and W. E. Hillis. 1965. *Wear 8*:238.
241. MacKenzie, I. A. and H. E. Street. 1970. *J. Exp.
      Bot. 21*:824.

242. MacKinney, A. L. and L. E. Chaiken. 1935. *U.S.D.A. Appalachian For. Expt. Sta. Tech. Note No. 18.*
243. MacLean, N. and J. A. F. Gardner. 1958. *For. Prod. J. 8*:107.
244. McMichael, B. L., W. R. Jordan, and R. D. Powell. 1972. *Plant Physiol. 49*:658.
245. Narayanamurti, D. 1957. *Holz als Roh- u. Werk-stoff 15*:370.
246. Něcesaný, V. 1966. *Holzforschung Holzverwert. 18*:61.
247. Něcesaný, V. 1968. *Holzforschung Holzverwetung 20*:49.
248. Něcesaný, V. 1973. *Holzforschung 27*:73.
249. Neely, D. 1970. *J. Am. Hort. Sci. 95*:536.
250. Nekrasova, G. N. 1973. *Biologicheski Nauki 1*:67.
251. Nelson, N. D. 1975a. *Can. J. For. Res. 5*:291.
252. Nelson, N. D. 1975b. *In* Proc. Third North Amer. For. Biol. Workshop (C. P. P. Reid and G. H. Fechner, eds.), pp. 381-382. Colorado State Univ., Fort Collins CO.
253. Nelson, N. D. 1976a. Gross influences on heartwood formation in black walnut and black cherry trees. *USDA For. Serv. F.P.L. Res. Paper.*
254. Nelson, N. D. 1976b. Abstr. *Plant Physiol.* (Annual Supplement).
255. Nelson, N. D., R. R. Maeglin, and H. E. Wahlgren. 1969. *Wood & Fiber 1*:29.
256. Nelson, N. D. and W. E. Hillis. Unpublished.
257. Nemec, S. 1975. *Can J. Bot. 53*:2712.
258. Nicholls, J. W. P. 1965. *Nature 207*:320.
259. Nicholls, J. W. P. and A. G. Brown. 1974. *Silvae Genetica 23*:138.
260. Nicholson, J. E., G. S. Campbell, and D. E. Bland. 1972. *Wood Sci. 5*:109.
261. Nicholson, J. E., W. E. Hillis, and N. Ditchburne. 1975. *Can. J. For. Res. 5*:424.
262. Nikles, D. G. 1970. *Unasylva 24*:9.
263. Nobashi, K. and T. Yokota. 1975. *J. Japan Wood Res. Soc. 21*:315.
264. Nobuchi, T. and H. Harada. 1968. *J. Japan Wood Res. Soc. 14*:197.
265. Onishi, H. and T. Goto. 1971. *Bull. Fac. Agric. Shimane Univ. 5*:61.
266. Pakianathan, S. W. 1970. *Planter's Bull. Rubber Res. Inst. Malaya 111*:351.
267. Panshin, A. J., C. de Zeeuw, and H. P. Brown. 1952. Textbook of Wood Technology, Vol. 1 (Second Ed.), McGraw-Hill Book Co., New York.

268. Parameswaran, B. and J. Bauch. 1975. *Wood Sci.
      Technol.* 9:165.

268a. Parker, M. L., G. N. Barton, and L. A. Jozsa. 1974.
      Wood Sci. Technol. 8:229.

269. Paul, B. H. 1930. *Southern Lumberman, Nashville,
      Tenn., Issue of Oct. 1.*

270. Peck, E. C. 1953. USDA For. Serv., *For. Prod. Lab.
      Rept. No. D 768,* Madison, USA.

271. Pilz, 1907. *Allgemeine Forst-und Jagd-Zeitung*
      83:265.

272. Plomley, K. F., W. E. Hillis, and K. Hirst. 1976.
      Holzforschung 30:14.

273. Polcin, J. and W. H. Rapson. 1971. *Pulp Paper Mag.
      Can. 72*:84.

274. Posey, C. E. and D. W. Robinson. 1969. *TAPPI 52*:110.

275. Pratt, H. K. and J. D. Goeschl. 1969. *Ann. Rev.
      Plant Physiol. 20*:541.

276. Priestley, J. H. 1932. *Forestry 6*:105.

277. Raffael, E. and W. Rauch. 1974. *Holz. Roh-Werkstoff
      32*:182.

278. Reid, R. W. 1961. *Forest Chron. 37*:368.

279. Reid, M. S. and H. K. Pratt. 1972. *Plant Physiol.
      49*:252.

280. Resch, H. and D. G. Arganbright. 1968. *Forest Sci.
      14*:148.

281. Rhodes, M. J. C. and L. S. C. Wooltorton. 1971.
      Phytochemistry 10:1989.

282. Ridge, I. and D. J. Osborne. 1970. *J. Exp. Bot.
      21*:720.

283. Roberts, D. R., N. M. Joye, A. T. Proveaux, W. J.
      Peters, and R. V. Lawrence. 1973. *Naval Stores Rev.
      83(5)*:4.

284. Roberts, L. W. 1976. Cytodifferentiation in Plants;
      Xylogenesis as a Model System. Cambridge Univ.
      Press.

285. Robitaille, H. A. and A. C. Leopold. 1974. *Physiol.
      Plant. 32*:301.

286. Roff, J. W. 1965. *Mycologia 56*:799.

287. Rudman, P. 1965. *Holz u. Organismen 1*:151.

288. Rudman, P. 1966. *Nature 210*:608.

289. Sachs, I. B., J. C. Ward, and R. E. Kinney. 1974.
      *Proc. 7th Ann. Scanning Elect. Micro. Symp. (Part
      II),* p. 453.

290. Sachsse, H. 1965. *Holz Als Roh- und Werkstoff 23*:425.

291. Sakai, A. 1960. *J. Japan For. Soc. 42*:97.
292. Sakai, S. and H. Imaseki. 1971. *Plant & Cell Physiol. 12*:439.
293. Salamon, M. and A. Kozak. 1968. *Forest Prod. J. 18(3)*:90.
294. Sandermann, W., H. H. Dietrich, and M. Puth. 1960. *Holz. Roh-Werkstoff 18*:63.
295. Sandermann, W., H. Preusser, and W. Schweers. 1960. *Holzforschung 14*:70.
296. Sandermann, W. and R. Kohler. 1964. *Holzforschung 18*:53.
297. Sandermann, W. and M. Puth. 1965. *Fabre Und Lack 71*:13.
298. Sandermann, W., B. Hausen, and M. Simatupang. 1967. *Das Papier 21*:349.
299. Scheffer, T. C. and E. B. Cowling. 1966. *Ann. Rev. Phytopathol. 4*:147.
300. Schroeder, H. A. and G. J. Kozlik. 1972. *Wood Sci. Technol. 6*:85.
301. Scurfield, G., A. J. Michell, and S. R. Silva. 1973. *Bot. J. Linn. Soc. 66*:277.
302. Shain, L. 1967. *Phytopathology 57*:1034.
303. Shain, L. 1971. *Phytopathology 61*:301.
304. Shain, L. and W. E. Hillis. 1971. *Phytopathology 61*:841.
305. Shain, L. and W. E. Hillis. 1972. *Phytopathology 62*:1407.
306. Shain, L. and W. E. Hillis. 1973. *Can. J. Bot. 51*: 1331.
307. Shain, L. and J. F. G. Mackay. 1973. *Forest Sci. 19*:153.
308. Shain, L. and J. F. G. Mackay. 1973. *Can. J. For. Res. 51*:737.
309. Shannon, L. M., I. Uritani, and H. Imaseki. 1971. *Plant Physiol. 47*:493.
310. Shigo, A. L. and W. E. Hillis. 1973. *Ann. Rev. Phytopathol. 11*:197.
311. Shortle, W. C. 1970. *Phytopathology 60*:578.
312. Sioumis, A. A. and L. S. Lau. 1976. *Appita 29*:272.
313. Skene, D. S. 1965. *Aust. J. Bot. 13*:367.
314. Smith, J. H. G., J. Walters, and R. W. Wellwood. 1966. *Forest Sci. 12*:97.
315. Squillace, A. E. 1965. *Proc. Eighth South. Conf. Forest Tree Improvement*, Savannah, Ga. pp. 73-76.
316. Squillace, A. E. 1971. *Forest Sci. 17(3)*:381.
317. Stahel, J. 1968. *Holz. Roh-und Werkstoff 26*:418.

318. Stahman, M. A., B. G. Clare, and W. Woodbury. 1966.
      Plant Physiol. *41*:1505.
319. Stamm, A. J. 1964. Wood & Cellulose Science. The
      Ronald Press Co., New York, USA.
320. Stewart, C. M. 1966a. *Science 153*:1068.
321. Stewart, C. M. 1966b. *CSIRO Div. Forest Prod. Tech.
      Paper 43.*
322. Stewart, C. M. 1967. *Nature 214*:138.
323. Stone, J. E. 1973. *Proc. IUFRO Division V Meeting,*
      South Africa, p. 988.
324. Sullivan, J. D. 1967. *Forest Prod. J. 17*:25.
325. Swan, B. and I.-S. Åkerblom. 1967. *Svensk Papperstid.
      70*:239.
326. Swarbrick, T. 1926. *J. Pomol. Hort. Sci. 5*:98.
327. Tachibana, S., M. Sumimoto, and T. Kondo. 1976. *J.
      Japan Wood Res. Soc. 22*:34.
328. Taras, M. A. and J. R. Saucier. 1967. *Forest Prod.
      J. 17*:97.
329. Tarkow, H. and J. Krueger. 1961. *Forest Prod. J.
      11*:228.
330. Tarkow, H. and P. M. Seborg. 1968. *Forest Prod. J.
      18*:104.
331. Taylor, F. W. 1974. *Wood Science 6(4)*:396.
332. Threlges, B. A. 1968. *Phytochemistry 7*:1411.
333. Trendelenburg, R. and H. Mayer-Wegelin. 1955. *In*
      Dan Holz als Rohstoff, pp. 245-262. Hanser Verlag,
      Munich.
334. Tsoumis, G. 1969. Wood as a Raw Material. Pergamon
      Press, pp. 276.
335. Uprichard, J. M. 1963. *Holzforschung 17*:129.
336. Vermass, H. F. 1975. *South African For. J. (92)*:24.
337. Van Buijtenan, J. P. 1967. *Proc. 14th IUFRO Congress
      9*:243.
338. Wangaard, F. F. and L. A. Granados. 1967. *Wood Sci.
      Tecnol. 1*:253.
339. Ward, J. C., J. E. Juntz, and E. McCoy. 1969. *Phyto-
      pathology 59*:1056.
340. Wardell, J. F. and J. H. Hart. 1970. *Can. J. Bot.
      48*:683.
341. Wardrop, A. B. and G. W. Davies. 1961. *Holzforschung
      15*:129.
342. Wardrop, A. B. and J. Cronshaw. 1962. *Nature 193*:90.
343. Webb, L. J. 1953. *Nature 171*:656.
344. Wellwood, R. W. 1955. *Forest Prod. J. 5*:108.
345. Wellwood, R. W. and P. E. Jurazs. 1968. *Forest
      Prod. J. 18*:37.

346. Werker, E. and E. Fahn. 1968. *Nature* *218*:388.
347. Westfelt, L. 1966. *Acta. Chem. Scand.* *20*:2829.
348. Wilcox, W. W. and C. G. R. Schlink. 1971. *Wood & Fiber* *2*:373.
349. Wolter, K. E. and D. F. Zinkel. 1976. Personal communication.
350. Yamagishi, Y., N. Kawai, and S. Ono. 1972. *Mokuzai Kogyo* *27*:588.
351. Yazaki, Y. In press.
352. Yazawa, K. and S. Ishida. 1965. *J. Fac. Agr. Hokkaido Univ.* *54*:123.
353. Yazawa, K. and S. Ishida. 1965. *J. Fac. Agr. Hokkaido Univ.* *54*:137.
354. Yazawa, K., S. Ishida, and H. Miyajima. 1965. *J. Japan Wood Res. Soc.* *11*:71.
355. Yokota, T. 1972. *J. Japan Wood Res. Soc.* *18*:525.
356. Yokota, T., K. Nobashi, H. Takehisa, and N. Shiraishi. 1974. *J. Japan Wood Res. Soc.* *20*:83.
357. Yoshimoto, T., A. Shibata, and K. Minami. 1975. *J. Japan Wood Res. Soc.* *21*:381.
358. Yoshimoto, T. and K. Minami. 1975. *J. Japan Wood Res. Soc.* *21*:439.
359. Zelawski, W. 1960a. *Bull. de L'Academi Polonaise des Sciences* *8(9)*:507.
360. Zelawski, W. 1960b. *Bull. de L'Academi Polonaise des Sciences* *8(11)*:509.
361. Ziegler, H. 1963. *In* The Formation of Wood in Forest Trees (M. H. Zimmerman, ed.), 562 pp. Academic Press, New York.
362. Ziegler, H. 1967. *Holz Roh-Werkstoff* *26*:61.
363. Zimmermann, M. H. and C. L. Brown. 1971. Tree Structure and Function. Springer-Verlag, Berlin.
364. Zobel, B. J., R. Stonecypher, C. Browne, and R. C. Kellison. 1966. *TAPPI* *49(9)*:383.
365. Zycha, H. 1948. *Forstwiss. Cent.* *67*:80.

Chapter Eight

DEGRADATION OF POLYMERIC CARBOHYDRATES BY MICROBIAL ENZYMES

E. T. REESE

*Food Science Laboratory, U.S. Army Research
and Development Command, Natick, Massachusetts
01769*

THE ORGANISM AND ITS SUBSTRATE

Higher animals prepare solid foods for digestion by mastication, grinding and other means of disintegration. Microorganisms must proceed in another manner. Spores of wood rotting fungi are brought to their substrate by wind, water or insects. In a moist environment, they germinate and the hyphae advance along open spaces till they reach a barrier, which they proceed to penetrate by physical or chemical means. The hyphal tip liberates the enzymes required to catalyze hydrolysis of the wall components, and grows into the space so liberated (Fig. 1), maintaining close contact with the wall. Bacteria behave similarly but with less ability to penetrate and advance rapidly into a complex structure.

Induction of Enzymes. A germinated spore cannot continue growth long before its food reserves are depleted. In the absence of soluble foodstuffs, enzymes must be synthesized which can catalyze the hydrolysis of the insoluble substrate. How does the organism recognize the substrate on which it finds itself, so that it can produce the proper

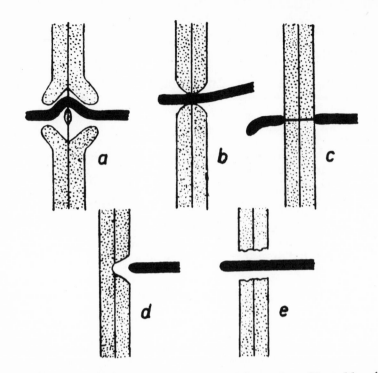

Figure 1. Hyphal penetration of wood cell-walls (a and
b) through pits, (c) constriction of hypha in wall, and
(d and e) enzymatic hydrolysis preceding penetration.[64]

enzyme? The fungus is probably producing all types of solubilizing enzymes, but in very minute amounts (Fig. 2, Table
1). Those enzymes which are secreted and find no substrate
send back no message. Those which do find a substrate form
soluble products which reenter the cell, notifying it to
turn on the proper machinery. For polysaccharases, these
inducing compounds are usually dimer and trimer products of
the enzyme action.

In a favorable environment, the organism consumes the
sugar products as rapidly as they are produced. That is,
there is no accumulation of product. In a less favorable
situation, sugars can accumulate. The fungus then runs
into the problem of catabolite repression of enzyme synthesis.
This is demonstrated when one supplies the inducer at high
concentration (0.5%) in shake flasks (Table 2). Here enzyme

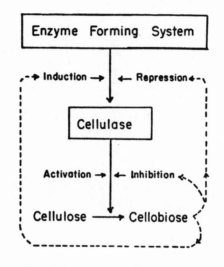

Figure 2.   The enzyme-forming system

Table 1.   Polysaccharase production by fungi

Constitutive	Induced
α-1,4-glucanase	α-1,6-glucanase (dextranase)
	α-1,3-glucanase
β-1,3-glucanase	β-1,4-glucanase
β-1,6-glucanase	β-1,2-glucanase
β-1,4-xylanase (?)	α-1,4-polygalacturonase
	β-1,4-mannanase
	β-1,4-chitinase
	β-2,1-fructanase
	β-2,6-fructanase

Table 2. Effect of modified inducers on enzyme yield[59]

Enzyme	Organism	Inducer (0.5%)	Yield
Cellulase EC 3.2.1.4	Trichoderma viride	Cellulose	22.5*
		Cellobiose	0.2*
		Cellobiose dipalmitate	4.8*
	Pestalotiopsis westerdijkii	Cellulose	35.9*
		Cellobiose	0.2*
		Cellobiose octaacetate	20.1*
	Pseudomonas fluorescens	Cellulose	514.
		Cellobiose (slow feeding)	430.
		Sophorose	397.
Dextranase EC 3.2.1.11	Penicillium funiculosum	Dextran	1080. *
		Isomaltose	2. *
		Isomaltose dipalmitate	1098. *
Invertase EC 3.2.1.26	Pullularia pullulans	Sucrose	1.3*
		Sucrose monopalmitate	108. *
Purine nucleosidase a EC 3.2.2.1	Aspergillus ambiguus	Adenosine	0
		Yeast RNA	57.
		Adenosine 5'PO$_4$	90.

*Unit values in International units/ml. Others as defined by authors.[59]

production is very limited. Consumption of a substance such as the parent polymer, or an ester of the inducer (which are acted upon by fungal enzymes to release the dimer inducer slowly) results in good yields of enzyme.[59] Sufficient examples of modified inducers have been described to lead us to believe that this is a procedure applicable to a wide range of enzymes. The best inducers of polysaccharases are modified *products* of the enzyme to be induced. But it is obvious (Table 2) that modified *substrates* are excellent inducers of other enzymes, e.g., glycosidases. Similarly in oxidative systems, a resistant substrate may induce an organism to produce much more enzyme than a highly susceptible one. Xylidine induced much more laccase in *Polyporus versicolor*[21] than did more rapidly consumed substrates of laccase.

In the above, the inducing substance is released slowly into the medium over a prolonged period. Slow feeding of inducers has a similar effect. Thus with *Ps. fluorescens*, very little cellulase is produced in the presence of 0.5% cellobiose (the inducer), but adding the cellobiose slowly over a long period results in the production of 100 times as much enzyme.[72] High concentration of inducer is not the only factor that is responsible for repression. *T. viride* produced no cellulase on 1% cellobiose unless Na oleate was added. The growth rate was inhibited by the addition of surfactant, but after consumption of cellobiose, large amounts of cellulase were produced. It would seem that the repression of enzyme formation is a function of the *rate* of consumption of cellobiose.

At times, it may appear as though the monomeric unit of the polymer is the inducing material. We, ourselves, thought at one time that glucose could induce cellulase. Much to our surprise, even "pure" glucose was found to contain a contaminating disaccharide, sophorose, in very small amounts; this turned out to be the inducer, a better one than cellobiose.[43] Whether inducers added to the medium are the agents acting within the cell to stimulate enzyme production is not established. In some instances, the dimer may be hydrolyzed as it enters the cell by β-glucosidase located on the surface of the hyphae. In others, the dimer may be modified within the cell to form a new but related compound which exerts its influence at the site of synthesis.

The ability of a fungus to speed up the production of those enzymes which it immediately needs greatly increases its efficiency. Less of the available substrates goes into enzymes not required.

For growth on a mixed substrate such as wood, fungi require that several enzymes be induced at the same time. Yet in laboratory experiments in shake flasks, one substrate is often consumed before the second is attacked, and there is a *sequence*[17] in the production and secretion of the enzymes needed to degrade the complex wall material (Fig. 3). These two observations seem to be contradictory. There appears to be an unexplained difference between growth in the natural habitat, and growth on the same substrate

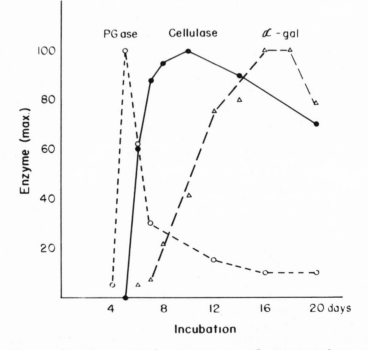

Figure 3. Sequential appearance of enzymes in a culture of *Colletotrichum* growing on *Phaseolus* cell walls. PGase, polygalacturonidase; α-gal, α-galactosidase. (After English *et al.*, 17.)

dispersed *in vitro*. In part, this may be due to differences
in the extent of adsorption of the enzymes on the solids in
the flasks.

   Synthesis and Release of Extracellular Enzymes. A
fungus having the genetic capability, and suitably induced,
turns on the synthesizing machinery and packages the enzyme
in minute vesicles as it is being formed[38] (Fig. 4). The
vesicles migrate to the cytoplasmic membrane, fuse with it,
and liberate enzyme outside the protoplast. If the cell
wall is freely penetrable to large molecules, this is equivalent to liberating enzyme into the surrounding environment.
If it is not, then the enzyme is cell-bound. Certain organisms tend to retain an appreciable part of the enzyme
between the protoplast and wall (yeasts; gram-negative
bacteria).

   If a fungus is 100% efficient, then all of the protein
it secretes should be the enzyme required for action on the
particular substrate on which it sits. This rarely turns
out to be the case, but a surprising amount of the extracellular protein can be the desired enzyme. One can evaluate
this by comparing the specific activity of the pure enzyme
with that of the crude filtrate. The lower the ratio, the
"purer" is the initial preparation (Table 3). An R value
of 7 indicates that 1/7 (14%) of the initial protein is the
desired enzyme. In some of the examples (e.g., β-glucosidase,
exo-β-1,3-glucanase), the major carbon source for growth is
starch, and amylase is the dominant enzyme. A high R value
(e.g., 400) indicates that the organism is *not* an effective
producer of the particular enzyme.

CONSTRAINTS ON THE SUSCEPTIBILITY
OF SOLID SUBSTRATES TO ENZYMATIC
HYDROLYSIS

   The most familiar enzyme systems are those in which
both the substrate and the enzyme are water soluble. In
these, the enzyme is generally much larger than its substrate,
and the substrate molecules diffuse to a suitable site on
the enzyme where the reaction takes place with the liberation of products. When the substrate is insoluble, it is
the enzyme which diffuses and "fits" itself to the substrate.
In the case of large molecules like cellulose and chitin,
the size relationship of enzyme and substrate is reversed

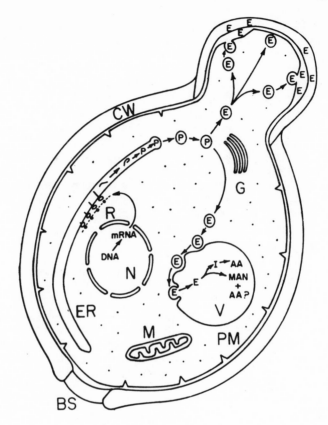

Figure 4. Schematic drawing of budding yeast cell
forming and secreting invertase. Production of invertase
protein (P) by ribosomes (R), and extrusion into the lumen
of the endoplasmic reticulum (E.R.). Transport of protein
in vesicles, and addition of mannan to form glycoprotein
(E). Fusion of vesicle with cytoplasmic membrane (PM)
with extrusion of enzyme into periplasmic space. (After
Lampen, J. O., 38). Reproduced with publishers permission.

(Fig. 5). Here the substrate greatly exceeds the enzyme in
size and it would appear necessary that the enzyme be free
in its environment if it is to function. This introduces
the problem of whether a cell-bound or immobilized enzyme
can ever attack an insoluble, polymeric substrate. It is
proposed[75] that enzymes bound on bacterial surfaces catalyze

Table 3. "Purity" of extracellular depolymerases

Enzyme	Source	R*	Ref.
Exo-β-1,3-glucanase	Basidiomycete QM 806	7	30
Exo-β-1,3-glucanase	Sclerotinia	8	15
Endo-β-1,3-glucanase	Rhizopus	165	45
β-1,4-Glucan cellobiohydrolase	Trichoderma	14**	7,23
β-Glucosidase	Trichoderma	50**	9
Endo-β-1,4-glucanase	Trichoderma	(10)	E. T. Reese (estimate)
Endo-β-1,6-glucanase	Rhizopus	400+	82
β-Glucosidase	Asp. phoenicis	7	E. T. Reese (estimate)

*R = Specific activity "pure" enzyme/specific activity
  "crude" enzyme.
**Based on commercial preparation.

cellulose hydrolysis. But it is difficult to visualize
how the intimate contact required for enzyme-substrate formation can be accomplished in such a situation. Some of the
factors that are peculiar to soluble enzyme-insoluble
substrate interaction are considered here.

Moisture. Cellulosic materials are protected from the
action of microorganisms if they are kept dry. The minimum
critical moisture level is characteristic of the material
and of the organism. For wood, this level is just above
the fiber saturation point, 24 to 32% of dry weight, and
for cotton about 10%. Moisture is required for its swelling
action on the substrate, and as a medium for movement of
extracellular enzymes.

Porosity. There are two types of capillary spaces in
cellulosic materials, the large lumina of cells, pores, and
apertures visible under the microscope (1 to 10 μm in
diameter), and the much finer spaces in cell walls between
microfibrils (0.5 to 4 nm). The total surface of large
capillaries[11] is about 2×10^3 cm^2/g of wood, several orders
of magnitude less than the surface of the cell wall capillaries,

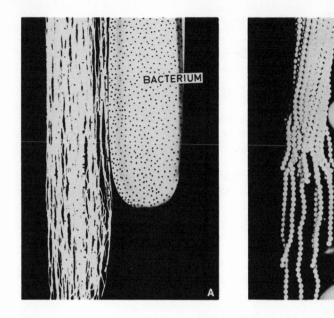

Figure 5.   (a) Model of a bacterium in contact with a bundle of microfibrils.   The dots represent cellulase molecules.   (b) Model of how enzyme molecules (large spheres) aligned on a cell surface may form enzyme substrate complexes facilitating the degradation of a microfibril.   Photographs provided by V. Hofsten.

3×10^6 cm^2/g.   The latter are, for the most part, too small to permit the entrance of large enzyme molecules (Fig. 6), and digestion is therefore limited to the surfaces of the large voids.   It is possible to increase the size of the wall capillaries by various types of swelling treatments, and by dissolving out inter-micellar materials (pulping). In nature, the fungus, itself, dissolves channels as it advances through the wall, thereby increasing the surface available to the enzyme.

The situation, however, is not as simple as it appears. The dimensions (Fig. 6) preclude the possibility of enzyme migration through the wall, yet there are definite indications that such migration occurs.   The residual cellulose changes, and this is the result of action at a distance

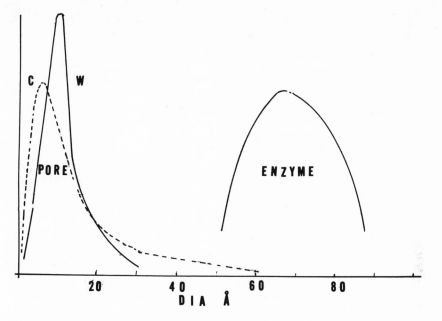

Figure 6. Comparison of pore diameter of cell wall capillaries of cotton (C) and spruce wood (W), with diameter of cellulase molecules. (After Cowling, 11.)

from the hyphae. One must conclude either that the porosity is greater than supposed, at least in the environment of the organism, or that some enzyme components are smaller than those that have been characterized, or both. Still another possibility is that the organism itself is able to open up the cell-wall pores by some unsuspected mechanism, such as the secretion of an active chemical of small size (see below) or by physical forces exerted in local areas. Our ignorance of these actions stems from the inability to detect small, local changes resulting from action of the organism on its substrate.

Steric Rigidity. By steric rigidity, we imply that certain constraints are put on a substrate molecule to restrict conformational change. By so doing, the ability of an enzyme to form an intimate association with the strained substrate is modified, and there is a concomitant

change in reaction rate.   For examples we must look at some
non-carbohydrate systems.

First System, Interfacial Association.   Small molecules
of limited water solubility form droplets as the concentration is increased.   Those molecules which remain free in
solution have maximum freedom of movement.   Those in the
droplet are under some degree of alignment and tension at
the interfaces.

Methyl butyrate is an example of such a molecule.   An
esterase (Fig. 7) hydrolyzes methyl butyrate, the rate being
a function of its concentration in the aqueous phase.   This
is the typical reaction involving a substrate under no
restraint.   Pancreatic lipase also hydrolyzes methyl
butyrate,[66] but the reaction is quite different.   It cannot
hydrolyze the ester in the aqueous phase, but it rapidly
hydrolyzes it when the ester exists as droplets.   Apparently
this enzyme requires that the ester linkages be oriented at

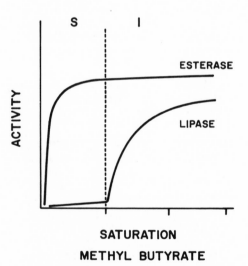

Figure 7.   Enzyme action on soluble (S) and on insoluble (I) substrate.   (After Sarda, 66.)

the interface, an orientation associated with tension. We
have, then, one enzyme whose action is inhibited by placing
the molecule under tension; and another enzyme that requires
such a state, and is inactive in the absence of such a
restraint.

Pancreatic phospholipase reportedly[13] behaves in a way
similar to pancreatic lipase. Its action on *dissolved*
molecules is extremely slow, as compared to its activity
on micelles. Furthermore, the rate of reaction varies
greatly with the *nature* of the aggregation in the micelle.
"Addition of detergents produced mixed organized structures
in which the molecules are spaced more widely, ... allowing
the enzyme to penetrate with subsequent more rapid hydrolysis."[13]

Second System, Soluble Polymers. An effect of strain
on hydrolysis rate was reported for α-amylase. Long substrate chains were hydrolyzed more rapidly than shorter
ones "primarily because of increased strain."[74]

Similar examples have been reported[79] for the action
of pullulanase on amylopectin. According to W. J. Whelan:
"We are all familiar with the idea that enzymes have a
three-dimensional structure, and that it is necessary to
retain the integrity of this structure for enzyme activity
to be displayed. We are equally familiar with the idea
that if we wish to break down a protein by enzyme action,
we must first denature the substrate. I think, however,
that we are less familiar with the concept that a polysaccharide also has a three-dimensional structure. If we
are to achieve complete breakdown, we must first denature
the substrate--the polysaccharide--just as we would have
to denature the protein. We have recently encountered an
example of this type involving the action of pullulanase.
We carried out an experiment in which starch was dissolved
in the cold in dimethylsulfoxide, and then the dimethylsulfoxide was dialyzed away. Now we had a solution of starch
dissolved by a very gentle method. We treated this with
pullulanase, and almost all the amylopectin was unattacked.
If the amylopectin is given a gentle treatment, which would
not break any chemical bonds, the amylopectin is totally
debranched by pullulanase. What we were seeing was that
it is necessary to destroy the three dimensional structure
of the substrate before complete enzyme attack can occur."[79]

On the other hand, another enzyme, e.g., isoamylase, *is* capable of *direct* action on the "ordered" amylopectin.

The next stage in the development of constraints is found with polymers which can exist both in solution and as gels. (Some polysaccharide conformations are shown in Fig. 8.) Agarase is highly active on agar solutions where the chains exist as random coils. It has little effect on agar gels, where the chains are "ordered" in hydrogen-bonded helices.[75]

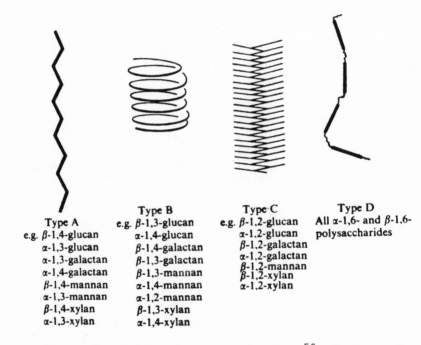

Type A	Type B	Type C	Type D
e.g. β-1,4-glucan	e.g. β-1,3-glucan	e.g. β-1,2-glucan	All α-1,6- and β-1,6-
α-1,3-glucan	α-1,4-glucan	α-1,2-glucan	polysaccharides
α-1,3-galactan	β-1,4-galactan	β-1,2-galactan	
α-1,4-galactan	β-1,3-galactan	α-1,2-galactan	
β-1,4-mannan	β-1,3-mannan	β-1,2-mannan	
α-1,3-mannan	α-1,4-mannan	β-1,2-xylan	
β-1,4-xylan	α-1,2-mannan	α-1,2-xylan	
α-1,3-xylan	β-1,3-xylan		
	α-1,4-xylan		

Figure 8.   Schematic representations[56] of the conformation of various polysaccharides: (A) Extended and ribbon-like, (B) flexible and helical, (C) crumpled and contorted, and (D) loosely jointed and extended. Reproduced with the publisher's permission.

Third System, Insoluble Polymers. Insoluble polymers
may vary in degree of order from gels to crystalline cellulose; i.e., from fully hydrated chains to structures into
which water can scarcely penetrate. The more rigid structures may owe their rigidity to hydrogen-bonding between
closely-packed chains, or to covalent cross-linkages.

Wool is one of the cross-linked polymers whose decomposition was investigated some years ago. An interesting
summary[41] has been presented by Linderstrøm-Lang: "You
are all familiar with the clothes moth, a nasty little
insect whose larvae eat keratin of the hair in the clothes
and thrive on it. Normally keratin is exceedingly stable
and is not attacked by enzymes; but when these larvae eat
the hair, it vanishes in their intestine and nice white
crystals of uric acid appear at the other end ... We set
out to investigate how the larvae do it ... The result of
our investigation showed that the clothes moth larva has a
very high alkalinity, pH 10, in its intestine, and secretes
an enzyme, a proteinase, which acts at high pH, and further
has the property of not being inhibited by SH groups. At
the same time, a reducing agent seems to be secreted which
reduces the hair so that the -S-S- bonds are turned into
-SH groups, whereby the protein is made soluble, and can be
cleaved by the proteinase." More recently, enzymes have
been found capable of cleaving the disulfide bond.

Covalent bonding, tying together unlike polymers, is
illustrated (Fig. 9) in a model proposed by Albersheim for
the sycamore cell wall.[32a] Only the cellulose molecules
themselves are free of cross linkages. Xyloglucan is tightly
bound to the micelles of cellulose by H-bonding, and to other
hemicelluloses by glycosidic linkages. The hemicelluloses
in turn form glycosidic linkages to pectic materials, and
to the wall protein via the hydroxyl groups of serine
residues. The cell walls of other plants may employ, in
addition, ester linkages between uronic acids and sugar
hydroxyls as a means of binding together adjacent layers.
Furthermore, in woods, lignin penetrates the entire structure, perhaps with occasional covalent bonding to other
components. The interconnecting materials may, then, be
looked upon as a single giant macromolecule. Cellulose
micelles, however, account for most of the wall substance
of *mature* cells.

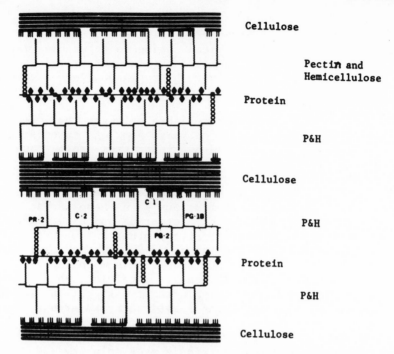

Figure 9. Model of sycamore cell wall showing cross-linking of the diverse components (except cellulose) to form one gigantic molecule occupying the space between cellulose microfibrils.[32a] Reproduced with permission of the publisher.

Based on a study of the invasion of kidney bean cell walls by *Colletotrichum*, it seems that the first enzyme to be secreted by the fungus--and to be required for penetration of the host--is endo-polygalacturonase (endo-PGase).[17] The random action of this enzyme liberates oligomers of polygalacturonic acid, many of which carry short glycan chains. The wall structure is "loosened" by this action, greatly increasing its susceptibility to cellulase, and to other enzymes. Without the prior action of the endo-PGase, cellulase is unable to reach its substrate.

Ordered structures have been investigated more vigorously in proteins than in polysaccharides, and we cite

another example from this group. Collagen is a triple-
stranded helix, a structure of high degree of order, in
which H-bonding plays an important role. Most proteases
cannot attack collagen, but there is a specific protease,
collagenase, which can (Fig. 10). The collagenases act
at a specific location yielding two fragments which spontaneously denature. The denatured products are readily
susceptible to further degradation by many proteases, but
not by collagenase.[44] Thus, the requirement of collagenase
is for a highly-ordered, constrained, substrate. In a
similar manner, there is a specific DNAse which acts on the
double helix of DNA, a polyester.[47]

   C_1-C_x _Concept Re-evaluated_.[58] The preceding examples
all lead up to the most highly ordered of the polymers,
the crystalline polymers, of which cellulose and chitin are
two well-known examples.

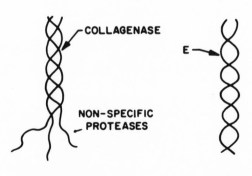

Figure 10. Action of enzymes on highly ordered structures. E = Enzyme acting on DNA.

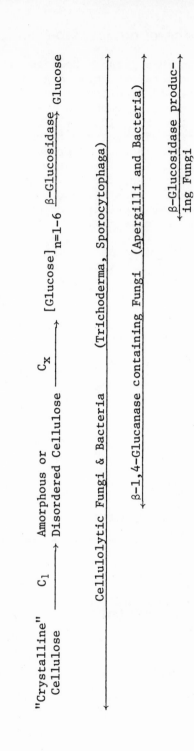

Twenty-six years ago it was shown that cellulase systems are made up of a complex of enzymes, one of which (C_1) appeared to be responsible for conversion of crystalline cellulose into a form accessible to normal hydrolytic endo-β-glucanases (C_x). The existence of C_1 was postulated on the fact that some organisms were able to degrade amorphous cellulose but were not able to degrade crystalline cellulose. Some organisms that degraded crystalline cellulose produced filtrates that were unable to degrade the crystalline material but could degrade carboxymethylcellulose (CMC) or amorphous cellulose. Grinding or swelling crystalline cellulose (processes which destroy crystallinity) made it suceptible to such filtrates. With the discovery of *Trichoderma viride* as an excellent source of active cellulases, the two components C_1 and C_x were separated in a number of laboratories. Strong synergistic effects were found for crystalline cellulose, but not for CMC, when the components were recombined. Further work showed the same two components to be present in other fungi, though usually the C_1 was in much smaller amounts, and the synergistic effects could be obtained by combining the C_1 of one fungus with the C_x of another.

Several workers have found an exo-enzyme in the C_1 fractions (from DEAE-Sephadex) which removes cellobiose units from the chain end to which is given the trivial name of cellobiohydrolase (CBH). Most investigators (the author excepted)[58] are in agreement that this is C_1. Since there are so few free ends in crystalline cotton, this interpretation has necessitated a change in our thinking concerning the sequence with which the two types of enzyme act. One of the endo-β-1,4-glucanase components is considered to act first, to form the free ends required for CBH activity-- quite the reverse of the mechanism originally proposed.

In the light of these developments, we have modified our original C_1-C_x concept.[58] Whereas we originally claimed only a disruption of hydrogen bonds by C_1, we now believe that a covalent linkage is split and that this act is accompanied by splitting of hydrogen bonds (as in the case of collagenase, above). Thus, C_1 becomes a member of the endo-β-glucanases (C_x). But it is a very special member having properties *not* possessed by most endo-glucanases, e.g., activity on *crystalline* cellulose; disruption of H-bonds; *lack* of action on CMC, and *inability* to act on

products of its own action (since it produces no soluble
products from crystalline cellulose). It would seem desirable, at least for the present, to maintain "C_1" as the
designation of this component. Other random acting components lack the special properties required for *initiation* of
the reaction, but are capable of catalyzing hydrolysis of
the products of C_1. They act (Fig. 11) on all partially
liberated chains. The exo-enzymes--glucohydrolase and
cellobiohydrolase--are limited in their action to *non*-reducing
chain ends. Endo-glucanases are required for the complete
removal of the liberated chains, a removal which is necessary
if C_1 is to act on underlying material.

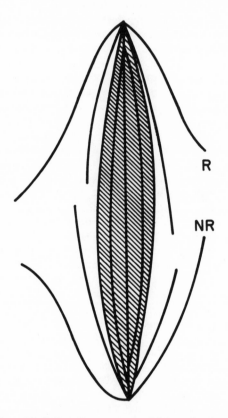

Figure 11. Model of a crystallite being opened up by
C_1. Shaded areas crystalline. R, reducing end; NR, non-
reducing end of liberated chain.

$$\text{Cotton} \xrightarrow{C_1} \begin{array}{c}\text{Swollen}\\\text{cellulose;}\\\text{alkali-}\\\text{soluble}\\\text{cellulose}\end{array} \left\{\begin{array}{l}\xrightarrow[\text{hydrolase}]{\beta\text{-1,4-glucan glucano-}} \text{oligomers}\\[2ex]\xrightarrow[\text{hydrolase}]{\beta\text{-1,4-glucan gluco-}} \text{glucose}\\[2ex]\xrightarrow[\text{biohydrolase}]{\beta\text{-1,4-glucan cello-}} \text{cellobiose}\end{array}\right.$$

The concept of successive actions by the various cellulase components has been questioned by some investigators--particularly those who work with bacteria (which secrete so little enzyme). Leatherwood[39] plated out a culture of *Ruminococcus* in roller tubes on cellulose agar. Some colonies formed clear zones (digesting the cellulose) around the colony; others did not. In addition, about one-half way between colonies of the two different types, there appeared a clear zone in the form of an arc. It seems that a diffusible factor from one colony, and another from tne second, were required for the clearing effect. Leatherwood's interpretation was that the two factors (an affinity factor and a hydrolytic factor) united to form one active enzyme. The data, however, are also consistent with the C_1 theory postulated above, involving *two* factors acting independently, and in an alternating fashion. von Hofsten[75] also favors simultaneous action in bacteria by the various bound components, these forming as many different types of enzyme-substrate complex as possible. This cooperative action explains the synergistic effects observed, the multiple binding tending to weaken the non-covalent bonds responsible for the ordered structure.

In this section, we have emphasized that enzymes exist in nature that are capable of initiating reactions on highly ordered biological structures, and that these enzymes *require* that their substrates be under some sort of constraint if action is to occur. As yet the number of examples is limited, probably because techniques for detecting minute local changes are inadequate. The *limiting* factor in degradation of crystalline polymers is this first step, the products of which are quickly solubilized by other components of the enzyme mixture.

ENZYME PROPERTIES

There are several features of the glycanases worth
noting, as they relate to their action in wood decomposition.

Physical Properties. *Stability*. The microbial glycanases are very stable enzymes. Fungal glycanases exhibit
optimum stability in the pH range 4 to 6. Usually they do
not require co-factors and generally they are resistant to
the levels of metals found in the environment.

Adsorbability. Like most enzymes, the glycanases can
be adsorbed on solid surfaces (including insoluble substrates)
and the extent of adsorption is a function of (1) the amount
of enzyme; (2) the available surface. (The solid may be
one which adsorbs proteins in general, or it may be a
substrate like cellulose with a special affinity for the
enzymes which act on it); (3) the physical properties of the
enzyme, e.g., charge, size, solubility; and (4) the environment, e.g., pH, salt concentration.

The source of enzyme determines how firmly cellulase
will be bound to cellulose. That of *T. viride* is very
strongly adsorbed (about 0.05 to 0.10 mg/mg cellulose);
and of the cellulase components of this fungus, cellobiohydrolase seems to be most firmly bound. Filtration of
very dilute cellulase solutions through filter paper may
remove most of the enzyme from them. The amount of free
enzyme in a cellulose digest is a function of the cellulose
concentration. Thus in a 1% cellulose suspension, most of
the *T. viride* cellulase is free in solution, but at 10%
cellulose over 90% of the enzyme is adsorbed.[42]

Many plants contain compounds which interact with
proteins to form enzymically inactive complexes, which may
or may not be soluble. Polymeric leucoanthocyanidins, and
to some extent, lignins, are good examples of protein inter-
actants. Some plant proteins also inhibit microbial polysaccharases.

Number and Nature of Components. The types of enzymes
that act on a *particular* linkage are few in number but
within each type there may be an appreciable number of
variants, detectable by their physical properties such as
size, charge, and solubility. In the cellulase systems of

fungi, the number of endo-β-1,4-glucanases (C_x) detectable
on various columns may vary from 1 to 10 or 12, and the C_1
component[16] from 1 to 3.   In other enzyme systems, there
are differences between cell-bound and liberated enzyme.
Usually these differences between the enzymes of a particular type are due to the amount of carbohydrate which is
covalently bound to it.   The cellobiohydrolases of *T. viride*,
for example, contain 10 to 15% carbohydrate.[23]   This may
imply an almost infinite number of components having the
same protein, but differing in carbohydrate content.
Actually the number of amino acids with which a glycosidic
linkage can be formed is limited, e.g., serine, threonine,
asparagine, and the size of the carbohydrate side chain is
usually--but not always--small.   (That of cellobiohydrolase
averages two units.)[23]   Multiple forms thus arise from
enzymatic additions or deletions of sugar units to or from
a "basic" structure.

The more important question about components is whether
they differ in their activities.   Endo-β-1,4-glucanases of
the same, and of different fungi, vary in amino acid composition, and the differences may account for the varied behavior
of the components.   Acting on carboxymethyl cellulose, the
components differ in the degree of randomness they exhibit,
as indicated by the slopes of the curves for reducing
sugar production vs fluidity change.   Other random acting
endo-glycanases (within a single type) vary in the limiting
size of the oligosaccharides on which they can act.   Some
have no activity on substrates smaller than 5 units, others
can act on tetramers, and still others on trimers.   Again,
there are some endo-β-1,3-glucanases that can split both
β-1,3 and β-1,4-bonds; and others that can split only
β-1,3-bonds.   The number of components based on activity
is much less than that based on physical properties.

Catalysis. *Enzyme Site*.   As already noted, most enzymes
are considerably larger than the substrate molecule on which
they act.   Glycanases are different.   If we accept 8 nm as
the diameter of a cellulase molecule, this would cover at
most 16 glucose units of a chain whose DP may be 1000 to
5000.   The *maximum* site size, then, would be about 16 units.
The *minimum* size of the site on a glycanase is for two
adjacent, unsubstituted units, based on the assumption that
branched units are excluded.   When, however, the substrates
contain more than one type of linkage, or of sugar, one
might expect that the site should be somewhat larger.

Current data (Table 4), however, indicate site size for
glycanases to be 4 to 10 sugar units. The most extensive
studies, made on lysozyme, indicate a site large enough for
6 units, a value which seems reasonable for a substrate
containing two different repeating units. It appears, from
the recorded values, that the magnitude of the site is *not*
a function of variability in the substrate, since homo-
polymers and hetero-polymers have the same size requirement.
Nor is it a function of enzyme type, the size for exo-
glycanases and endo-glycanases being about the same. The
data for the most part are based on the K_f values (K_f =
V/K_m) of an enzyme acting on oligomers of different sizes.
Perhaps the development of other approaches is needed to
confirm the currently accepted values.

Table 4. Enzyme site size in polysaccharases

Enzyme		Units in site	Refer- ence
Endo-glycanases			
α-Amylase			
Liquefying	*B. subtilis*	8 or 9	68
Saccharifying	*B. subtilis*	4	68
Soybean		8	68
Taka amylase	*(A. oryzae?)*	7	68
Porcine pancreatic		5	68
α-1,6-glucanases	*Penicillium*	5	32
	Arthrobacter	10	32
	Cytophaga	8	32
β-1,4-glucanases	*P. notatum*	5	54
	Myrothecium	5	81
Lysozyme		6	
Exo-glycanases			
β-Amylase	Soybean	7	68
	Wheat bran	5	68
Glucamylase	*Ryizopus delemar*	7	68

Molecular Activity. The rate of hydrolysis, i.e., the molecular activity, of polysaccharases acting on insoluble substrates is very low (10^2 to 10^4 bonds/min/enzyme molecule) compared to that of enzymes acting on small soluble molecules (10^7 to 10^8 bonds/min/enzyme molecule).

This no doubt reflects the difficulty of an enzyme diffusing to and attaching to the substrate, as compared to the more rapid movement of small substrate molecules to an enzyme site.

A low molecular activity necessitates a greater productivity of the required enzyme if an organism is to grow at its maximum rate. In general, the release of consumable product is limiting, and the growth rate on solid substrates consequently much less than that on a soluble sugar.

Enzyme Action and the Configuration of Products. Glucosides were originally considered to be α or β depending on the anomeric form of glucose which was produced on enzymatic hydrolysis. The assumption was made that the configuration would be retained. Although these observations helped to establish the configuration of these glycosides, their role was fortuitous because it was realized only later that the anomer liberated by some enzymes-- notably β-amylase--is the inverted form rather than that in which configuration is retained. The fact that inversion can take place has become understandable with the finding that the hexosyl-oxygen bond rather than the glycosidic oxygen-aglycone bond is cleaved in hydrolysis. Retention of configuration then may be accounted for as the result of two successive inversions.

Configurational relationships such as these had been examined almost invariably by polarimetry. A number of additional methods are now used. Earlier results have been confirmed and extended by direct gas-liquid chromatographic measurement of trimethylsilyl ethers of the monomeric sugar in its α- and/or β-forms,[50] proton magnetic resonance, and use of enzymes specific for one anomer: e.g. glucose oxidase for β-glucose,[77] and xylose isomerase for α-xylose.[33]

Results permit some broad generalizations regarding the configuration of the product resulting from enzyme action (Table 5). Random splitting, or endo-glycanases, retain configuration, i.e., from a β-glycan are obtained

Table 5.   Enzyme action and configuration of products

Enzyme Class	Retention or Inversion
Endo-acting enzymes	R
Exo-acting enzymes	I
Glycosidases	R
Synthetases:	
UDP-α-glucose→ α-glucans	R
UDP-α-glucose→ β-glucans	I
Phosphorolytic enzymes	I
	(R)*

*Inversion (I) is more common in phosphorolytic systems
 than retention (R); examples of both types have been
 reported.

oligomers having the β-anomeric configuration (β-1,4-
glucan[81]), and from α-glycans, the α-configuration (α-1,4-
glucan, α-1,3-glucan,[26] pullulan). Similarly, glycosidases
act in such a way that the products retain the configuration
of the substrates, while exo-glycanases act in such a way
that the configuration is inverted (Tables 6, 7).

      "Inversion by an *exo*-glucanase may be depicted as
involving strong association both with the glycosyl residue
and the adjacent residue (OR) to which C-1 is bonded in
such a way that the enzyme is well disposed to assist in
the departure of the OR portion (Fig. 12). Cleavage of the
glycosyl carbon-to-oxygen bond may then be accompanied by
backside attack of water at C-1, yielding the anomer in
which the hydroxyl group possesses the inverted configuration.

      "By contrast, retention of the configuration in the
product could result from two consecutive inversions. A
nucleophilic group of the enzyme, such as a carboxyl anion,
attacks C-1 to yield a glycosyl-enzyme product of inverted
configuration (Fig. 12). The direction of the subsequent
approach by water to C-1 of the glycosyl unit is such as
to lead to overall retention. Such a pathway shows less
obvious need for participation of the OR moiety than does
that envisaged above for the *exo*-glycanases, which is in
harmony with the specificity characteristics of glycosidases."

Table 6.   Action of enzymes on disaccharides as related to acceptor

Anomer	Donor	Acceptor	Enzyme	Product	Configuration*	Bond**
α	Maltose (etc)	H_2O	α-glucosidase	α-glucose	R	
	α-Glucosyl-F	H_2O	α-glucosidase	α-glucose	R	
β	Cellobiose (etc)	H_2O	β-glucosidase	β-glucose	R	
	β-Glucosyl-F	H_2O	β-glucosidase	β-glucose	R	
β	Sucrose	H_2O	Invertase	β-fructose	R	
β	Thioglucoside (Sinigrin)	H_2O	Myrosin	β-glucose	R	
α	Maltose	H_2O	Exo-α-1,4-glucanase	β-glucose	I	$1{\rightarrow}4$
	α-Glucosyl-F	H_2O	Exo-α-1,4-glucanase	β-glucose	I	
β	Laminaribiose	H_2O	Exo-β-1,3-glucanase	α-glucose	I	$1{\rightarrow}3$
α	Maltose	PO_4	Phosphorylases	β-G-1-P	I	$1{\rightarrow}4$
α	Trehalase	PO_4	Phosphorylases	β-G-1-P	I	$1{\rightarrow}1$
β	Cellobiose	PO_4	Phosphorylases	α-G-1-P	I	$1{\rightarrow}4$
β	Laminaribiose	PO_4	Phosphorylases	α-G-1-P	I	$1{\rightarrow}3$
α	Sucrose	PO_4	Phosphorylases	α-G-1-P	R	
α	Sucrose	$(\alpha\text{-}1,6\text{-}G)_n$	Glycan sucrase	$(\alpha\text{-}1,6\text{-}G)_{n+1}$	R	$1{\rightarrow}6$
α	Sucrose	$(\alpha\text{-}1,4\text{-}G)_n$	Glycan sucrase	$(\alpha\text{-}1,4\text{-}G)_{n+1}$	R	$1{\rightarrow}4$
β	Sucrose	$(\beta\text{-}2,1\text{-}F)_n$	Glycan sucrase	$(\beta\text{-}2,1\text{-}F)_{n+1}$	R	$2{\rightarrow}1$
β	Sucrose	$(\beta\text{-}2,6\text{-}F)_n$	Glycan sucrase	$(\beta\text{-}2,6\text{-}F)_{n+1}$	R	$2{\rightarrow}6$

*Retention of configuration, R; inversion of configuration, I.
**Nature of bond formed in oligo- or poly-saccharide.

Table 7. Classification of exoglycanases based on anomeric configuration

Type of exo-glycanase	Source	Reference
A. Inversion of product		
Transfer of a monomeric unit		
α-Linked substrates		
α-1,4-glucan + H_2O →	Fungal	50
...β-D-glucose	Glucoamylase	20,25
α-1,3-glucan + H_2O →	*Cladosporium*	76
...β-D-glucose	*resinae*	
α-mannan + H_2O →	Fungi	19
β-mannose		
α-glucosyl fluoride →		5
β-glucose		
β-Linked substrates		
β-1,3-glucan → α-D-	Fungi	50,20
glucose	*Euglena*	6
β-1,4-glucan → α-D-	Bacteria	10
glucose	Fungi	40
Transfer of dimer unit		
α-Linked		
α-1,4-glucan → β-maltose	β-amylase of	37
	higher plants	
β-Linked		
β-1,4-glucan → α-cello-	Bacteria	70
biose	Fungi	18
B. Retention		
Transfer of monomeric unit		
α-1,6-glucan → α-glucose	*Streptococcus*	77
Transfer of dimeric unit		
α-1,6-glucan → α-isomaltose	*Arthrobacter*	67

Such schemes for classification are usually quite helpful in trying to understand enzyme reactions; but they are not quite as "neat" as one would like. Exceptions to the rule soon appear. The major exceptions fortunately, have something in common. They are all enzymes that act

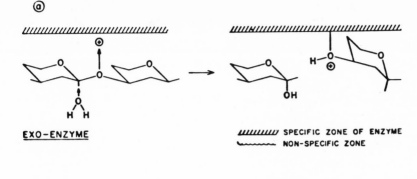

EXO-ENZYME

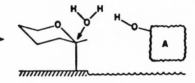
SPECIFIC ZONE OF ENZYME
NON-SPECIFIC ZONE

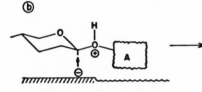

GLYCOSIDASE OR ENDO-ENZYME

Figure 12. Reactions of (a) exo-glycanase and of (b) glycosidase with their substrates.[20] Reproduced with permission of the publisher. Aglycone, A.

on α-1,6-glucans, and they do exactly the opposite from what we had come to expect. The endo-α-1,6-glucanases lead to inversion[35,67] rather than to retention.

The exo-α-1,6-glucanases removing a single glucose, and the exo-α-1,6-glucanases removing a dimeric unit both lead to retention (Table 7). We have confirmed the results, leaving little room to doubt the nature of the action. One needs now some basis for understanding why the α-1,6-glucan-ases behave differently. Coupling of sugar units in a chain from C1 of one unit to a secondary OH group of the next, permits rotation about two bonds: C1-O, and the O-C4 (or O-C3); while coupling to a primary OH (i.e., C6) adds a third bond (C6-C5) about which rotation can occur.[56] (Is it possible that a terminal α-1,6-glucose unit resembles an α-glucoside more closely than it resembles a terminal α-1,4 or α-1,3-glucose unit?) While this feature separates the 1,6-glucans from the 1-4, 1-3 and 1-2, it does not shed light on the mechanism. It suggests that enzymes acting

on β-1,6-glucans might more closely resemble those acting
on the α-1,6-glucans, than those acting on the more familiar
β-1,4- and β-1,3-glucans. As yet, no results have been
reported on the anomeric nature of products of β-1,6-glucan-
ase actions.

Another difference between α-1,4-glucans and α-1,6-
glucans is in the difference in the free energy change
($\Delta G°$) associated with bond formation.[67] Formation of an
α-1,6-bond is exergonic ($\Delta G° = -1200$ cal/mole); formation
of an α-1,4-bond is endergonic ($\Delta G° = +1200$ cal/mole).

Having isolated the exceptions to the rule associated
with α-1,6-glucanases, there remain other "exceptions" to
be considered. These fall chiefly in the category: glycosidases vs exo-glycanases. The difficulty arises because
these two groups have substrates in common. Other criteria
must be brought into play to establish whether an enzyme is
to be considered a glycosidase or an exo-glycanase. If we
are content to say that an enzyme that acts on the glucose
trimer, for example, with retention of configuration is a
glucosidase; and another that acts on the same trimer with
inversion is an exo-glycanase then there are no exceptions.
The definition takes care of the differences. (Indeed, we
have placed the α-1,3-glucanase example (Table 7) under the
exo-glycanases simply on this basis.)

Specificity.[46] *Glycosidases Versus Exo-glycanases*
(Table 8). Several criteria have been suggested for differentiating glycosidases from exo-glycanases.[60] We prefer to
use "glycosidase" in the narrow sense, i.e., to indicate
enzymes which act on "simple" glycosides. Those that act
only on *aryl* glycosides are not a problem. It is those that
act on dimers and trimers of sugars that are confused with
exo-glycanases. The first criterion--retention vs. inver-
sion--has been discussed above. The second is the effect
of the degree of polymerization of the oligomer on its
susceptibility to hydrolysis. Dimers and trimers are
readily hydrolyzed by glycosidases, but the rate falls
rapidly as the chains get longer. Exo-glycanases show the
opposite effect, long chains are rapidly hydrolyzed, and
dimers are quite resistant.

A third criterion is bond-specificity. β-Glucosidases
readily hydrolyze all β-linked dimers of glucose (Table 9);
α-glucosidases are not quite as general, the 1,1- and 1,2-

Table 8.   Criteria for the characterization of glucosidases and exo-glucanases[60]

Criterion	Glucosidases		Exo-glucanases	
	α	β	α	β
Effect of DP on hydrolysis	2 > 3 >> 4		6 > 5 >4> 3> 2	
Configuration of product	Retention		Inversion	
Effect of inhibitors				
Nojirimycin (R I/S 50)*	.002	.0002	.50	.05
Transfer	+	+	?	?
Dimer specificity	Less specific		More specific	

*R I/S 50 = ratio of inhibitor to substrate required to inhibit the enzyme reaction by 50%.

Table 9.   Specificity of glucosidases and exo-glucanases for action on dimers of glucose (relative rates)

Dimer (Linkage)	α-Gluco-sidase A. *niger*	Exo-α-1,4-Glucanase* A. *niger*	β-Glucosidase P. *melinii*	Exo-β-1,3-Glucanase Basidiomy-cete 806
1→1	0	< 0.1	78	7
1→2	6	NT	84	12
1→3	21	4	100	100
1→4	100	100	45	1
1→6	40	2.5	69	0

*Abdullah et *al.*, 1.

dimers being somewhat more resistant to hydrolysis. In fact, the α-1,1-linked dimer has its own enzyme, α,α-trehalase, but β,β-trehalose is hydrolyzed by β-glucosidase, and no specific enzyme has been reported for it.

Exo-enzymes show greater specificity for dimer linkage than do glucosidases. Exo-β-1,3-glucanases act on laminari-biose, have a slight action on β,β-trehalose and on sophorose

(1→2), and no activity on cellobiose (1→4) or on gentiobiose
(1→6). Similarly, the exo-α-glucanase, glucamylase, acts
primarily on maltose (1→4) with little or no action on the
the other α-linked dimers. Such enzymes should prove useful
in digesting a specific dimer from a mixture of dimers.

The high specificity of the exo-β-1,3-glucanase is
further shown in its action on β-trimers of glucose. The
rate of hydrolysis of laminaritriose (1→3, 1→3) is high;
but the hydrolysis of the 1→3, 1→4 trimer is quite low,
even though the action in both cases is on the 1→3 linkage
at the non-reducing end of the molecule. No action occurs
on β-1,4- or on β-1,6-linked trimers.

The greater specificity of the exo-enzymes for a particular linkage has interesting implications. The exo-β-1,3-
glucanases are highly specific for β-1,3-linkage; the exo-
α-glucanases (glucamylases) prefer the α-1,4, but can act
more slowly on the α-1,6, particularly in larger oligomers.
At the dimer level, however, the preference is nearly 100-
fold in favor of a particular linkage. This suggests that
perhaps the specific "maltases", "isomaltases", etc.,
reported in the literature are in reality exo-α-1,4-glucan-
ases (glucamylase); exo-α-1,6-glucanases, etc.

A fourth criterion for distinguishing glucosidases from
exoglycanases is based on the relative inhibitory effect of
various compounds.

Specific competitive inhibitors should be useful as
tools to characterize enzymes. Δ-Gluconolactone is an
excellent inhibitor of β-glucosidases and is almost without
effect on exo-β-1,3-glucanase.[62] For the β-glucosidases,
one molecule of the lactone competes successfully with over
100 molecules of the substrate. One thousand times as
much lactone is needed to give a similar degree of inhibition of exo-β-1,3-glucanase. A variety of substrates was
used in these experiments, including a substrate common to
both enzymes. The lactone is much less effective in inhibiting α-glucosidases (than β-), and cannot be used to
differentiate them from exo-α-glucanases. However, a new
competitive inhibitor of glucosidases, nojirimycin, has
been discovered[48] which is not only much more stable than
the lactone, but also more potent. It is an analog of
glucose in which the ring oxygen is substituted by an NH
group. This inhibitor has been used to differentiate

β-glucosidase from exo-β-1,3-glucanase, and also α-glucosidase from exo-α-1,4-glucanase.[62]

Like gluconolactone, nojirimycin is specific for glucosidases and ineffective against β-mannosidase, β-galactosidase, β-thioglucosidase, invertase, and β-xylosidase. Perhaps even more unusual is its ineffectiveness against α,α-trehalase. This, and the ineffectiveness of α-glucosidases on trehalose suggest that the active site of α,α-trehalase is quite different from that of other α-glucosidases.

The fifth criterion is the relative ability to transfer the glucosyl unit to an acceptor other than water. Again the "relative" nature must be emphasized, especially in the light of recent experiments of Hehre[29] and his collaborators which show transfer by enzymes where no such action was previously known. In general, glucosidases are able to transfer much more readily than exo-glycanases (or than endo-glycanases), i.e., transfer products can be detected at lower levels of acceptor. If, however, the acceptor level (10 to 30% solutions), and the enzyme concentration are increased, transfer products can be produced even by exo-glucanases and endo-glucanases. The substrate is often also the acceptor in many of these experiments, with trimers and tetramers arising from action of enzyme on the dimer. (We have observed that a mixture of β-linked trimers is produced when *Trichoderma viride* grows on 3% cellobiose, and perhaps also when cellulase acts on cellulose.) It must be recognized that these are unusual conditions, and that in nature such products rarely occur, since the sugars are consumed as rapidly as they are produced.

The transfer products undergo changes in concentration with time (Fig. 13). The action of β-glucosidase[11] (*A. niger*) on cellobiose results in transfer of a β-glucosyl moiety from a cellobiose unit to the sugar whose concentration is highest, and mainly to the C6-carbon. At zero time, the reaction favors addition to cellobiose to form β-1,6-glucosyl cellobiose. As cellobiose decreases and glucose increases, transfer to glucose yields β-1,6-glucosyl glucose (gentiobiose). Further incubation leads to hydrolysis of the intermediates, and to the accumulation of glucose.

Whereas the nature of the reaction is usually determined in the hydrolysis direction, it can also be observed in the

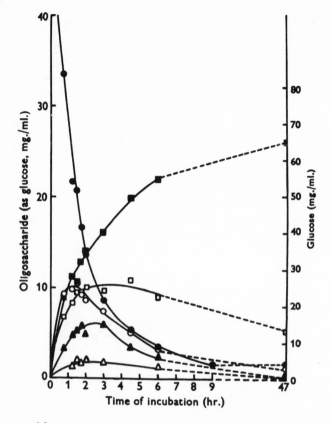

Figure 13. Action of β-glucosidase (*A. niger*) on
cellobiose (●) to form glucose (■), the β-1,6-, β-1,4-
glucose trimer (O), gentiobiose (□), and other products
(△,▲) less well-characterized. (After Crook and Stone, 12).

reverse direction.[29] Here, again, the glycosidases not
only trasfer more rapidly, but they show less specificity
for formation of a particular bond. However, on long incubation, the dominant reversion product is usually the 1→6
dimer. Thus, β-glucosidase acts on the β-anomer of glucose
to yield the β-1,6-dimer (gentiobiose). The exo-enzymes
which show high bond specificity in cleavage, also show
high bond specificity in synthesis. β-Amylase (Fig. 14)
acts on β-maltose to synthesize α-1,4-maltosyl maltose
(maltotetraose), and glucamylase (Fig. 15) on β-glucose to

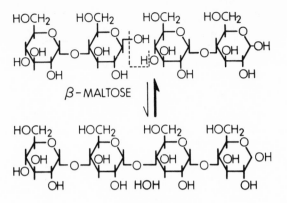

β-MALTOSE

MALTOTETRAOSE

Figure 14. β-Amylase: Synthesis and hydrolysis of maltotetraose with inversion of configuration. (Hehre *et al.*, 29.)

form α-1,4-glucosyl glucose (and to a lesser extent α-1,6-glucosyl glucose). Cellobiohydrolase also forms reversion products.[9] Thus, exo-enzymes act by inversion, and glycosidases by retention, when involved in synthetic reactions, as they do in hydrolysis.

The five criteria were based on examination of a limited number of examples. As indicated above, exceptions have been reported and these require explanation. For the most part, only a single criterion has been used by most investigators. Before the recognition of exo-glycanases, an enzyme that hydrolyzed maltose was called an α-glucosidase. Now it must also retain configuration to be so considered. Yet there are gradations, matters of degree, that indicate variability within each category. The ability of exo-α-1,4-glucanase to attack an α-1,6-linkage is one example. The susceptibility of the α-1,6-linkage is much less than that of the α-1,4 and practically disappears when the α-1,6-links occur in sequence. The exo-β-1,3-glucanase also has a peculiarity. When acting on a β-1,3-glucan that has β-1,6-linked glucose branches, it removes glucose units up to the branch point and then skips

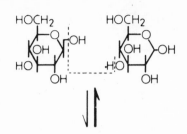
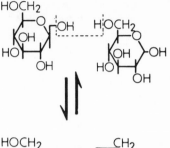
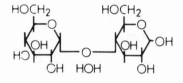
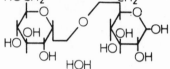

MALTOSE   ISOMALTOSE

most rapidly synthesized   more slowly synthesized
  "   "   hydrolyzed   "   "   hydrolyzed

Figure 15.   Glucamylase: Synthesis and hydrolysis of
disaccharides with inversion of configuration (Hehre *et al.*,
29).

$$- 3G - 3G - 3G - 3G ---$$

$$\begin{array}{ccc} 6 & & 6 \\ | & & | \\ G & & G \end{array}$$

over the new terminal 1-6 linkage, to attack the next 1-3,
forming gentiobiose and glucose as products. Here the enzyme
is specific for the 1,3 bond, but that bond is not necessarily terminal.

A recent exception amongst glycosidases[33] has been
reported. The β-xylosidase of *B. pumilus* acts on an aryl
β-xyloside by *inversion* to form α-xylose. This enzyme has
almost no activity on xylo-oligosaccharides, and thus has no
claim to being an exo-β-xylanase. It is apparent that classification is a man-made effort to bring order to the myriad
bits of data. Like all such schemes, this classification
relies on incomplete information and on conflicting data, and
may rapidly be replaced by a new insight involving as yet
unexplored approaches.

Synthesis by Reversion

Glycosidase

2 β-Glucose $\xrightarrow{\text{β-glucosidase}}$ $\begin{array}{l}\text{β-1,3-}\\\text{β-1,4-}\\\text{β-1,6-}\end{array}$ dimers of glucose

Exo-glycanases

2 β-Glucose $\xrightarrow{\text{exo-α-1,4-glucanase}}$ α-1,4-glucosyl glucose[29]

2 β-Maltose $\xrightarrow{\text{β-amylase}}$ α-1,4-maltosyl maltose[29]

2 α-Isomaltose $\xrightarrow{\text{exo-}(G_2)\text{-1,6-glucanase}}$ α-1,6-isomaltosyl isomaltose[69]

The best known glycanases are those which remove a mono-meric or dimeric unit from the non-reducing end of the chain (Table 7). There are reports, however, that trimers or longer groups may similarly be removed. Thus, cellotriose was reported to be the dominant product formed by the action of a termite cellulase.[55] However, the trimer was characterized only by its R_G value in paper chromatograms. The substrate used was a cellodextrin prepared with sulfuric acid, a system we had examined some years ago in which a similar product was observed and which turned out to be a sulfate ester. This example, then, seems to be an artifact. The second case of trimer removal was the action of an exo-enzyme on dextran[71] to form isomaltotriose as the dominant product. This product, too, requires confirmation of structure.

A tetramer (G_4) was the dominant product from α-1,4-glucans acted on by a *Pseudomonas* enzyme.[63] A pentamer was similarly obtained on hydrolysis of β-1,3-glucan[14] by an *Arthrobacter* enzyme. These, too, need re-examination to con-firm that such large pieces are removed by action of an exo-enzyme, and to see whether inversion takes place.

(Trimers and tetramers may be dominant products of hydro-lysis of glycans of mixed linkage by *endo*-glycanases. In the above discussion, we are limiting ourselves to glycans con-taining but a single sugar, and a single type of linkage.)

Endo-glycanases. [*] The most common enzymes involved in depolymerization are the random–acting enzymes known as endo-glycanases. With these the ultimate products of hydrolysis are small molecules which retain the various linkages present in the initial polymer. Acting on homo-polymers the enzyme usually produces dimers and trimers, and, less frequently,

```
        E         E
        ↓         ↓
- A - A - A - A - A - A - A -
```

larger oligomers (pentamers, etc., by α–amylase of *B. subtilis*) The variability in the product size results from differences between endo-glycanases in their ability to hydrolyze the shorter chains. The products that remain are, obviously, those that resist attack. Since resistance is relative, the products of short digestion times tend to be larger than those of longer incubation periods. Endo-glycanases "prefer" not to act on terminal linkages. Linkages of dimer and trimer are in this category—with the result that little monomer is produced. The anomeric linkage of the product is the same as that of the substrate. In other words, these enzymes act with retention of configuration. But mutarotation soon produces a mixture of the α– and β–anomers.

There are a large number of endo-polysaccharases. Different ones are required for the hydrolysis of each of the various homopolymers. Thus, there are glucanases, galactanases, chitinases, etc.; and different enzymes are necessary in each group because of the diversity of linkage types. Thus, of eight possible glucan types, seven are known to occur in nature, and a different endo-glucanase is needed for the hydrolysis of each. The high degree of specificity of these enzymes is sometimes overlooked, and the occasional claim is made that a cellulase acts on xylan, or vice versa. Such claims probably arise from inadequate separations, and insufficient information. As separation methods improve, the degree of specificity is found to increase. Indeed, within each enzyme type there may be several "isozymes," each acting on the same substrate, but having such different physical properties that they are separable by techniques currently available.

----

[*]In all generalized formulae, the reducing sugar is considered to be on the right-hand side.

Hetero-polymers may be acted upon by two or more different endo-enzymes. There are a number of polysaccharides composed of two sugars alternating with each other.

$$E_1 \qquad\qquad E_2$$
$$\downarrow \qquad\qquad\quad \downarrow$$
$$- A - B - A - B - A - B - A - B - A - B -$$

E_1 acts only at the C1-O bond of sugar A; E_2 only at the C1-O bond of sugar B. The resulting products are:

$$E_1 \quad B-A, \quad B-A-B-A \quad \text{etc.}$$
$$E_2 \quad A-B, \quad A-B-A-B \quad \text{etc.}$$

Trimers and pentamers are usually absent, but tetramers, hexamers, etc. are more prominent than they are in hydrolysates of homo-polymers.

Homo-polymers of mixed linkage may be acted upon by two or more different enzymes:

$$E_3 \quad E_4 \qquad\qquad E_3 \quad E_4$$
$$\downarrow \quad\ \downarrow \qquad\qquad\quad \downarrow \quad\ \downarrow$$
$$- 4A - 3A - 4A - 4A - 3A - 4A - 4A$$

These enzymes may be capable of acting also on homo-polymers of a single linkage type.[51,53] This introduces a new view regarding the specificity of enzymes. The above example may represent lichenin, a β-glucan containing 1→3 and 1→4 units as shown.

E_4 (cellulase of *Streptomyces*) acts on 1→4 linkages of both cellulose (β-1,4-glucan) and on lichenin. However, on lichenin, it acts on only one of the two possible 1→4 linkages (i.e., it does not hydrolyze at the 1→4 site shown for E_3). Even more unusual is the action of E_3 (β-1,3-glucanase of *Rhizopus*). On laminarin, E_3 hydrolyzes 1→3 linkages (the only linkages present); but in lichenin it hydrolyzes a 1→4 linkage, but again only one of the two available. These conclusions are based on the products obtained.

Common to all of the cellulase products is a cellobiose (1→4) moiety at the reducing end of the oligosaccharide; common to all β-1,3-glucanase products is a laminaribiose (1→3) moiety. The new interpretation regarding specificity is that enzyme E3 has a site for the glucose dimer (A-3A) and splits

E_4	E_3
Cellulase	β-1,3-Glucanase (endo-)

from cellulose	G1→4G̲	from laminarin	G1→3G̲
	G1→4G1→4G̲		G1→3G1→3G̲
			G1→3G1→3G1→3G̲

from lichenin	G1→4G̲	from lichenin	G1→3G̲
	G1→3G1→4G̲		G1→4G1→3G̲
	G1→4G1→3G1→4G̲		G1→4G1→4G1→3G̲

the adjacent linkage whether it be A-3A or A-4A. Thus the
emphasis is on the enzyme-substrate site, rather than on the
particular bond hydrolyzed. Similarly E4 has a site for the
1→4 dimer, and splits the adjacent linkage, which in both
cellulose and lichenin is 1→4. More recently it was shown[69]
that an endo-β-1,6-glucanase, E_5, acts on the β-1,3-linkage
of the tetramer depicted in Figure 16, thus further supporting
the interpretation given above. This enzyme has a site for
the 1→6 dimer, splitting the adjacent 1→3 linkage in the tetramer, and the 1→6 linkage in its normal substrate, β-1,6-glucan.

In addition, there are enzymes specific for mixed-linkage
substrates (i.e., lichenin) which have no action on the 1→3,
or 1→4 homo-polymers. These act at the site E_3 (above), and
tend to produce predominantly the trimer (A-4A-3A) (and to this
extent resemble the β-1,3-glucanase). As yet, there seems to
be no report of an enzyme specific for the 1,3-linkage of this
mixed-linkage polymer, lichenin.

The fact that lichenin is a substrate for two different
enzymes (E_3 and E_4) makes it possible to investigate its
hydrolysis in the presence and absence of these enzymes. The
nature of the products (above) reveals the extent to which each
enzyme is involved. The two enzymes exhibited little difference in hydrolytic power when used independently, but the
action of the β-1,3-glucanase on lichenin was *completely inhibited* in the presence of only half its weight of cellulase.
On the other hand, when β-1,3-glucan was the substrate, there
was *no* interference by the cellulase. The results indicate
that the two different substrate sites on lichenin overlap
appreciably, that there is a competition for these sites by

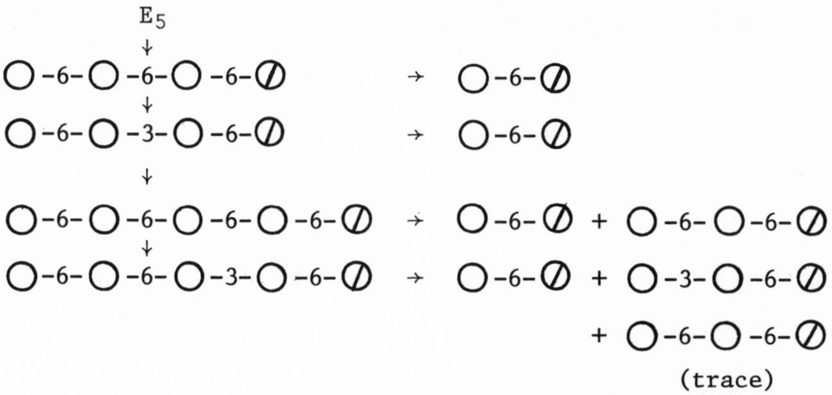

$$E_5$$
$$\downarrow$$
$$\bigcirc-6-\bigcirc-6-\bigcirc-6-\oslash \qquad \rightarrow \qquad \bigcirc-6-\oslash$$

$$\bigcirc-6-\bigcirc\overset{\downarrow}{-3-}\bigcirc-6-\oslash \qquad \rightarrow \qquad \bigcirc-6-\oslash$$

$$\downarrow$$

$$\bigcirc-6-\bigcirc-6-\bigcirc-6-\bigcirc-6-\oslash \quad \rightarrow \quad \bigcirc-6-\oslash \; + \; \bigcirc-6-\bigcirc-6-\oslash$$

$$\bigcirc-6-\bigcirc-6-\bigcirc-3-\bigcirc-6-\oslash \quad \rightarrow \quad \bigcirc-6-\oslash \; + \; \bigcirc-3-\bigcirc-6-\oslash$$

$$+ \; \bigcirc-6-\bigcirc-6-\oslash$$

(trace)

Figure 16.   Action of endo-β-1,6-glucanase on β-1,6-
and on β-1,3-linkages (Shibata 69). O = glucose unit; ∅ =
glucose unit at reducing end of oligomer.

the two enzymes, and that cellulase is the more successful
competitor. Drastic reduction in the proportion of cellulase again permits action by the β-1,3-glucanase.

In other homo-polymers, two linkages may alternate, as
in mycodextran, an α-glucan:

$$E_6$$
$$\downarrow$$
- 3A - 4A - 3A - 4A - 3A - 4A - 3A - 4A -

The only known enzymes capable of hydrolyzing this glucan are
specific for the α-1,4-linkage.[61]   (No enzymes are known to
act on the α-1,3-linkage of this compound). The products are
the dimer, nigerose (A-3A) and the tetramer (A-3A-4A-3A), of
which the tetramer is dominant. Enzymes of this type are specific for the 1→4 link of the mixed glucan, and cannot act on
a homo-glucan where all linkages are 1→4. Nor can the α-
amylases, which act on an α-1,4-linkage in α-1,4-glucan, act
on the 1→4 linkage of the mixed glucan.

For some time we thought that mycodextranase[61] might be
an α-1,3-glucanase, by analogy with the endo-β-1,3-glucanase
specificity noted above. However, when these enzymes became
available, they were found to be ineffective on the mixed
linkage glucan. And, conversely, mycodextranase was without
effect on the α-1,3-homo-polymer.

Another α-glucan, pullulan, has a sequence of two α-1,4-linkages followed by an α-1,6, and is usually depicted as:

$$
\begin{array}{l}
\quad\quad\quad - A \\
\quad\quad\quad\quad \downarrow \\
E_7 \rightarrow 6 \\
\quad\quad A - 4A - 4A \\
\quad\quad\quad\quad\quad \downarrow \\
\quad\quad\quad\quad\quad 6 \\
\quad\quad\quad\quad A - 4A - 4A \\
\quad\quad\quad\quad\quad \uparrow \quad\quad\quad 1 \\
\quad\quad\quad\quad\quad E_8
\end{array}
$$

Two different endo-enzymes have been reported acting on this polymer:

pullulanase (E_7) → A - 4A - 4A   maltotriose[80]

isopullulanase (E_8) → A - 4A - 6A   isopanose[65]

Both enzymes are specific for the mixed linkage polymer, having no effect on the corresponding homo-polymer. A search for an enzyme that will act on the third linkage to form panose (A - 6A - 4A) was unsuccessful. (These two known pullulanases should be investigated to see whether one of them may inhibit the action of the other when acting on the same mixed linkage substrate.)

Some polysaccharides contain both a mixture of sugars and a mixture of linkages. Agarose is one of these in which the two sugars (and linkages) alternate. The sugars are anhydro-L-galactose (A) and D-galactose (B):

$$
\begin{array}{c}
\quad\quad E_9 \quad\quad\quad E_{10} \\
\quad\quad \downarrow \quad\quad\quad\quad \downarrow \\
A-\underline{\alpha 3}B-\underline{\beta 4}A-\underline{\alpha 3}B-\underline{\beta 4}A-\underline{\alpha 3}B-\underline{\beta 4}A-\underline{\alpha 3}B \\
\quad\quad \uparrow \\
\quad\quad E_{11}
\end{array}
$$

Enzymes E_9 and E_{10} act on the same linkage of the octomer but E_9 acts preferentially at the middle of the molecule, and E_{10} at the bond nearest the reducing end. A different enzyme (E_{11}) acts on the α-linkage. Tetramers often form an appreciable part of the products.

In the hydrolyses of mixed-linkage polymers, there is a difference between the α-polymers and the β-polymers. For the α-series no enzyme has been found which is capable of acting on both uniformly linked, and on mixed-linked glycans (as known for cellulases and β-1,3-glucanases of the β-series).

The next step in complexity is the addition of a side group to a homo-polymer backbone. The substituent may be a simple sugar linked glycosidically, a side chain, or a non-carbohydrate linked through ester or ether linkages to the main chain. The backbone of *branched* polymers is hydrolyzed by the same enzymes that hydrolyze unbranched polymers of the same sugar-types and linkages. The branch, however, interferes with action of the enzyme in the vicinity of the branch point. As a result, the *products* of hydrolysis generally contain the branch on a non-terminal unit. For example

$$
\begin{array}{c}
E_1 \qquad\qquad E_1 \\
\downarrow \qquad\qquad\quad \downarrow \\
- A - A - A - A - A - A - A - A \\
| \\
B
\end{array}
$$

typically yields the products $A - A - A$, $A - A - A - A$, etc.
$$
\begin{array}{ccc}
& | & \qquad | \\
& B & \qquad B
\end{array}
$$

However, as branching becomes more frequent, the resistance to hydrolysis increases, and the maximum branching that permits endo-enzyme action is that in which two adjacent units are unsubstituted.

$$
\begin{array}{c}
E_1 \\
\downarrow \\
- X - X - X - X \quad . \\
| \qquad\qquad | \\
A \qquad\qquad A
\end{array}
$$

There is no hydrolysis of
$$
\begin{array}{c}
- X - X - X - \quad . \\
| \qquad\quad | \\
A \qquad\quad A
\end{array}
$$

Perlin[52] found that the products of hydrolysis of arabino-xylans have *un*substituted xylose units at both the reducing and non-reducing ends of the oligomer. In other investigations, however, end-units appear to be substituted.[34] When this substitution is at the non-reducing end of the molecule, the presence of a contaminating glycosidase may be suspected, the action of which would be to remove the non-reducing *un*-substituted unit of the primary product. An alternative explanation is that the *location* of a substituent may affect the extent to which it can inhibit the enzyme approach to its site. Thus, a substituent at C6 would be less likely to block than one at any other location. Where a substituent is found on an end unit of the product, it is usually at the non-reducing end.

There are other enzymes, however, which can act *only* at the sugar unit containing the branch. These enzymes cannot hydrolyze the same polymer after the branch has been removed, or, if they do so, it is at a greatly reduced rate,[57] e.g.

$$- A - A \overset{\underset{\displaystyle\downarrow}{E_{12}}}{-} A - A - A - \quad \text{and} \quad - A - A \overset{\underset{\displaystyle\nrightarrow}{E_{12}}}{-} A - A - A - \, .$$

$$ || $$

$$ X X$$

The best known enzyme of this type is lysozyme.

It is easy to understand that addition of a substituent might interfere with the action of an enzyme on its substrate. It is less easy to see that the removal of a branch from a susceptible substrate similarly interferes with enzyme action (the "hole" concept). Some examples:

(1) Xylan is a cellulose lacking a $-CH_2OH$ "branch." Most cellulases cannot act on xylan.

(2) Chitosan is a chitin lacking a $-\overset{\displaystyle O}{\overset{\displaystyle \|}{C}} -CH_3$ "branch." Chitinases cannot act on chitosan.

(3) Lysozyme is unable to hydrolyze its natural sub-strate when acetyl groups are removed from the N-acetyl glucosamine moiety.   (This, however, is a deletion from the sugar one unit away from the site of hydrolysis.)

There are also examples of this effect among enzymes that hydrolyze simple molecules. Thus, in β-glucosidases (A = glucose),

	Substrate	Non-substrate
E_{13}	$A-\beta \rightarrow 4A$	$PO_4-6'A-\beta \rightarrow 4A$
E_{14}	$PO_4-6'A-\beta \rightarrow 4A$	$A-\beta \rightarrow 4A$.

E_{13} which hydrolyzes cellobiose cannot act on the $6'-PO_4$ of cellobiose, because of the bulky phosphate substituent. E_{14} hydrolyzes cellobiose-$6'-PO_4$, but cannot act on cellobiose,[49] presumably because of the unfilled hole left in the enzyme site, due to deletion of the phosphate.

To explain these results, we assume that an enzyme possesses a groove into which a portion of the polymer fits, the portion varying perhaps from the space required for a dimer to that for a hexamer. Those enzymes which attack at a branch point require an additional space in the enzyme for the branch. In the absence of a branch, this hole is un-filled. There is a looser fit of polymer section to enzyme, and this, in turn, interferes with the catalytic action. In other words, removal of a substituent can be--for these enzymes--just as serious an obstacle as the addition of a substituent to the substrate of those enzymes previously considered.

Endo-glycanases are not generally considered to be "trans-ferases." Transfer has, however, been detected provided that the concentrations of enzyme, "donor" and "acceptor" are very high.

Depolymerization by Non-hydrolytic Enzymes. **Eliminases**. One occasionally needs to be reminded that hydrolysis is not the only enzymatic means of depolymerization of polysaccharides and that the products of such reactions may be quite different from those of hydrolysis. An interesting and unusual action is that of the recently[2] discovered "eliminases" where chain scission does not involve the addition of water. As a result of a β-elimination reaction, a product is formed having unsaturation in the non-reducing sugar moiety between C4 and C5 (Fig. 17).

At present, eliminases seem to be restricted in their action to polymers containing a uronic acid group: hyalu-ronic acid, pectic acid, pectin and alginic acid (i.e., poly-mers containing glucuronic, galacturonic or mannuronic acids).

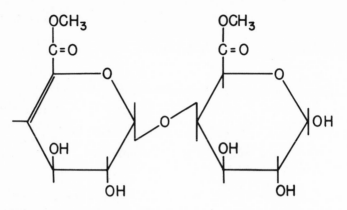

Figure 17. Unsaturated dimer produced from pectin by bacterial eliminase.

The linkage may be α or β. The hydrogen on C5 may be *cis* or *trans* to the oxygen on C4. These enzymes bring about depolymerization, the products having reducing properties and lowered viscosity. As a result--unless the products are further characterized--one is not likely to suspect that the action is anything but hydrolytic. There is, however, an easy way to detect the "eliminase," namely, the appearance of a strong absorption by the unsaturated products at 232 nm. Another difference from hydrolases is in the position of cleavage. In eliminases the split is between C4 and the bridge oxygen, whereas in hydrolytic enzymes, it is between C1 and oxygen. While most of the reported eliminases are endo-enzymes, exo-polygalacturonic acid-*trans* eliminases have been described.[27,28] They differ from exo-polysaccharases in that they act from the reducing end of the chain.[28]

The products of the eliminase reaction contain an unsaturated galacturonic acid at the non-reducing end of the oligosaccharide. The modification should effectively inhibit the subsequent action of exo-PG-ases. An enzyme from *A. niger*[27] has been found, however, which can remove this modified terminal group. The rate, as one might expect, is about 3% of that in which the terminal group is unmodified. The product obtained is not the expected 4,5 unsaturated galacturonic acid, but a re-arrangement product thereof, e.g., 4-deoxy-5-keto-D-glucuronic acid.

The number of different enzymes acting on pectic substances appears to be much greater than on any other natural
polymer. They include:

hydrolases, endo-
               exo- dimer removing; monomer removing

eliminases, endo-
               exo- dimer removing, monomer removing

methyl esterases

separate enzymes for pectins (vs. polygalacturonic
   acid).

Eliminases are the principal enzymes involved in *bacterial*
breakdown of pectic substances, and are somewhat less common
in fungi.

Intramolecular Transferases. Recently a depolymerase of
inulin (β-2,1-fructan) has been reported[73] which produces
the anhydride of the fructose dimer.

$$\text{--- F2 - 1F2 - 1F2 - 1F ---} \longrightarrow \quad \begin{matrix} \overset{\frown}{2 \quad 1'} \\ F \quad F \\ \underset{\smile}{3 \quad 2'} \end{matrix}$$

This dimer may be considered a diglycoside, since both C2
positions are involved. This reaction may be essentially
like that of the *B. macerans* enzyme which forms cyclic dex-
trins of 6 to 8 units from starch. They are specialized
examples of carbohydrate transfer where the donor and acceptor are part of the same molecule. The dimer anhydride would
represent the smallest ring arising from such a reaction, and
by its very formation indicate that in the parent polymer C3
of one fructose unit occurs in a favorable position relative
to C2 of another.

Phosphorolytic Enzymes. Another depolymerization reaction involves the transfer of the terminal sugar of a chain
to phosphate. Phosphorolytic enzymes are involved in the
depolymerization of α-1,4-glucans, cellobiose, and β-1,3-
oligoglucans. These behave like exo-glycanases in that they
act by inversion. The continual appearance of new enzymes
(such as the above) should alert us to the possibility that
still other mechanisms of depolymerization may exist.

Oxidative Enzymes. The idea that oxidative actions may be involved in the biological degradation of cellulose has been put forth from time to time. Recent data (Table 10) show that such oxidative enzymes indeed occur.[18] These experiments show for a variety of organisms that cell-free filtrates solubilize crystalline cellulose much more rapidly under *aerobic* than under *anaerobic* conditions. These data were supported by results showing a consumption of oxygen during the solubilization, and the presence of glucuronic and gluconic acids as products in the hydrolysate, although neutral sugars predominated. It was suggested that the enzyme oxidizes cellulose by forming uronic acid moieties, thus breaking H-bonds between chains and resulting in a swelling action that permits other enzymes to bring about the hydrolysis. The oxidative enzyme thus resembles C_1 in its presumed disruption of hydrogen bonds of the crystalline structure.

Most investigators of cellulolytic systems have observed only neutral sugars in the hydrolysates. The oxidative theory (above) does not require that a large number of modified glucose units be present, and it is just possible that such sugar acids could have been missed. But then, too, one wonders why the effect of oxygen on rate has been overlooked, and again it may be argued that the amount of oxygen required is very low. (The writer has been unable to confirm the presence of this oxidative enzyme. Re-examination by others will be necessary.)

Table 10. Degradation of cotton cellulose by "cellulases" of four different cellulose degrading fungi in the presence and absence of oxygen*

Organism	Cellulose degradation weight loss %	
	O_2-Atmosphere	N_2-Atmosphere
Sporotrichum pulverulentum	52.1	21.5
Polyporus adustus	42.6	18.0
Myrothecium verrucaria	33.6	17.0
Trichoderma viride	20.6	10.0

*Eriksson *et al.*, 18.

The Coupling of Lignin and Cellulose Degradation in White Rot Fungi.[78] It has long been known that white rot fungi produce oxidative enzymes, such as polyphenol oxidase and laccase, capable of oxidizing phenolic compounds. Since lignins are polymeric phenols, the implication has been that such enzymes are involved in lignin breakdown.[3] There is a good correlation between ability to produce these enzymes and ability to degrade lignin, but certainly not *all* organisms that produce these enzymes degrade lignin. Laccase acts on lignin to produce *larger* molecules (dominant) as well as smaller ones. As lignin degradation is much more rapid when other carbon sources (cellulose) are present, it may be that the metabolism of the carbohydrates yields products which favor the degradative rather than synthetic action of the polyphenol oxidases. Thus, while these oxidases are ineffective *by themselves* in solubilizing lignin, a coupling with some other system may give the desired result.

Recently, an extracellular enzyme was reported from *Polyporus versicolor*, and from *Chrysosporium pulverulentum* (both cause white rot) which catalyzes the oxidation of cellobiose with simultaneous reduction of quinones. As the enzyme utilizes a carbohydrate derived from cellulose, and an oxidized phenol derivable from lignin, the suggestion was made that this enzyme may play an important role both in lignin and in cellulose breakdown. The enzyme has been called cellobiose:quinone oxidoreductase, or cellobiose dehydrogenase (Fig. 18).

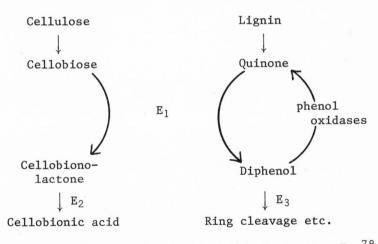

Figure 18.   Action of cellobiose dehydrogenase, E_1.[78]

The high degree of specificity of this enzyme for hydrolysis products of cellulose (Table 11), and the presence of
a lactonase in these cultures (to hydrolyze the cellobiono-
lactone produced), lend credence to this hypothesis. The
non-specific nature of the quinone indicates that a variety
of lignin products might be used. The enzyme is induced by
cellulosic materials, but not by quinones.

At present, it is reasonable to suggest a role for
cellobiose dehydrogenase in the *further decomposition* of the
products of digestion of cellulose and lignin. No evidence
has been produced to couple these actions with the *initial*
attack on either the cellulose or the lignin. In the absence
of a better scheme, further work is necessary.

Non-enzymatic Depolymerization. While many fungi produce and secrete highly active enzyme systems, there are
others which secrete very little or none. Yet both groups
may degrade cellulose at the same rate. The explanation for
this is not at all clear. The activity of enzymes located
at bacterial surfaces has been mentioned above. Still, there

Table 11. Specificity of cellobiose dehydrogenase

Substrate (2 μmole/3 ml)		Oxidation Rate (relative)
Cellobiose	G–β→4G	100
Cellopentaose	G–β→4G–β→4G–β→4G–β→4G	100
Cellulose	$[G–β→4G]_n$	0
Lactose	Gal–β→4G	40
Mannosyl glucose	M–β→4G	0
Glucosyl mannose	G–β→4M	36
Mannobiose	M–β→4M	0
Xylobiose	X–β→4X	0
Maltose	G–α→4G	0

[*]Reaction mix (pH 4.5; 25 C) contains 2 μmole carbohydrate,
1 μmole 3-methoxy-5-*tert*-butyl-benzoquinone, and enzyme in
3 ml (Westermark & Eriksson, 78).

is a strong belief by many that other mechanisms (as yet undetected) are involved. One of these suggests that the organism secretes a highly reactive molecule capable of bringing about the depolymerization reaction. Enzymes are involved in producing the reactive moiety, but these are one step removed from a direct action on the polymeric substrate.

$$S \xrightarrow{E} P + R^*,$$

where E is an enzyme other than cellulase, acting on its substrate (S) to yield two products, P and R^*, R^* being a highly reactive molecule. Then

$$(G)_n + R^* \longrightarrow (G)_{n/2} + R_1$$

Cellulose   Degraded Cellulose

According to this mechanism (an oxidative cleavage), the living organism carries on some process which yields a reactive product, such as H_2O_2[24],[36] and the superoxide[22] anion radical $(O_2^- \cdot)$. Many enzymes are known which can do this. (The superoxide radical, however, does not leave the microbial cell, and thus can not bring about changes in the *extracellular* polymers.) The mechanism has been suggested particularly to account for degradation on ligno-celluloses (woods), where the lignin components represent (S) in the above equations, and H_2O_2 is R^*. An examination of many wood decay fungi (basidiomycetes) reveals that they do indeed produce H_2O_2 when grown on wood, but the reaction is not limited to rot organisms. Many fungi that grow on simple substrates produce enzymes which yield H_2O_2, e.g., those that oxidize sugars to sugar lactones. In this latter case, both products are reactive, the lactone spontaneously hydrolyzing to acid.

The reactive product (R^*) must be able to degrade cellulose. The superoxide anion radical[22] can do this. The reaction is not specific, i.e., it can depolymerize most polysaccharides, and modify other compounds. H_2O_2 has a rather feeble effect unless it is coupled with Fe^{++} (Fenton's reagent). Then it is highly reactive. If this process is to occur in wood, there must be present (a) suitable substrates for the H_2O_2-producing enzymes, and (b) a source of Fe^{++}. Glucose, xylose and phenolics are present in combined forms, and their release must be mediated by other fungal enzymes.

Iron, too, is present in sufficient amount in woods, but requires conversion into the active (Fe^{++}) state. It is not now clear how or whether the decay organism accomplishes this.

Assuming the availability of H_2O_2 + Fe^{++} in sufficient quantities to degrade cellulose, one asks how this reaction compares with that of cellulolytic enzymes, and of an organism growing on wood. In reactions with Fenton's reagent, the initiating agent is believed to be the hydroxyl radical, a radical which can be produced by ionizing radiation, as well as by chemical and biological means. Cellulose undergoes random oxidative cleavage to yield much shorter molecules, and as such this resembles the hydrolytic cleavage accomplished by random-acting cellulases. Both systems result in lower DP, a consequent increase in alkali solubility, an increase in alkali swelling, and eventually an increase in water solubility as the chains become very short. The systems differ in that all products of the hydrolytic system are sugars readily utilized by the growing fungus, while the products of the oxidative system are less readily metabolizable.

The H_2O_2 + Fe^{++} action seems most closely related to the brown rots in the results it achieves, and in the requirements it imposes. Brown rot fungi produce sufficient H_2O_2 to account for the depolymerization and loss in strength observed when these fungi grow on wood.[36] Both the fungal effects and the reagent effects are those associated with a widespread--rather than local--attack, indicating extensive diffusion of active agents (R^*; or enzyme) from the brown rot organism. White rot fungi and fungi growing on textiles produce localized effects, completely consuming cellulose in isolated areas, and leaving the bulk of the cellulose unchanged. As a result, these residues show little change in DP, alkali solubility, or strength loss.

While the H_2O_2 + Fe^{++} mechanism offers an alternate explanation of depolymerization, the hypothesis is not proven by any means. The refractory nature of the ultimate products, as far as their utilization by fungi is concerned, requires examination. Unfortunately such products are diverse in nature and as yet poorly characterized. Amongst them may be compounds that are *toxic* to living things.

The wood decay fungi produce at least the C_x (random splitting) component of cellulase. Usually the yields are not great, but the amount is appreciable. (They produce little or none of the C_1.) Why should they do so if they rely on an alternate method of decomposition? It may be that the H_2O_2 + Fe^{++} system is involved only in the *early* stages of cellulose breakdown in wood, i.e., in bringing about the changes which one observes in DP and strength losses. Four successive random cleavages would reduce a cellulose molecule of DP 2000 to about DP 100, yet modify less than 1% of the glucose units. If further depolymerization then occurs *via* endo-β-1,4-glucanase action, the final products would be predominantly glucose and cellobiose, thus overcoming the objection stated above relative to the refractory nature of the products.

An examination of the primary products of the H_2O_2 + Fe^{++} reaction on *cotton* shows that these are indeed susceptible to the subsequent action of endo-β-1,4-glucanases (Table 12). Considering that the pretreatment had such a great effect on the alkali solubility of the product (26%), the *increased* digestibility by endo-glucanases is very little (0.8%). The big difference between the *P. westerdijkii* and *T. viride* enzymes is in the C_1 component; the H_2O_2 + Fe^{++} pretreatment definitely does not supplant the C_1 action. Thus, *P. westerdijkii* action on pretreated cotton gave only 1.1% decomposition,

Table 12. Enzymatic hydrolysis of cotton and of cotton pretreated with Fenton's reagent

Enzymes of	Properties of enzyme*		Alkali solubility of substrate		Digestion** by enzymes	
	C_x	C_1	Control	Pretreat.	Control	Pretreat.
Pestalotiopsis westerdijkii QM381	170	0.1	1	26	0.3	1.1
Trichoderma viride QM9414	220	10.0	1	26	8.9	9.7

(%)

* Arbitrary enzyme units.
** Digestion, 2 hr at 50 C (E. T. Reese, unpublished).

while *T. viride* (containing C_1) gave 8.9% decomposition of
the *untreated* cotton. Both C_1 and the H_2O_2 + Fe^{++} systems
greatly increase the alkali solubility of crystalline cellulose, but the products formed are quite obviously different
in their susceptibility to subsequent action of the random-
acting glucanase. These results do not support the hypothesis that H_2O_2 + Fe^{++} can substitute either for the whole
cellulase system, or for the C_1 part of it. It may be, however, that in wood the structures are opened up by the oxidative action in such a way as to permit more facile penetration of the enzymes. The hypothesis certainly merits
continued support.

ACKNOWLEDGMENTS

        I thank Drs. Mary Mandels, G. Dateo, and D. Ball for
their assistance in the preparation of this manuscript.

REFERENCES

1.   Abdullah, M., I. D. Fleming, P. M. Taylor, and W. S.
     Whelan. 1963. *Biochem. J.* *89:*35p.
2.   Albersheim, P., H. Neukom, and H. Deuel. 1960. *Helv.
     Chim. Acta 43:*1422.
3.   Ander, P. and K. E. Eriksson. 1975. *Svensk. Papper.*
     *78:*643.
4.   Araki, Y., T. Nakatani, H. Hayashi, and E. Ito. 1971.
     *Biochem. Biophys. Res. Commun. 42:*691.
5.   Barnett, J. E. G. and W. T. S. Jarvis. 1967. *Biochem.
     J. 105:*9p.
6.   Barras, D. R. and B. A. Stone. 1969. *Biochim. Biophys.
     Acta 191:*342.
7.   Berghem, L. and L. G. Pettersson. 1973. *Euro. J. Bio-
     chem. 37:*21.
8.   Berghem, L. and L. G. Pettersson. 1974. *Euro. J. Bio-
     chem. 46:*295.
9.   Brown, R. 1975. Personal communication.
10. Charpentier, M. and D. Robic. 1974. *C. R. Acad. Sci.
     Paris 279 (D):*863.
11. Cowling, E. 1975. Biochem. & Bioeng. Symp. No. 5, 163
     (C. R. Wilkie, ed.). John Wiley & Sons, N. Y.
12. Crook, E. M. and B. A. Stone. 1957. *Biochem. J. 65:*1.
13. de Haas, G. 1973. *J. Biol. Chem. 248:*4023.
14. Doi, K., A. Doi, and T. Fukui. 1971. *J. Biochem. 70:*711.

15. Ebata, S. and Y. Santomura. 1963. *Agr. Biol. Chem.*
    27:478.
16. Emert, G. H., E. K. Gum, J. A. Lang, T. H. Liu, and
    R. D. Brown. 1974. Adv. Chem. Ser. #136, p. 79.
    ACS, Washington, D.C.
17. English, P. D., J. B. Jurale, and P. Albersheim. 1971.
    Plant Physiol. 47:1.
18. Eriksson, K. E., B. Pettersson, and U. Westermark.
    1974. *FEBS Letters 49*:282.
19. Eveleigh, D. 1969. Personal communication.
20. Eveleigh, D. E. and A. S. Perlin. 1969. *Carbohyd.*
    Res. 10:87.
21. Fahraeus, G. 1967. *Acta Chem. Scand. 21*:2367.
22. Fridovich, I. 1972. *Acc. Chem. Res. 5*:321.
23. Gum, E. 1974. Ph.D. Dissertation. Dept. Biochemis-
    try, VPI, Blacksburg, Va.
24. Halliwell, G. 1965. *Biochem. J. 95*:35.
25. Hamauzu, Z., K. Hiroma, and S. Ono. 1965. *J. Biochem.*
    (Japan) *57*:39.
26. Hasegawa, S., J. H. Nordin, and S. Kirkwood. 1969.
    J. Biol. Chem. 244:5460.
27. Hatanaka, C. and J. Ozawa. 1969. *Ber. Ohara Inst.*
    Landwirt. Biol. 15:25.
28. Hatanaka, C. and J. Ozawa. 1970. *Agr. Biol. Chem.*
    (Tokyo) *34*:1618.
29. Hehre, E. J., G. Okada, and D. S. Genghof. 1969.
    Arch. Biochem. Biophys. 135:75.
30. Huotari, F., T. Nelson, F. Smith, and S. Kirkwood.
    1968. *J. Biol. Chem. 243*:952.
31. Hutson, D. H. and H. Weigl. 1963. *Biochem. J. 88*:588.
32. Jansen, J. C. 1972. Ph.D. Dissertation. Acta Univ.
    Upsaliensis, Uppsala, Sweden.
32a. Keegstra, K., K. W. Talmadge, W. D. Bauer, and P.
    Albersheim. 1973. *Plant Physiol. 51*:188.
33. Kerstens-Hilderson, H., M. Claessens, E. Van Doorslaer,
    and C. K. De Bruyn. 1976. *Carbohyd. Res. 47*:269.
34. Klop, M. and P. Kooiman. 1965. *Biochem. Biophys. Acta*
    99:102.
35. Kobayashi, T. 1954. *J. Agr. Chem. Soc.* (Jap.) *28*:352.
36. Koenigs, J. W. 1975. *In* Cellulose as a Chemical and
    Energy Source (C. R. Wilkie, ed.), p. 151. John Wiley
    & Sons, New York.
37. Kuhn, R. 1924. *Ber. 57*:1965.
38. Lampen, J. O., S. C. Kuo, F. R. Cano, and J. S. Tkacz.
    1972. *Proc. NIFS Ferment. Techn. Today 819-24*.

39. Leatherwood, J. M. 1969. *Advan. Chem. Ser. 95*:53.
40. Li, L. H., R. M. Flora, and K. King. 1965. *Arch. Biochem. Biophys. 11*:439.
41. Linderstrom-Lang, K. 1952. Proteins and Enzymes. Stanford Univ. Press, Stanford, California.
42. Mandels, Mary, J. Kostick, and R. Parizek. 1971. *J. Polymer Sci. (C) 36*:445.
43. Mandels, Mary, F. W. Parrish, and E. T. Reese. 1962. *J. Bacteriol. 83*:400.
44. McCroskery, P. A., S. Wood, and E. D. Harris. 1973. *Science 182*:70.
45. Marshall, J. J. 1974. *Carbohyd. Res. 34*:289.
46. Marshall, J. J. 1974. *Adv. Carbohyd. Chem. & Biochem. 30*:257.
47. Marx, J. L. 1973. *Science 180*:482.
48. Niwa, T., S. Inouye, T. Tsuruoka, Y. Koaze, and T. Niida. 1970. *Agr. Biol. Chem.* (Toyko) *34*:966.
49. Palmer, R. E. and R. L. Anderson. 1972. *J. Biol. Chem. 247*:3420.
50. Parrish, F. W. and E. T. Reese. 1967. *Carbohyd. Res. 3*:424.
51. Perlin, A. 1963. *In* Advances in Enzymic Hydrolysis of Cellulose and Related Materials. Pergamon Press, N.Y.
52. Perlin, A. S. 1965. *Biochem. Biophys. Res. Commun. 18*:538.
53. Perlin, A. and E. T. Reese. 1963. *Can. J. Biochem. Physiol. 41*:1842.
54. Pettersson, G. 1969. *Arch. Biochem. Biophys. 130*:286.
55. Potts, R. C. 1974. *Comp. Biochem. Physiol. (B) 47*:327.
56. Rees, D. A. 1975. *In* Carbohydrates (G. O. Aspinall, ed.). Butterworths, England.
57. Reese, E. T. 1968. *In* Organic Matter and Soil Fertility, p. 535. Pontif. Acad. Sci., Rome, Italy.
58. Reese, E. T. 1975. *In* Biological Transformation of Wood by Microorganisms (W. Liese, ed.), p. 165. Springer-Verlag, Heidelberg, Germany.
59. Reese, E. T., J. E. Lola, and F. W. Parrish. 1969. *J. Bacteriol. 100*:1151.
60. Reese, E. T., A. H. Maguire, and F. W. Parrish. 1968. *Can. J. Biochem. 46*:25.
61. Reese, E. T. and Mary Mandels. 1964. *Can. J. Microbiol. 10*:103.
62. Reese, E. T., F. W. Parrish, and M. Ettlinger. 1971. *Carbohyd. Res. 18*:381.
63. Robyt, J. and R. J. Ackerman. 1971. *Arch. Biochem. Biophys. 145*:105.

64. Rypacek, V. 1975. *Drevarsky Vysksm. 20:*1.
65. Sakano, Y., N. Masuda, and T. Kobayashi. 1971. *Agr. Biol. Chem.* (Tokyo) *35:*971.
66. Sarda, L. and P. Desneulle. 1958. *Biochim. Biophys. Acta 30:*513.
67. Sawai, T. and Y. Niwa. 1975. *Agr. Biol. Chem.* (Tokyo) *39:*1077.
68. Shibaoka, T., K. Miyano, and T. Watanabe. 1974. *J. Biochem.* (Japan) *76:*475.
69. Shibata, Y. 1974. *J. Biochem.* (Japan) *75:*85.
70. Storvick, W. O., F. E. Cole, and K. W. King. 1963. *Biochemistry 2:*1106.
71. Sugiura, M., A. Ito, and T. Yamagachi. 1974. *Biochim. Biophys. Acta 350:*61.
72. Suzuki, H., K. Yamane, and K. Nisizawa. 1969. *Advan. Chem. Ser. 95:*60.
73. Tanaka, K., T. Uchiyama, and A. Ito. 1972. *Biochim. Biophys. Acta 284:*248.
74. Thoma, J. 1971. *J. Biol. Chem. 246:*5621.
75. von Hofsten, B. 1975. *In* Enzymatic Hydrolysis of Cellulose (M. Bailey, T. M. Enari, and M. Linko, eds.), p. 281. SITRA, Aulanko, Finland.
76. Walker, Gwen. 1976. Personal communication.
77. Walker, Gwen and A. Pulkownik. 1973. *Carbohyd. Res. 29:*1.
78. Westermark, U. and K. E. Eriksson. 1975. *Acta Chem. Scand. (B) 28:*204.
79. Whelan, W. J. 1971. Proc. Intnl. Symp. Conversion and Manufacture of Food Stuffs by Microorganisms, p. 87. Kyoto, Japan.
80. Whelan, W. J., G. S. Drummond, E. E. Smith, and H. Tai. 1969. *FEBS* Meeting Abstract 102.
81. Whitaker, D. R. 1954. *Arch. Biochem. Biophys. 53:*436.
82. Yamamoto, S., R. Kobayashi, and S. Nagasaki. 1974. *Agr. Biol. Chem.* (Tokyo) *38:*1493.
83. Young, K. K., C. Hong, M. Duckworth, and W. Yaphe. 1971. Proc. 7th Intnl. Seaweed Symposium, p. 469. Univ. Tokyo Press, Japan.

Chapter Nine

ADVANCES IN UNDERSTANDING THE MICROBIOLOGICAL DEGRADATION

OF LIGNIN

T. KENT KIRK, W. J. CONNORS

Forest Products Laboratory, USDA, Forest Service, Madison, Wisconsin   53705

J. G. ZEIKUS

Department of Bacteriology, University of Wisconsin, Madison, Wisconsin   53706

INTRODUCTION

From the earlier chapters the reader can gain an up-to-date concept of the chemical and physical structure of lignocellulosic cell walls, and the biosynthesis of their structural components. About half of the earth's carbon that is fixed annually via photosynthetic reactions is incorporated into these constituents.[3] The carbon, of course, does not remain fixed, but is released ultimately as volatile products to complete the carbon cycle that is central to life. The carbon is released primarily as CO_2 by the activities of microorganisms.

Most microbes which are able to degrade isolated wood carbohydrates by processes such as those described in the preceding chapter cannot attack woody tissues unless they are able to decompose the lignin which physically protects the carbohydrates, or unless the tissue is reduced to small enough particles that the lignin barrier is overcome, as is the situation with termites. Thus the microbial degradation of lignin is a most important aspect of the carbon cycle, not only because lignin is a major repository of reduced carbon, but also because it protects wood carbohydrates from decomposition by the majority of carbohydrate-degrading microbes. Lignin is perhaps the most serious impediment to the development of successful bioconversion processes for waste cellulosics.

This chapter is an interpretative summary of some of the recent findings in the study of the microbial degradation of lignin. Because the senior author has already reviewed the literature through 1970,[35] this chapter will emphasize work since that time. In the six years since that review, progress has been made in understanding the complex process of lignin degradation, and the authors feel that the next few years will see substantial increases in our understanding. A greatly increased impetus to research in this area has resulted from the interest in having lignin-degrading enzymes and/or organisms available for possible exploitation in a variety of scientific and industrial applications.

METHODOLOGY

One of the major problems in studying lignin biodegradation has been the lack of a definitive and sensitive assay. This problem has been largely solved by the development of assays based on ^{14}C-labeled synthetic lignins[28,40] and [lignin-^{14}C]-lignocellulosics.[15,28] In the two laboratories describing syntheses of ^{14}C-lignins, the immediate precursor of conifer lignin, coniferyl alcohol, was chemically synthesized with the labeled carbon in the aromatic ring, in the propyl side chains, or in the methoxyls, and oxidatively polymerized *in vitro* in a peroxidase/H_2O_2-catalyzed reaction. In one laboratory, the resulting polymer termed *dehydrogenative polymerizate* (DHP), was shown[40] by various chemical and spectroscopic methods to be similar in subunit structure to spruce lignin. Although the relative proportions of the various substructures differed from those in the natural lignin, the synthetic polymer was demonstrated to be qualitatively representative, and to be decomposed to

CO_2 by lignin–degrading organisms. The availability of these model lignins has already permitted progress to be made in studies of the microbiological aspects of lignin degradation, as described in the following sections.

Labeled natural lignocellulosics for biodegradation studies have also been prepared recently by feeding lignin precursors to lignifying plant tissues.[15,28] Such materials from cattail, oak, and maple were shown[15] to contain at least 90% of their ^{14}C in non–acid–hydrolyzable (mainly lignin) components. Corresponding tissues labeled in the cellulose fraction also have been prepared (R. Crawford, personal communication). These materials will also aid greatly in microbiological studies, particularly in screening for lignin–degrading organisms, and in defining the relationships between lignin and cellulose degradation.

Another serious impediment to studies of lignin bio-degradation has been--and remains--the difficulty in isolating substrate quantities of lignin from natural tissues. The most trusted preparation, milled wood lignin (MWL),[5] has served well for 20 years in various analytical chemical studies, but its preparation is tedious, involving grinding, pre-extractions to remove extraneous components, ball-milling to a very fine powder, extraction of the lignin, and multi-step purification. Yields are low. In a variation developed by Pew and Weyna,[54] the ball-milled woody tissues are treated with polysaccharidases and the resulting "cellulolytic enzyme lignin, CEL," collected. Yields are nearly quantitative. A recent advance was made by Chang et al.,[11] who showed that purified CEL and MWL are essentially equivalent, which should allay the concern of some that MWL is not adequately representative of the total lignin in wood, or that CEL has been altered in the polysaccharidase enzyme treatment.

Attempts to prepare useable lignin in aspen tissue cultures met with failure due to contamination of the lignin by polyphenolic materials.[63] However, more recent studies of the biochemistry and physiology of lignification in tissue cultures[20] should point the way for utilizing such cultures for large-scale production of radiolabeled natural lignins and lignocellulosics with very high specific activities.

Progress has also been made in methods for characterizing lignins. Miksche and co-workers[21] have refined and greatly extended the methods for oxidatively degrading lignins after methylation, and for identifying and quantifying the products and interpreting the results in terms of lignin structure. Similarly, Lundquist[51] and Higuchi[29] have further refined procedures for analyzing low molecular weight products formed on treatment of lignin with HCl in aqueous dioxane ("acidolysis"). Nimz[49] has developed the procedures for applying carbon-13 nmr spectroscopy to lignin characterization, and has completed the spectral analysis of several lignins by comparisons with numerous substructure model compounds. These various procedures can aid greatly in investigations of the integrity of lignin after microbial attack in various environments, as for example in defining the contributions of lignin to humus formation.

ORGANISMS AND ENVIRONMENT

Some progress has been made in recent years in defining the groups of microorganisms active in lignin degradation, and in describing the factors favoring degradation in nature. Table 1 lists those groups of microorganisms which have been shown, beyond a reasonable doubt, to cause structural alterations in, or to completely decompose, sound (unmodified) lignins. The relative importance of these particular groups in cycling the carbon of lignin in nature, however, is far from being known. It would appear on the basis of our present, limited, knowledge that the white-rot fungi and the closely related litter-decomposing fungi play a predominant role in the complete decomposition of lignin, and that the brown-rot fungi are mainly humifiers, having been shown to cause only limited degradation of lignin.[37] Soft-rot fungi apparently degrade lignin quite slowly,[22] and perhaps incompletely; their participation in nature would seem therefore to be limited, but they probably contribute to humification and, in some cases, to complete conversion to CO_2. Whether other groups of fungi, including molds, soil fungi, and yeasts, decompose or partially degrade lignin is not known.

Even though Table 1 lists only fungi, it is possible that bacteria also play a predominent primary role in nature. Their ability to attack unmodified lignin is not known.

Table 1. Groups of organisms that degrade lignin[a]

Group of organisms[b]	Substrates	Assay method	References
Basidiomycetes			
White-rot fungi (e.g. *Coriolus versicolor*)	Various wood tissues, synthetic lignins	Gravimetric,[c] radioactivity, various chemical and spectroscopic.	9,13,38,39 40
Brown-rot fungi (e.g. *Gloeophyllum trabeum*)	(As above)	(As above)	9,13,36,40
Litter-decomposing fungi (e.g. *Collybia butyracea*)	Forest litter	Gravimetric	48
Ascomycetes			
White-rot fungi (e.g. *Hypoxylon deustrum*)	Various woody tissues	Gravimetric	9,10
Soft-rot rungi (e.g. *Chaetomium globosum*)	Various woody tissues	Gravimetric, radioactivity	22,47
Fungi imperfecti			
Soft-rot fungi[d] (e.g. *Papulospora sp.*)	Various wood tissues	Gravimetric	22

[a]Only those organisms are included which have been shown beyond a reasonable doubt to cause structural alterations in, or complete decomposition of, unaltered lignins.

[b]The white-rot, brown-rot, litter-decomposing, and soft-rot fungi are not taxonomically classified as such. Their delineation is based solely on the types of decay caused.

[c]Gravimetric analyses for lignins, based on acid-insolubility,[55] are neither fully accurate nor specific, and can be used as criteria for lignin decomposition only when the depletion is great enough to leave no reasonable doubt, and when care is taken to rule out spurious results.

[d]These soft-rot fungi are probably Ascomycetes, but their perfect stages are not known, and they are placed in Fungi Imperfecti.

There have been several reports of bacterial degradation
of lignin (see ref. 35), but none of these presents evidence
that is "beyond a reasonable doubt." Bellamy[4] recently
demonstrated that the lignin in alkali-treated fibers from
manure is degraded by a thermoactinomyces, but whether
unaltered lignin is also a substrate was not reported.
Bacteria in the genus *Pseudomonas* have recently been shown
to degrade lignin substructure model compounds.[17,18,33]
The primary features of the pathway for degradation of one
such model, 1-(3,4-dimethoxyphenyl)-2-(2-methoxyphenoxy)
propane-1,3-diol, by *P. acidovorans*, were elucidated by
Crawford.[17,18]

Attempts to obtain lignin-degrading bacteria and fungi
through the usual microbiological enrichment procedures
have not been successful. The reason for this may be
that a growth substrate is always required by those organisms
capable of attacking unmodified lignin; such a requirement
has recently been demonstrated in white-rot fungi (see
p. 384).

A distinction may of course be made between those
primary organisms, such as white-rot fungi, which are able
to attack the unaltered lignin polymer, and those organisms
that may further degrade the lignin after it has been
altered by primary degraders. Bacteria are probably
important in various soil and water environments in a
secondary role whether or not they are primary decomposers.

Recent studies have given some indication of the rates
of degradation of lignin by the mixed microflora in soils
and other natural materials, and of the factors that affect
degradation in such materials. Crawford and Crawford,[15]
and Crawford *et al.*[16] studied the decomposition of ^{14}C-
lignin-labeled oak, maple, and cattail tissues in compost,
humus, soil, and creek water samples. Rates of degradation
varied substantially with the inoculum source. In 30 to 35
days up to about 45% conversion to $^{14}CO_2$ was observed in
one soil sample, and over 30% conversion to $^{14}CO_2$ was
demonstrated in one water sample. Hackett *et al.*[27] examined
lignin biodegradation in a variety of natural materials
using synthetic ^{14}C-lignins labeled in the side chains,
aromatic rings, or methoxyl groups. Natural materials
included soils, lake sediments, silage, animal bedding, and
rumen contents. Both aerobic and anaerobic conditions
were employed. No degradation to labeled gaseous products

occurred under anaerobic conditions. Aerobic degradation
varied greatly with the type of material employed, site,
soil type and horizon, and temperature. The greatest
degradation to $^{14}CO_2$ occurred in a soil from Yellowstone
National Park, and exceeded 42% in a 78-day period, although
degradation to $^{14}CO_2$ in most of the samples was considerably
slower. Mineralization of the ^{14}C-lignins in Wisconsin
soils was significantly correlated with organic carbon,
organic nitrogen, nitrate nitrogen, exchangeable calcium,
and exchangeable potassium.

CHEMISTRY AND PHYSIOLOGY OF
FUNGAL DEGRADATION

Virtually all of the detailed studies of the chemistry
and physiology of lignin biodegradation have been with wood-
destroying fungi of the white-rot type, such as *Coriolus
versicolor* (=*Trametes*, *Polyporus*, *Polystictus versicolor*),
Phanerochaete chrysosporium (=*Sporotrichum pulverulentum*),
Polyporus anceps, *Pleurotus ostreatus*, and *Polyporus
dichrous*.

Several factors make these studies difficult. One of
these is that analogies between the degradation of lignin
and the degradation of other biopolymers are only superficially relevant. In contrast to proteins, polysaccharides,
and nucleic acids, lignin does not contain repeating units
connected by easily hydrolyzable bonds. Because it is
polymeric, lignin remains outside microbial cell walls and
must be attacked by extracellular enzymes, as are other
common biopolymers. Hydrolases, the kind of enzyme usually
involved with other biopolymers, are clearly of minor
importance in lignin degradation. A unique kind of extracellular system is indicated. Another factor making the
study of lignin biodegradation difficult is that degradation
is relatively slow. Connected with this and with the
heterogeneous structure of lignin is a third factor:
significant amounts of identifiable, informative, low
molecular weight degradation products are not found. Two
strictly methodological problems have already been mentioned:
(a) large amounts of representative lignin are still difficult to obtain, and (b) until recently there was no sensitive and definitive assay for biodegradation. Despite
these difficulties some progress has been made in defining
the overall chemical changes that comprise degradation,[38],[39]

and some speculation has been offered as to the nature of the suspected enzymes.[36] Elucidation of the specific structural changes is of course necessary before the enzymatic aspects can be studied.

One approach to elucidating the chemical changes in polymer biodegradation is to interrupt the process, and compare the partially decomposed residual polymer with the unattacked material. This approach, if applied to polysaccharide degradation by fungi, would reveal in most cases that the partially degraded polymer is simply shorter in chain length and that the end groups are the same as in starting material. This would define the nature of the enzymes involved. In lignin degradation, the partially degraded polymer differs from starting material in a large number of ways (Table 2). These and more detailed differences,[39] point clearly to a predominantly oxidative mode of degradation. The most important conclusion from the interpretation of the detailed results of these studies is that aromatic moieties are oxidatively cleaved while they are still in the polymer.[39] This conclusion is based on several points of indirect evidence which indicate the presence of a substantial amount of oxidized aliphatic residues, the most likely origin of which is through oxidative degradation of aromatic nuclei.[39] An unprecedented type of reaction is indicated: extracellular cleavage of aromatic rings. If the classical modes of aromatic ring cleavage[19] are functioning in the system, mono-oxygenases and their coenzymes would prepare the rings for subsequent cleavage by di-oxygenases. This preparation of the rings for cleavage would entail demethylation of methoxyl groups and/or other aryl-ether cleavages, aromatic hydroxylation, or a combination of these reactions. How these coenzyme-requiring reactions can be catalyzed extracellularly is an intriguing question. However, that demethylation of aryl-methoxyl groups and hydroxylation of aromatic rings occur in the lignin polymer has been firmly established through studies of lignin in brown-rotted wood.[37]

Other degradative reactions indicated by the comparisons of sound and partially white-rotted lignins include oxidation of α-carbinol groups, revealed by spectroscopic studies, and oxidative shortening of propyl side chains, as revealed by chemical and spectroscopic methods.[39]

Table 2. Some changes in general properties of lignin in spruce wood during decay by white-rot fungi[38,39]

Property	Method of analysis[a]	Change[b] Increase	Decrease
Carboxyl content	C, UV, IR, NMR	+	
Hydroxyl content			
Total	C		+
Aliphatic	C, NMR		+
Aromatic	C, UV, NMR		+
Carbonyl content	UV, IR, NMR	+	
Hydrogen/C content	C		+
Oxygen/C content	C	+	
Methoxyl/C content	C		+
Total yield of methoxy-lated aromatic acids on oxidative degradation after methylation[c]	C		+
Yield of principal acidolysis products[d]	C		+

[a] C = Various chemical methods; UV, IR, and NMR = ultraviolet, infrared and nuclear magnetic resonance spectroscopy, respectively.

[b] Degraded lignins were compared to sound lignins.

[c] The dominant product is 3,4-dimethoxybenzoic acid. Some of the individual products were higher and some lower in the degraded lignin than in the sound lignins, but total yields were considerably lower.

[d] The dominant product is 3-hydroxy-1-(4-hydroxy-3-methoxyphenyl)-2-propanone.

These results, together with what is known from studies of the biodegradation of low molecular weight aromatics, suggest the following pattern of degradation of lignin by the white-rot fungi.

Phenolic units in various parts of the polymer are converted to diphenolic units by demethylation and perhaps by aromatic hydroxylation. This apparently is a rate-limiting step since no new phenolic hydroxyl groups are detected. The diphenolic units are cleaved to produce aliphatic, carboxyl-rich residues. These are degraded,

probably via both hydrolytic and oxidative reactions, with
release of low molecular weight aliphatic products which
enter the fungal hyphae. New phenolic units are gradually
freed for further attack by these reactions. In the process,
some low molecular weight aromatics are released and metabolized intracellularly. While this attack on aromatic
regions is progressing, some terminal side chains are being
oxidized to aromatic carboxyl (C_6-C_1) residues, and other
side chains are being oxidized in the α-positions. The
kinds of oxidized residues that the above speculative
pattern of attack could produce are illustrated in Figure 1.
The structures of ring cleavage fragments in the lignin
are not known; for lack of a better structure, Figure 1
shows a substituted muconic acid residue even though such
structures apparently are not formed, or if they are
formed, do not persist.[39]

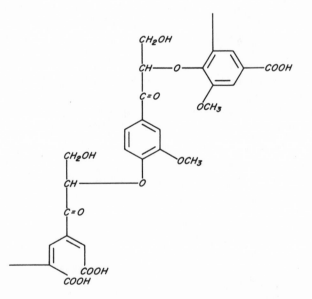

Figure 1. Hypothetical structure of a portion of
lignin after attack by white-rot fungi. Known consequences
of attack include decreases in H; OCH_3; aromaticity; phenolic,
total and aliphatic hydroxyl; yield of aromatic products
on chemical degradations; and increases in COOH; O; C=O;
α,β-unsaturated COOH; and aryl COOH.[38,39]

Unfortunately the studies of degraded lignins, while allowing the above speculation about the pattern of degradation, have not given enough information to allow meaningful predictions of the specific structures of intermediates or the specific enzyme reactions involved. Other approaches are necessary.

Two approaches are currently being pursued in our laboratory. In the first, cultures actively degrading ^{14}C-lignin to $^{14}CO_2$ are challenged with simple model compounds representing specific lignin substructures, and transformation products are identified. This will be followed by cell-free studies and will, if successful, result in a definition of specific reactions. With this approach care must be taken that the lignin-degrading enzyme system is what is acting on the models, rather than an irrelevant system induced by the model compounds. As pointed out below, studies with such models in the absence of demonstrated lignin degradation suggest that certain transformations indeed may not be relevant to lignin degradation. Our preliminary results with this approach indicate very low levels of activity of enzymes acting on the models used.

A second approach involves attaching substructure models to a biologically inert polymeric support, incubating this product in a culture actively degrading lignin, re-isolating the polymer with attached degradation fragments, and cleaving off, purifying, and identifying the fragments. This approach has two advantages: (1) it circumvents the above-mentioned possibility of irrelevant intracellular modifications of the models, and (2) it circumvents the anticipated problems connected with low levels of enzymes. Lignin degradation is a slow process under laboratory conditions, and the immobilized substrate approach should allow intermediate product accumulation over a period of several days. Preliminary results showed that model compounds which were degraded by a lignin-degrading fungus were not degraded when attached to a cross-linked polystyrene-divinylbenzene copolymer (Connors and Kirk, unpublished results). Further study suggested that the compounds are inaccessible when attached to this support, and other supports are now being investigated.

Some of the conversions of lignin-related compounds by lignin-degrading fungi are interesting despite the fact that they have been difficult to relate to the process of lignin degradation. These conversions in any case show some of the capacities of these fungi for modifying aromatic compounds.

The transformations of a number of 4-alkyl ethers of vanillic acid (4-hydroxy-3-methoxybenzoic acid), of 3-ethoxy-4-hydroxybenzoic acid and of syringic acid (3,5-dimethoxy-4-hydroxybenzoic acid) by the lignin-degrading fungus *Polyporus dichrous* in glucose-containing liquid cultures were studied in an attempt to learn more about the specific reactions of lignin degradation.[44] Identification of intermediate products suggested the function of a relatively nonspecific system for oxygenating the 4-alkyl substituents (Fig. 2). The 4-alkyl groups were cleaved from most of the compounds, presumably as a result of hydroxylation of the alkyl carbon atom in the ether linkage. However, syringic acid ethers and the β-glycerol ether of 3-ethoxy-4-hydroxybenzoic acid were not cleaved, and the compounds were not metabolized by the fungus. The absence of degradation of these compounds questions the relevance of all of these reactions to lignin degradation,[44] because lignins contain both syringyl ether-type units and aryl-β-glyceryl ethers.[56]

The parent acids, produced on cleavage of 4-alkyl groups, were completely degraded. Detailed studies with vanillic acid showed that it was degraded via oxidative decarboxylation to methoxyhydroquinone (1,4-dihydroxy-methoxybenzene) by *Polyporus dichrous*, an unusual pathway.[43] In contrast, *C. versicolor* had been reported to degrade vanillic acid via demethylation to protocatechuic (3,4-dihydroxybenzoic) acid.[8,24] Very few studies have been made of the pathways of complete catabolism of aromatic compounds by lignin-decomposing fungi, although aromatic catabolism, particularly in bacteria, has been studied quite extensively.[19]

The possible role in lignin degradation of extracellular phenol-oxidizing enzymes, which are produced copiously by most white-rot fungi, has probably been studied more than any other aspect of the lignin biodegradation problem. In the earlier review[35] it was pointed out that the phenol oxidases theoretically can bring about limited oxidation and degradation of lignin, but that their function, direct or indirect, was obscure. Research in the intervening

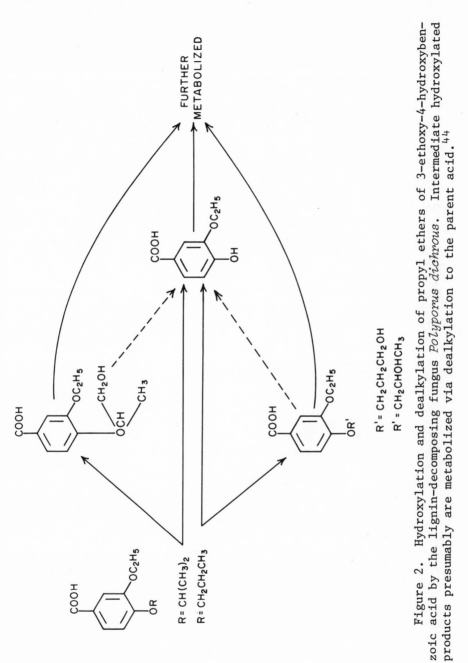

Figure 2. Hydroxylation and dealkylation of propyl ethers of 3-ethoxy-4-hydroxybenzoic acid by the lignin-decomposing fungus *Polyporus dichrous*. Intermediate hydroxylated products presumably are metabolized via dealkylation to the parent acid.[44]

period has confirmed the proposal that phenol oxidases can cause limited degradation, and also has indicated that these oxidases are required for lignin degradation, although their true rule still has not been defined.

Purified laccase (E.C. 1.10.3.2) from *C. versicolor*, in the presence of oxygen, has been shown to alter isolated lignins. Ishihara and Miyazaki[30] demonstrated a simultaneous condensation (coupling between phenolic units) and depolymerization of maple milled wood lignin (MWL). Ferm[23] reported that condensation of a sodium lignosulfonate, but not depolymerization of the lignosulfonate or of a MWL, occurred on treatment with laccase and O_2. Konishi *et al*.[46] and Konishi and Inoue[45] demonstrated a laccase/O_2-catalyzed condensation of a conifer MWL, and showed a concommittant formation of carboxyl groups. Evidence was presented that part of the carboxyl groups were in glyceric acid-β-aryl ether residues, which presumably resulted from side-chain cleavage reactions. Formation of analogous residues via phenol oxidase action had been demonstrated earlier with model compounds.[42,53] Some evidence also indicated that the laccase/O_2 treatment of MWL produced γ-COOH moieties.[45] The proportion of the lignin that resisted settling on centrifugation from water (15,000g, 20 min) was diminished by about 75% by the enzyme treatment, and the rate of adsorption onto fungal mycelia was markedly increased.[45] These effects, thought to reflect COOH formation,[45] might be expected to enhance the rate of degradation, although condensation reactions presumably would have the opposite effect. However, Westermark and Eriksson[61] suggested that phenol oxidase-catalyzed condensation reactions in lignin may be prevented during white-rot wood decay by the reduction of phenoxy radicals and quinones via the action of their newly discovered enzyme cellobiose dehydrogenase.

Limited demethoxylation of lignin by laccase/O_2 (<1.5% of the methoxyl content) has been reported.[31] Methanol was the product from the methoxyl group, and related studies with model compounds showed that the demethoxylation reaction resulted in o-quinone residues.[32] Treatment of some model compounds with peroxidase (E.C. 1.11.1.7) and H_2O_2 caused substantial demethoxylation.[12] However, demethoxylation of lignin by phenol oxidase action has not been demonstrated to be a quantitatively important reaction. Even though removal of methoxyl groups is suspected to be a key reaction,[36],[39] the role of phenol oxidases in this regard is unclear.

Gierer and Opara[26] studied the effects of the peroxidase/H_2O_2 and laccase/O_2 systems on monomeric and dimeric lignin-related phenols with the guaiacyl (4-hydroxy-3-methoxyphenyl) type of substitution. Because of the limited amount of H_2O_2 employed (1 eq/eq of phenolic hydroxyl) in the peroxidase reaction, and the short duration of the laccase/O_2 reaction, they found only products resulting from carbon-to-carbon and carbon-to-oxygen coupling reactions. Methoxyl groups and other aryl-alkyl ether groups remained intact. In accord with the accepted action of the enzymes, no demethylating or other effect was observed with non-phenolic compounds, reactions which had been reported by Trojanowski et al.[60] for the peroxidase/H_2O_2 system.

Using genetic manipulation, Ander and Eriksson[2] obtained strong evidence that phenol oxidases are required for fungal lignin degradation, perhaps in a regulatory role. A non-phenol-oxidase-producing mutant of *Phanerochaete chrysosporium* (=*Sporotrichum pulverulentum*) was screened from among normal phenol-oxidase producing colonies obtained from irradiated spores. A revertant strain was also obtained. The phenol-oxidase negative mutant grew normally on a glucose medium, and degraded cellulose and xylan, but not wood. The fact that lignin was not attacked presumably accounted for the lack of ability to degrade wood. However, in the presence of phenolic compounds, degradation of cellulose and xylan was almost completely inhibited, although growth on glucose was unaffected. The deleterious effect of the phenolics was apparently due to inhibition of production of the polysaccharideases.[2] Addition of purified laccase to cultures of the mutant restored its capacity to degrade kraft lignin and to degrade cellulose and xylan in the presence of phenolics. The wild-type strain and the revertant strain were alike; both degraded kraft lignin, all components of wood, and cellulose and xylan in the presence or absence of phenolics. These findings led to the suggestion[2] that the phenol oxidase, presumably peroxidase in this case, functions somehow in regulating production of various wood-degrading enzymes. The exact role of phenol oxidases in lignin degradation, however, remains cloudy.

In order to optimize the degradation of lignin under defined culture conditions, a number of studies have been completed in our laboratory, using as the assay ^{14}C-lignin $\rightarrow$ $^{14}CO_2$. Synthetic lignins labeled in the aromatic rings, in the β- and γ-positions of the propyl side chains, or in the methoxyl groups were used.[40] The fungus *Phanerochaete chrysosporium* was chosen for these studies from among many examined saprophytic fungi because it grows very rapidly, has a relatively high optimum temperature, degrades lignin relatively rapidly, and forms copious asexual spores. These studies have provided some of the first definitive information about the factors influencing lignin biodegradation.

Results have established that lignin cannot serve as a sufficient carbon and energy source for its own catabolism.[41] Two species of lignin-decomposing fungi, *P. chrysosporium* and *C. versicolor*, required a suitable growth substrate such as cellulose, xylose, or glucose. In cultures containing no growth substrate, but containing substrate quantities of spruce milled wood lignin, no growth was detected, and no production of $^{14}CO_2$ from included synthetic ^{14}C-lignin (ring-labeled) was observed. Addition of growth substrate to such cultures, however, resulted in production of $^{14}CO_2$, the amount depending on the amount of growth substrate provided (Fig. 3). The basis for this requirement is not clear but it might be that the energy required to degrade the lignin to molecules that can enter the hyphae is not compensated for by the energy recovered in (intracellular) respiration of the products.

It has also been established (Schultz *et al.*, unpublished) that the level of nutrient nitrogen provided in the culture medium exerts a profound effect on the rate of lignin degradation by *P. chrysosporium* (Fig. 4). Degradation ceased above approximately 10 mM reduced N, although growth at the higher levels appeared to be as good as, or better than, at lower levels and the rate of growth substrate depletion was shown to be as rapid at 30 mM as at 3 mM N. Of several tested N-sources that maintained good growth, casamino acids or a mixture of an ammonium salt and L-asparagine were best for lignin catabolism. The basis for the unexpected effect of nitrogen level is unknown, but is consistent with the fact that

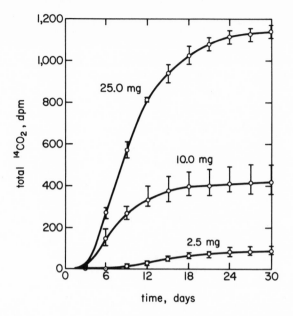

Figure 3.   Relationship between level of growth
substrate and lignin decomposition by the fungus *Phanerochaete chrysosporium*.   Shown here is the effect of addition
of various levels of cellulose on decomposition of synthetic
[ring-^{14}C]-lignin to $^{14}CO_2$ in a defined medium containing
substrate quantities of (unlabeled) spruce lignin.[41]
Reproduced with permission.

lignin-degrading fungi normally grow in woody tissues,
which are often very poor in nitrogen.[14]

Studies were made of the rate of conversion to $^{14}CO_2$
of the ^{14}C-lignins in a defined medium under various oxygen
tensions (Schultz *et al.*, unpublished).   As with nutrient
nitrogen, a very marked effect was noted (Fig. 5).   Best
degradation occurred in an atmosphere of 100% oxygen
(2 to   times faster than in air), and degradation did
not occur at all in an atmosphere of 5% O_2/95% N_2.

Table 3 gives the composition of the culture medium
that has given relatively high rates of lignin
decomposition by *P. chrysosporium*.   Because
an infinite number of comginations and levels of

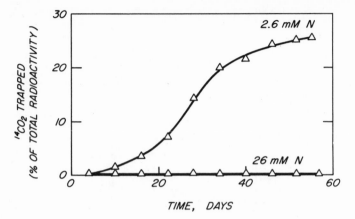

Figure 4.   Influence of the initial level of nutrient nitrogen in a defined culture medium (Table 3) on decomposition of synthetic [ring-^{14}C]-lignin to $^{14}CO_2$ by *Phanerochaete chrysosporium*.

Table 3. Culture medium for lignin degradation by *Phanerochaete chrysosporium*

Component	Amount
	g
Glucose	10.0
NH_4NO_3	.05
L-Asparagine	.10
KH_2PO_4	.20
$MgSO_4 \cdot 7H_2O$	.05
$CaCl_2$	.01
Thiamine·HCl	.0025
Trace metals[a]	< .001 of each
Lignin	.01–3.0
Buffer	---[b]
H_2O	1 liter

[a]Mn^{+2}, Na^{+1}, Fe^{+2}, Co^{+2}, Zn^{+2}, Cu^{+2}. Al^{+3}, BO_3^{-3}, MoO_4^{-7}.

[b]*o*-Phthalate, 0.01 M final concentration, pH 4.6.

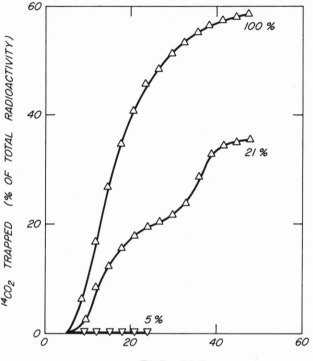

Figure 5.   Influence of O_2 level on decomposition of
synthetic [ring-^{14}C]-lignin to $^{14}CO_2$ by *Phanerochaete
chrysosporium* in a defined culture medium (Table 3).

medium constituents is possible, further improvements are
certainly expected. Typically the fungus is grown in the
medium in shallow standing cultures, which are incubated
at the optimum growth temperature of 40°, and in an atmosphere of essentially 100% O_2, maintained by periodic
flushing. Agitation of the cultures from the time of
inoculation greatly slows lignin degradation. However,
degradation is improved by shaking if growth is allowed to
become established first in stationary culture.

The effect of dilution of the ^{14}C-lignins with various levels of natural lignin (spruce milled wood lignin), and with unlabeled synthetic lignin, on degradation rate and extent was also investigated (Schultz *et al.*, unpublished). Both the unlabeled synthetic lignin and the spruce lignin formed a common metabolic pool with the synthetic ^{14}C-lignins, although the rate and extent of degradation was substantially greater than would be obtained by simple dilution of label. This result provides biological evidence augmenting the chemical and spectroscopic evidence[40] that the synthetic and natural lignins are very similar. The fact that simple dilution of label did not occur also shows for the first time that at least some part of the lignin-degrading enzyme complex is inducible.

INDUSTRIAL LIGNIN BIODEGRADATION

Modified lignins are produced in excess of 18 million tons annually in the United States as byproducts of chemical pulping processes.[1] These "industrial lignins" are mostly burned to recover pulping chemicals and to provide fuel. Interest in their biodegradability by various microorganisms stems from a need to understand their polluting potential and their potential for possible bioconversions to more valuable products.

Industrial lignins produced as byproducts of the sulfite pulping processes are termed lignosulfonic or lignin sulfonic acids. During pulping, water solubilization accompanies depolymerization and the introduction of sulfonic acid groups. In 1971 Pandila[52] reviewed the literature on the microorganisms degrading lignosulfonic acids. He concluded that various bacteria and fungi can effect limited degradation, and that mixed cultures may be more effective than pure cultures. In accord with this latter point, Sundman and Näse[59] demonstrated a synergistic action of various pairs of fungi in causing limited degradation of a lignosulfonate. Also, Selin and Sundman[57] showed that a mixture of bacteria and protozoa could degrade the low molecular weight components of a commercial lignosulfonate.

Ferm[23] and Selin *et al.*[58] showed that lignosulfonates were polymerized by extracellular phenol oxidases of white-rot fungi. Selin *et al.*[58] found also that the lignosulfonates

are utilized by these fungi to a limited extent; in fact,
C. versicolor used a low molecular weight fraction, but
not higher molecular weight fractions, as sole carbon and
energy source. Wojtaś-Wasilewska and Trojanowski[62] reported
that lignosulfonates were converted in part to biomass by
wood-decaying fungi in a glucose-containing medium.

   Lundquist *et al.*[50] prepared ^{14}C-labeled lignosulfonates by acid sulfite pulping of synthetic ^{14}C-lignins, and
studied their biodegradation by *Phanerochaete chrysosporium*.
Approximately 30% of the aromatic carbon and 23% of the
side chain carbon in the synthetic lignosulfonates were
converted to CO_2. Chromatographic evidence showed that
the residual radiolabeled material extracted from spent
cultures had also been substantially altered. A portion
of the polymeric, radiolabeled material extracted from
the spent cultures was shown to be further converted to
CO_2 in fresh cultures, although results suggested that a
portion of these samples was resistant to degradation by
this fungus.

   Most of the world's chemical pulp is produced by the
kraft, or sulfate, process,[1] but kraft lignin biodegradation has been studied much less than lignosulfonate biodegradation. Ganczarczyk[25] obtained evidence that most
of the relatively small amount of kraft lignin that enters
activated sludge treatment facilities, where it is removed,
is polymerized and adsorbed rather than being biodegraded.
However, Kawakami *et al.*[34] reported that the aerobic
bacterium *Pseudomonas ovalis* caused significant degradation
of kraft lignin.

   Bouveng and Solyom[6,7] compared the stability of
industrial lignins (both kraft lignins and lignin sulfonates) in a model aquatic ecosystem. Kraft lignins were
more readily degradable than lignin sulfonates. They
found that waste lignins contained two fractions, a readily
biodegradable fraction and a more recalcitrant component.
Biodegradation of 96% of the total organic component of
kraft lignin required 40 weeks in their model system.

   More definitive information about the biodegradability
of kraft lignin has recently been obtained by the use of
radiolabeled lignins. In our laboratory, radiolabeled
kraft lignins were prepared from synthetic ^{14}C-lignins,
and their degradation by *P. chrysosporium* was studied.[50]

Over 40% of the radiocarbon in the aromatic rings and 30% of that in the side chains was converted to $^{14}CO_2$ in standing liquid cultures. As with the lignosulfonates, chromatographic evidence indicated that the residual radio-labeled material in the spent cultures had been substantially altered during incubation. A portion of the residual polymeric ^{14}C-material extracted from the spent cultures was converted to $^{14}CO_2$ in fresh cultures. The total proportion of the kraft lignin converted to CO_2 was greater than with the lignosulfonates; in fact, whether or not any of the kraft lignin was completely resistant to degradation by *P. chrysosporium* was not clear.

Crawford *et al.*[16] prepared uniformly radiolabeled kraft lignins by pulping lignin-[^{14}C]-lignocellulosics from hardwoods. The isolated lignins were incubated under aerobic conditions with soil and water samples from a variety of sites, and evolved $^{14}CO_2$ was trapped and measured. Extents of degradation varied substantially among various soil and water samples; up to 27% of the added radiolabel was converted to $^{14}CO_2$ in one water sample. It was demon-strated with one soil sample that the low molecular weight fractions of the kraft lignin was more rapidly degraded to CO_2 than were higher molecular-weight fractions.

CONCLUSION

In recent years the study of lignin biodegradation has become an increasingly active research area. Important progress has been made, particularly in methodology, with the development of sensitive and definitive assays for biodegradation based on ^{14}C-lignins, and with further refinements in the techniques for characterizing lignins. A big gap in methodology remains the difficulty in preparing large quantities of unaltered lignin for detailed investigations.

The availability of the ^{14}C-lignin assays has already made possible rapid progress in elucidating various micro-biological and environmental aspects; for example, it has been found that anaerobic biodecomposition of lignin to gaseous products apparently does not occur. Use of ^{14}C-lignin assays for biodegradation in elucidating the cultural and environmental factors affecting fungal degradation has revealed: (a) a requirement for a growth substrate during lignin degradation, (b) a critical sensitivity to the level of nutrient nitrogen, (c) a dramatic influence of the level of O_2, and (d) an induction of lignin-degrading enzymes by lignin. Studies with the ^{14}C-assays have also demonstrated that kraft lignins and lignosulfonic acids are substantially biodegradable.

Other investigations have demonstrated a requirement for phenol-oxidizing enzymes in lignin degradation by fungi; these enzymes probably function in a regulatory role.

Chemical and spectroscopic comparisons of sound and fungus-degraded lignins have provided evidence for a unique, oxidative, mode of biodegradation. New approaches, however, must be pursued to decipher the detailed chemistry and biochemistry of the process.

ACKNOWLEDGMENTS

The research from our laboratories cited here has been supported in part by grants GB-41861 and BMS-7301195 from the National Science Foundation.

REFERENCES

1. American Paper Institute. 1975. Wood Pulp and Fiber Statistics, Book 1. *Am. Paper Institute, N.Y.* 119 p.
2. Ander, P. and K.-E. Eriksson. 1976. *Arch. Mikrobiol.* (in press).
3. Bassham, J. A. 1975. *Biotechnol. Bioeng. Symp. No. 5*:9.
4. Bellamy, W. D. 1974. *Biotechnol. Bioeng. 16*:869.
5. Björkman, A. 1956. *Sven. Papperstidn. 59*:477.
6. Bouveng, H. O. and P. Solyom. 1973. *Sven. Papperstidn. 76*:26.
7. Bouveng, H. O. and P. Solyom. 1976. *Sven. Papperstidn. 79*:224.
8. Cain, R. B., R. F. Bilton, and J. A. Darrah. 1968. *Biochem. J. 108*:797.
9. Campbell, W. G. 1952. *In* Wood Chemistry (L. E. Wise and E. C. Jahn, eds.), pp. 1061-1116. Reinhold, N.Y.
10. Campbell, W. G. and J. Wiertelak. 1935. *Biochem. J. 29*:1318.

11. Chang, H.-M., E. B. Cowling, W. Brown, E. Adler, and
     G. Miksche. 1975. *Holzforschung* 29:153.
12. Connors, W. J., J. S. Ayers, K. V. Sarkanen, and J. S.
     Gratzl. 1971. *Tappi* 54:1284.
13. Cowling, E. G. 1961. *USDA Tech. Bull. 1258.* 79 p.
14. Cowling, E. G. and W. Merrill. 1966. *Can. J. Bot.*
     44:1533.
15. Crawford, D. L. and R. L. Crawford. 1976. *Appl.
     Environ. Microbiol. 31:714.*
16. Crawford, D. L., S. Floyd, A. Pometto, and R. L.
     Crawford. 1977. *Appl. Environ. Microbiol.* (in press).
17. Crawford, R. L., T. K. Kirk, J. M. Harkin, and E.
     McCoy. 1973. *Appl. Microbiol.* 25:322.
18. Crawford, R. L., T. K. Kirk, and E. McCoy. 1975.
     Can. J. Microbiol. 21:577.
19. Dagley, S. 1971. *Advan. Microbiol. Metab.* 6:1.
20. Ebel, J., B. Schaller-Hekeler, K.-H. Knobloch, E.
     Wellman, H. Grisebach, and K. Hahlbrock. 1974.
     Biochim. Biophys. Acta 362:417.
21. Erickson, M., S. Larsson, and G. E. Miksche. 1973.
     Acta Chem. Scand. 27:9.
22. Eslyn, W. E., T. K. Kirk, and M. J. Effland. 1975.
     Phytopathology 65:473.
23. Ferm, R. 1972. *Sven. Papperstidn.* 75:859.
24. Flaig, W. and K. Haider. 1961. *Arch. Mikrobiol.* 40:
     212.
25. Ganczarczyk, J. 1973. *J. Water Poll. Control Federa-
     tion* 45:1898.
26. Gierer, J. and A. E. Opara. 1973. *Acta Chem. Scand.*
     27:2909.
27. Hackett, W. F., W. J. Connors, T. K. Kirk, and J. G.
     Zeikus. 1977. *Appl. Environ. Microbiol.* (submitted).
28. Haider, K. and J. Trojanowski. 1975. *Arch. Mikrobiol.*
     105:33.
29. Higuchi, T., M. Tanahashi, and A. Sato. 1972. *Mokuzai
     Gakkaishi* 18:183.
30. Ishihara, T. and M. Miyazaki. 1972. *Mokuzai Gakkaishi*
     18:415.
31. Ishihara, T. and M. Miyazaki. 1974. *Mokuzai Gakkaishi*
     20:39.
32. Ishihara, T. and M. Ishihara. 1975. *Mokuzai Gakkaishi*
     21:323.
33. Kawakami, H. 1975. *Mokuzai Gakkaishi* 21:629.
34. Kawakami, H., M. Sugiura, and T. Kanda. 1975. *Japan
     Tappi* 29:33.
35. Kirk, T. K. 1971. *Ann. Rev. Phytopathology* 9:185.

36. Kirk, T. K. 1975. *Biotechnol. Bioeng. Symp. No. 5*: 139.

37. Kirk, T. K. 1975. *Holzforschung 29*:99.

38. Kirk, T. K. and H.-M. Chang. 1974. *Holzforschung 28*:217.

39. Kirk, T. K. and H.-M. Chang. 1975. *Holzforschung 29*:56.

40. Kirk, T. K., W. J. Connors, R. D. Bleam, W. F. Hackett, and J. G. Zeikus. *Proc. Nat. Acad. Sci. (U.S.A.) 72*:2515.

41. Kirk, T. K., W. J. Connors, and J. G. Zeikus. 1976. *Appl. Environ. Microbiol.* (in press).

42. Kirk, T. K., J. M. Harkin, and E. B. Cowling. 1968. *Biochim. Biophys. Acta 165*:145.

43. Kirk, T. K. and L. F. Lorenz. 1973. *Appl. Microbiol. 26*:176.

44. Kirk, T. K. and L. F. Lorenz. 1974. *Appl. Microbiol. 27*:350.

45. Konishi, K. and Y. Inoue. 1971. *Mokuzai Gakkaishi 17*:255.

46. Konishi, K., Y. Inoue, and T. Higuchi. 1974. *Mokuzai Gakkaishi 18*:571.

47. Levi, M. P. and R. D. Preston. 1965. *Holzforschung 19*:183.

48. Lindeberg, G. 1955. *Z. Pflanzenernähr. Düng. Bodenk. 69*:142.

49. Lüdemann, H.-D. and H. Nimz. 1974. *Makromol. Chem. 175*:2409.

50. Lundquist, K., T. K. Kirk, and W. J. Connors. 1977. *Arch. Mikrobiol.* (in press).

51. Lundquist, K. and R. Lundgren. 1972. *Acta Chem. Scand. 26*:2005.

52. Pandila, M. M. 1973. *Pulp Paper Magazine Can. 74*:T78.

53. Pew, J. C. and W. J. Connors. 1967. *Nature 215*:623.

54. Pew, J. C. and P. Weyna. 1962. *Tappi 45*:247.

55. Ritter, G. J. and J. H. Barbour. 1935. *Ind. Eng. Chem., Anal. Ed. 7*:238.

56. Sarkanen, K. V. and C. H. Ludwig. 1971. Lignins. Wiley Interscience, N.Y.

57. Selin, J.-F. and V. Sundman. 1971. *Suomen Kem. Tied. 80*:11.

58. Selin, J.-F., V. Sundman, and M. Räihä. 1975. *Arch. Mikrobiol. 103*:63.

59. Sundman, V. and L. Näse. 1972. *Arch. Mikrobiol. 86*: 339.

60. Trojanowski, J., A. Leonowicz, and M. Wojtaś. 1967.
     Acta Microbiol. Polon. 15:215.
61. Westermark, U. and K.-E. Eriksson. 1974. *Acta Chem.
     Scand. B28*:204.
62. Wojtaś-Waselewska, M. and J. Trojanowski. 1975. *Acta
     Microbiol. Polon. 7*:77.
63. Wolter, K. E., J. M. Harkin, and T. K. Kirk. 1974.
     Physiol. Plant. 31:140.

Chapter Ten

THE NON-SPECIFIC NATURE OF DEFENSE IN BARK AND WOOD DURING
WOUNDING, INSECT AND PATHOGEN ATTACK[1,2]

D. BIR MULLICK

Pacific Forest Research Centre
Canadian Forestry Service
Department of the Environment
506 West Burnside Road
Victoria, British Columbia

---

[1]Studies of Periderm, IX. See reference 35 for Part VIII

[2]Part of the research reported here was carried out under a University of British Columbia and Canadian Forestry Service cooperation project while the author was seconded (1967-1973) to Faculty of Forestry, UBC.

INTRODUCTION

Lack of knowledge of the nature of a host response when attacked by a pathogen* is the major factor contributing to the inadequate understanding of disease. In this chapter, evidence is presented to show that whenever the functioning of tissues essential to a tree is affected, regardless of cause, non-specific autonomous processes are triggered: of the three processes described here, two involve restoration of vital lateral meristems, namely, the phellogen and the vascular cambium, and the third, blocking of conductive sapwood. These processes involve dynamic metabolic and anatomical alterations, and once triggered, occur automatically in pre-existing totipotent cells.

Understanding the sequence of these alterations would be greatly enhanced if it were possible to compare them, by direct observation, with the normal chemical state of cells. Fluorescence, commonly used in identification procedures in chromatography, was a convenient choice to begin such investigations. We were able to avoid losses and changes of chemicals associated with usual histological techniques by freezing tissues under appropriate conditions, cutting frozen sections and examining them while still frozen by fluorescence and other optical microscopy techniques.[27]

Cork cambium (phellogen) is a tissue essential to trees because it accommodates circumferential growth through seasonal renewal of the impervious outer covering, the phellem. Phellogen is the first living tissue affected during penetration by pathogens. The cryofixation technique[27] showed that whenever phellogen becomes non-functional, regardless of cause, the autonomous process of phellogen restoration, constituting the host component in host-pathogen interactions, is initiated. The process of phellogen restoration entails the formation of periderms of the necrophylactic category.[31] Wound periderm has been assigned only a passive role in defense[48] because the process of its formation before the establishment of phellogen had remained little understood due to technological limitations. According to our findings, phellogen restoration, involving necrophylactic periderm

---

*Pathogen is defined as any agent (microbe, insect, parasitic plant) which causes chronic physiological disorders (pathogenesis) in the host.[28,46]

formation, is an active process in defense interactions, since it is triggered as soon as the functionality of the first living layer of cells is affected.

All three processes are non-specific as to incitant. The processes are triggered singly or collectively, depending upon the site of interaction. Failure to trigger the processes or degree of interference with the processes should determine the degree of resistance or susceptibility. The specific effects of a pathogen, for example, on the process of phellogen restoration may be determined by using, as a control, the process of phellogen restoration in the absence of injury and pathogens, e.g. at rhytidome (bark scales).

Evidence and arguments are presented to show that phytoalexins in general and other known biochemical and structural factors in defense reactions are components of these autonomous processes. The question as to whether the concept of defense is anthropocentric or real is considered. These facts and concepts are essential for understanding the biochemical pathways involved in these processes, and how they are affected by injuries and by environmental and edaphic variations. How these variations in biochemical pathways affect and are affected by specific pathogens is also considered.

Although the three non-specific processes described here were discovered in the absence of pathogenic interactions, their involvement in *Abies* defense reactions against balsam woolly aphid, *Adelges piceae* (Ratz.), was confirmed. In many respects, these host-aphid interactions resemble those of microbial pathogens. The involvement of these host processes against *Phellinus* (Poria) *weirii* (Murr.) Gilbertson, causing root rot in Douglas-fir, *Pseudotsuga menziesii* (Mirb.) Franco, and against *Cronartium ribicola* J. C. Fischer ex. Rabenh., causing blister rust in white pine, *Pinus monticola* Dougl., has been observed.

BALSAM WOOLLY APHID, A MODEL PATHOGEN FOR STUDY OF
HOST-PATHOGEN INTERACTIONS

Our interest in defense reactions arose through research on the microscopic insect, balsam wooly aphid (BWA), which damages only true firs, *Abies* spp., by feeding on the living bark. A number of similarities exist between the interactions of BWA and the microbial pathogens that attack living trees.

BWA completes at least two generations per year on the coast of British Columbia.[20] Immature stages of the aphid are difficult to see, but adults (in North America, partheno-genetic females only) are readily spotted because of the copious white waxy "wool" that covers their bodies (Fig. 12, color plates). The aphid has a high reproductive potential and infested stems often appear white from the density of attack. Crown attack results in gouting of leader and branch tips (Fig. 2), causing deformation of apical buds, loss of foliage and dieback.

After hatching, BWA undergoes three moltings to become an adult female. Only the first instars (Fig. 11, color plates), prior to settling and entering diapause, are motile and are referred to as crawlers. Second and third instars and adults are sessile. The stylets in all instars are ca. 1.5 mm long; their 3 μm diameter is thinner than plant cell walls (Fig. 1).[16]

The crawlers insert their stylets into the bark, develop fringes of wool (Fig. 11) within a few days, and go into diapause which lasts 3 to 8 weeks in summer and approximately 28 weeks in winter (overwintering), depending on local wea-ther and other factors. After termination of diapause, the first instars feed for a few days and then undergo three moltings at intervals of about 10 days, giving rise to second and third instars and adults, all of which insert stylets at the original feeding zone.

Penetration of the host by the stylet is generally inter-cellular and occasionally intracellular. The depth of the feeding zone is limited by the length of the stylet, but the salivary effects may reach vascular cambium and sapwood.

BWA is spread by wind dispersal of eggs and crawlers. Host selection is not involved. The crawlers, wherever they land, may insert their stylets into anything penetrable, but they die in the absence of the specific host, *Abies*. BWA penetrates and feeds on both *Abies* and other conifers, but it dies before completion of its life cycle on resistant hosts and on non-hosts. The similarities between BWA and microbial pathogens are numerous and include dispersal patterns, and the conditions needed for successful development.

The salivary secretion of aphids and other insects con-sists of various enzymes and metabolites, such as a

polyphenoloxidase, pectin polygalacturonase (PGU), other
pectinases, and probably cellulases that are associated with
intercellular penetration by the stylets.[21] In addition,
several amino acids, 3-indoleacetic acid (IAA) and gibberellic
acids found in saliva of various insects are likely involved
in pathogenic interactions. BWA is too small for ready analysis of its saliva. Adams and McAllan[1] demonstrated PGU in
extracts of BWA. As suspected by earlier investigators,[6] BWA
saliva shows auxin activity.[34]

Balch,[5] Oeschssler[33] and Saigo[38] have described and
reviewed the nature of BWA-host plant interactions. Briefly,
cell walls near the stylet thicken and nuclei move toward the
stylets. The crawlers fully insert their stylets before
entering diapause and cause abnormal growth (hypertrophy) of
bark tissues (Fig. 3) before feeding commences after termination of diapause.[5] Feeding is carried out by repeated partial
withdrawal and re-insertion of the stylets in a new direction,
giving rise to an approximately spherical pocket of tissue
forming the feeding zone. In gout formation, salivary secretions stimulate the apical meristem. Hypertrophy of the cells
around the stylet is believed to be induced by auxin-like compounds in the salivary secretions,[6] and results in the formation of giant cells of irregular shape, with diameters 6 or
7 times those of normal cells. Secretions also affect the
sapwood, causing the formation of "rotholz" or red wood (Fig.
4), which is impervious to water and is weak in fibre strength.
Certain microbial pathogens also induce cell wall thickenings,
movement of nuclei and the ubiquitious hypertrophy.

Ten years ago, we began an investigation to determine why
BWA damages only true firs and not related indigenous conifers, and whether variations in the bark chemistry might
account for the variation in susceptibility. However, considerable work on the chemistry of other host-pathogen interactions indicated that the problem was complex; a compound
responsible for susceptibility still remained to be isolated.[41]

The emphasis shifted from the comparative chemistry of
bark to a study of the nature of defense using BWA. While
individual trees may succumb to BWA attack, others of the
same age and species, on the same site and, at times, with
foliage intermingling with that of a heavily infested tree
may remain 'resistant'. Where environmental and edaphic conditions are constant, the variation in susceptibility is probably
due to genotypic variations, or to differential stress from

Figure 1.   Electron micrograph of a cross-section of the crumena of BWA containing the looped stylet bundle.   Each central duct contains 3 dentrites.   CD, central duct; FdC, food canal; MdS, manibular stylet; MxS, maxillary stylet; SC, salivary canal (from Forbes and Mullick[16]).

Figure 2.   Gouted tips of an *A. amabilis* twig with all lateral growth stopped.

Figure 3.   Cryofixed radial section of *A. grandis* bark through the feeding site of a BWA first instar, 56 days after settling.   The stylets (tailless arrows) are traceable to the circular zone of hypertophied cells, the boundary of which is marked with tailed arrows.   The relatively light appearance of cells adjacent to the stylets is indicative of modifications in these cells.   fep, first exophylactic periderm; vc, vascular cambium.

Figure 4.   Cross-section of an *Abies* species stem that had been heavily attacked by BWA.   The *rotholz* or red wood (rw) below the bark (b) is sharply contrasted with the normal wood (nw).

Figure 5.   Cross-section through the stem of a mature *A. amabilis*.   The deeply fissured rhytidome (r) layers are separated from one another and from the living bark (lb) by reddish-purple necrophylactic periderm (np).   vc, vascular cambium; x, xylem.

Figure 6.   The same section as that in Figure 26 after thawing.   Most distinguishing characteristics of all tissues including NIT are lost under these conditions.   All cell contents are gone and all cell walls would fluoresce more or less strongly.

Figure 7.   A scanning electron micrograph of a bordered pit from unaffected conducting sapwood.   The pit membrane appears as an open unobstructed network.

Figure 8.   A scanning electron micrograph of a bordered pit from the non-conducting sapwood below a deep injury.   The entire pit is heavily encrusted.

other less apparent primary causes, such as root diseases.
In the absence of other primary causes, the genotypic variation should be the determinant of susceptibility, because the
apparent constancy of environmental factors would still dif-
ferentially affect gene expressivity in resistant and susceptible genotypes. What is the basis for this tree to tree
variation and how should it be studied?

      Patterns of population buildup vary in stem-attacked
trees. On the west coast of North America, BWA may heavily
infest the stem from base to a certain height, past which
the bark remains free of the aphid. What is the basis of
these physiological differences in the bark? Furthermore,
some heavily infested trees succumb, while others recover
after years of attack. What is the basis of recovery?

      Lack of understanding of the nature of disease fre-
quently continues because of lack of knowledge of the host
component in host-pathogen interactions. Understanding
entails isolation of the nature of the host response when
attacked by a pathogen.

CHEMISTRY AND LOCALIZATION OF REDDISH-PURPLE PIGMENTS
OF NECROPHYLACTIC PERIDERMS

      A search for characteristics of host response in the
'non-induced' state began with procurement and examination
of BWA feeding sites in the stem bark of *Abies amabilis*
(Dougl.) Forbes and *Abies grandis* (Dougl.) Lindl. The microscopic size of the stylets made delineation and large-scale
procurement of feeding sites difficult. Reddish-purple pigments in the periderm surrounding the feeding sites[5] were
chosen as markers. Feeding sites, usually ovoid, are readily
seen because of the conspicuous pigmentation in slices through
the gouted leader and branch terminals. Slicing through the
feeding site showed that the reddish-purple wound periderm
surrounded a yellowish-brown interaction zone (Fig. 15, color
plates).

      Reliable delineation of feeding sites on the stem, using
pigments as markers, posed problems because reddish-purple
pigments were also found at varied sites in bark unrelated to
BWA attack. Observations were undertaken to clarify the basis
of reddish-purple pigmentation in the two *Abies* species mentioned above, and in *Thuja plicata* Donn and *Tsuga heterophylla*

(Raf.) Sarg. In these species, the pigments were normally
found at sites of abscission, e.g. needle scars and cone
scars, as long as first periderm persisted. They were also
found abutting old resin blisters in *Abies* species. The
major sites of pigmentation were in areas of rough and dead
outer bark; here pigments were found abutting bark scales,
i.e., rhytidome. In addition, they were found at healed
sites of bark diseases, regardless of the cause.

Some of the foregoing observations suggested injury as
the cause of pigmentation. Accordingly, the bark was mechanically injured to the depth of the vascular cambium in early
summer. The pigments were found three months later on the
surface of healed-over tissues in the four species. Initial
attempts at precise localization of the pigments by the conventional histological techniques in the healed-over tissues
were unsuccessful, because the usual kill-fixes and tissue
dehydrants destroyed or extracted the pigments.

A large array of chemical compounds are known to give
rise to reddish-purple coloration. The literature supported
the view that wound periderm, and therefore its pigmentation,
could have been induced under the specific influence of BWA
salivary secretions or their breakdown products. Oechssler[33]
reported that periderm around the BWA feeding sites was
induced largely under mass attack and rarely under spot attack.
She concluded that sufficient buildup of salivary toxins in
the feeding zone was necessary for induction of wound periderm. It has also been reported that wound periderms around
feeding sites of scale insects which, like BWA, are sessile
and feed at a single site, arise in response to salivary
toxins.[10,11] Similarly, wound periderms around fungal and
other microbial lesions were believed to be induced in response to microbial toxins[2,17] or to death of cells or their
breakdown products resulting from mechanical injuries.[9] Thus
for reliable procurement of feeding sites and for understanding whether the response resulting in pigmentation was specific to causal agents, it was essential to determine if the
pigments at varied sites were or were not chemically identical.

Chemical Analysis of Reddish-Purple Pigments. The chemical nature of reddish-purple pigments from the bark of several
conifers has been investigated extensively (see references
cited by Mullick[23]). It was believed that the pigments,
referred to as phlobaphenes before our studies,[23] consisted,

within a species, of a single complex polymeric compound.
The pigment from *Abies magnifica* A. Murr. was believed to be
a low molecular weight polymer of cyanidin; that from *T.
heterophylla* was either identical or closely related. Some
investigators concluded that the pigmentation of *T. heterophylla* was not attributable to cyanidin as such, or to related anthocyanins. The pigment of *T. plicata* was reported
to be a heterogeneous polyphenolic-polyester containing an
organic cation and an organic anion. It reportedly occurred
as sheets in the bark of *T. plicata* and in phelloderm tissues
of *T. heterophylla*.

We were interested in a chromatographic comparison of
the pigments from varied sites within a species. By development of new solvent systems, the pigments were successfully
resolved on thin layers of microcrystalline cellulose.[23] The
pigments constituted a new class of *non-anthocyanic* reddish-
purple compounds. Contrary to established beliefs, they
consisted not of a single compound, but of mixtures of several
distinct compounds. *A. amabilis* contained at least six of
these new compounds, *T. plicata* at least eight and *T. heterophylla* at least nine; the nature of pigments varied among
species.

Although the major portion of the pigments was non-
anthocyanic, a minor portion consisted of anthocyanidins,
occurring *as such* in nature.[24,25,26] The reddish-purple pigments, non-anthocyanic and anthocyanic, at varied sites of
their occurrence within a species, regardless of the cause,
were identical;[31] thus their induction at feeding sites was
not associated specifically with salivary secretions of BWA.
The identity of the pigments at the sites of wounding or
diseases with those in the absence of wounding and diseases,
e.g. old resin blisters and around rhytidome, merited special
attention. It meant there was an endogenous process of pigment formation that could occur in the presence or absence of
injuries. It was, therefore, essential to determine, unambi-
guously, the localization of the pigments in order to understand the nature of the physiological process with which the
pigment formation was associated and why it took three months
to develop. Since pigments were found in periderms, a brief
description of periderm, as it was understood at the time
this research was initiated, follows.

Prior Concepts of Periderms. Our understanding of periderms was largely descriptive and resulted from histological

studies. The physiology of periderm formation was virtually
unknown. In general, periderm replaces the epidermis, usually
after the first year's growth, in most woody plants possessing
secondary growth. Periderm usually consists of three tissues,
phellem (cork), phellogen (cork cambium) and phelloderm,
which may or may not be present in all plants. The theory
of periderm formation was advanced about 125 years ago.[45]
According to this theory, both phellem and phelloderm are
produced *de novo* by phellogen. The mechanics of cell division in periderm formation are illustrated in Figure 9.

        In general, phellem cells outnumber phelloderm cells in
most plants. Phellem cells die at maturation, while phellogen and phelloderm remain alive. Phellem walls develop suberin
in the course of maturation, making them impervious to water.
The mechanics of wound periderm formation are similar, except
that, in some cases, the wounded surfaces may develop suberin
and wound gums before wound periderm formation.[4,7,8,48]

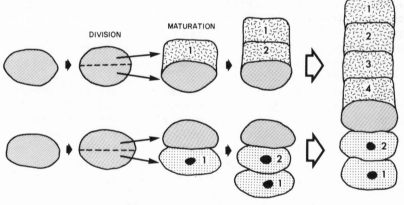

PHELLEM FORMATION

PHELLODERM FORMATION

        Figure 9. Formation of periderm by the phellogen. After
division either the external or the internal cell remains
meristematic. In the former case, the remaining (internal)
cell differentiates into a phelloderm cell, in the latter
the remaining (external) cell differentiates into a phellem
cell. Phellem production is generally greater than phelloderm production.

The periderm may arise, depending upon the plant species, either in the epidermis, in cells immediately below the epidermis or in cells several cell-layers below the epidermis. This periderm was termed *first periderm*.

The bark of trees generally stays smooth as long as the first periderm persists. The first periderm in most plants with secondary growth is replaced by "sequent" periderms, which arise at varying intervals of time in successively deeper layers of bark. The tissues external to the sequent periderm die and are referred to as rhytidome. This results in the rough appearance of bark (Fig. 5). The causal factors in the origin of periderms, in spite of extensive studies, were not understood. However, the first and sequent periderms were believed to be alike, differing only on the basis of time or origin rather than differences in anatomy, chemistry or site of origin. The differences in the site of origin pertained only to the first periderm.[14,15,45]

Wound periderms were believed to differ from 'natural' periderms mentioned above, because they arose under the stimulus of injury as well as factors other than those responsible in the origin of 'natural' periderms.[2,7,8,9,45] Wound periderms, sometimes referred to as pathological periderms, were believed to be induced under the specific stimulus of microbial toxins and insect saliva.[2,11,17,33] Bloch[9] cautioned that induction of wound periderms could have resulted from chemical substances released from death, or from breakdown products of host cells alone in the host-pathogen interaction zones. Periderms developing under normal physiological stimulus, for example, abscission of plant parts such as leaves and branches,[14,18] were variously referred to as secondary protective layers, cork or simply periderm. It was, however, believed that physiological factors in the origin and development of wound periderm as well as periderm around leaf-fall differ from those in natural periderms (i.e., first and sequent periderms) (see review by Srivastava[45]).

New Concepts of Periderms. The chemical identity of reddish-purple pigments at varied sites provided the stimulus for their localization in tissues. A cryofixation technique[27] was used for localization of the pigments. It depends upon fixation of bark tissues by freezing and microscopic examination of frozen cryostat sections in frozen state by fluorescence and other modes of optical illumination. It does not require pretreatments or staining of tissue sections. It showed that whereas the phellem contents of the first periderm were dark brown, those of the sequent periderm were reddish-purple. Furthermore, the pigments occurred in all four species in the phellem of wound and pathological periderms, regardless of the causal agent of injury, biotic or abiotic. Additionally, the pigments occurred in sequent periderms, in secondary protective layers at the site of abscissed organs and beneath old resin blisters. The fifteen parameters provided by the cryofixation and chemical techniques for characterizing reddish-purple periderms were identical within a species at all these sites.[30,31]

On the basis of cryofixation characteristics and chemical analysis, we established conclusively that within a species the normal sequent periderm, wound and pathological periderms of any origin, the secondary protective layer at abscission scars and periderms beneath old resin blisters were all identical. Thus, the usage in the literature of the terms "sequent," "wound" and "pathological periderms" obscures their identity with one another. These periderms constituted a single category and were named necrophylactic (*necrus*, dead; *phylaca*, a guard) periderms because they were found next to dead tissues and protected living bark from adverse effects associated with death of cells and diseased tissues.[31]

Contrary to the common belief that first and sequent periderms were similar, we found, based on chemical and cryofixation characteristics, two distinct kinds of sequent periderms, one with reddish-purple phellem and another with dark-brown phellem.[27] The dark-brown sequent periderm was identical with the first periderm, also dark brown, thereby making the latter term incomplete. The two brown periderms constituted a second category of periderms, the exophylactic (*exo*, external; *phylaca*, a guard) periderms,[31] which are distinguished as first exophylactic periderms and sequent exophylactic periderms. The latter develops only abutting necrophylactic periderm (NP). It appears to be associated with exfoliation or sloughing of dead and diseased tissues, and develops only rarely in trees with adhering dead bark.

A. amabilis, *T. plicata* and *T. heterophylla* were studied in detail to establish the concepts of exophylactic and necrophylactic periderms. However, prior to this work, differences in the natural pigmentation of the first and sequent periderms of 40 species of conifers belonging to 13 genera were examined. Periderm pigmentation varied among species. Within a species,

the pigmentation of the two categories of periderms was distinct.[27] In addition, examination of the pigmentation of
wound periderms, which were available in most of the 40 species, showed that within a species it was identical with
pigmentation of the sequent periderm. This survey was, therefore, instrumental in generating a belief in the general
validity of our concepts. Doppelreiter[13] and Schellenburg[39]
validated these concepts in *Pinus monticola* Dougl. and *Picea
sitchensis* (Bong.) Carr., respectively, and Soo[44] in representatives from all four families of conifers, as well as
several deciduous trees.

   *Significance of Periderm Concepts in Host-pathogen Inter-
actions*. The above findings led to the conclusion that NP
and, therefore, the reddish-purple pigments at healed sites
of insect-plant, pathogen-plant and parasitic plant-plant
interaction zones arose not as a specific response induced
by the toxins of the agent of injury, but as an inherent,
non-specific response of the host, presumably against any
agent causing cellular damage or death.

   Cellular damage as a common causal factor in the origin
of NP was a reasonable hypothesis for injuries, diseases and
even abscission scars. However, the validity of this hypothesis in the origin of NP at rhytidome was questionable,
because wounding and disease agents were not involved and
because of the consistent statement in the literature that
periderm was the cause and not the result of death of external bark tissues.[14,45] We felt that there must be a common
cause in the origin of all NP's, and that stimuli resulting
from injury, death of cells, breakdown products of cells,
wound hormones, or specific effects of microbial and insect
toxins (see reviews by Bloch,[7,8,9] Lipetz,[19] Akai,[2] and
Carter[10]), either singly or collectively, might be triggering
a common mechanism. The answer we felt lay in understanding
the reason for NP development at rhytidome. We knew that NP
initiation occurs three to four weeks after injury in summer
and that maturation of NP phellem and appearance of reddish-
purple pigments in the phellem takes about another six weeks.
There must be a stepwise sequence of events in the process
of NP formation. Awareness of the process at rhytidome would
provide us with knowledge of the process in a "non-induced,"
pathogen-free state. It became clear that we had started
looking at the end stage of the process, and we would have to
backtrack to determine when and how it was initiated.

DISCOVERY OF A NON-SUBERIZED IMPERVIOUS TISSUE: A
LANDMARK IN UNDERSTANDING THE PROCESS OF NECRO-
PHYLACTIC PERIDERM FORMATION

A discovery that greatly aided our pursuit of common
causality in the origin of NP was that of a non-suberized
impervious tissue (NIT). We observed in the course of ultrastructural studies that osmium tetroxide fixative failed to
penetrate the BWA feeding zones, which are surrounded by NP,
and, in some instances, those zones where periderm was absent
or incompletely formed. Could this non-periderm impermeability be responsible for cutting off BWA diseased tissues from
living bark and therefore be associated with a defensive response of the host? If so, was it a non-specific response of
the host, or was it specific to interactions resulting from
BWA saliva with the host tissues?

In the summer of 1968, we found that wounding induced
non-periderm impermeability. Briefly, shallow mechanical
injuries were made on the stem bark of A. grandis, A. amabilis, T. heterophylla, and T. plicata; 3-cm diameter disks of
bark, with an injury in the center, were removed with an arch
punch on alternate days to study the time required for the
impermeability development. The impermeability was detected
by an F-F test[28] which is based on the premise that radial
diffusion of test solutions through the bark stops at the
boundary of the impervious zone. The test entails permeation
with a 2% $FeCl_3$ solution, followed by permeation with 4%
$K_3Fe(CN)_6$, resulting in blue coloration of the permeated zone.
The apparatus of F-F testing by diffusion through the cambial
surface towards the periderm or through a wound surface towards
the vascular cambium is shown in Figure 10.

For the first two years of research, the F-F test was
carried out only through the cambial surface, and resulted
in blue coloration of the entire sample, including the injured
zone, until the development of impermeability, when the test
solutions stopped at a concave zone some distance from the
injury (Fig. 16, color plates). The impermeability around
injuries in A. amabilis, A. grandis, T. plicata, and T. heterophylla developed in three to four weeks in summer.[28] Cryostat sections through such injuries showed that the stain had
stopped at the internal boundary of a zone of enlarged cells.
Critical examination failed to reveal any special feature of
the enlarged cells ascribable to impermeability (Fig. 6). In
general, they did not show suberization but occasionally

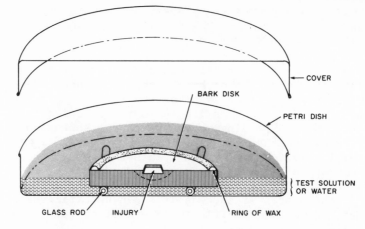

Figure 10. A sectional view of the arrangement for F-F testing bark samples through either the wound or the cambial surface. When testing through the wound surface, the test solutions are pipetted onto the injured area. When testing through the cambial surface, the test solutions are placed in the petri dish (from Mullick[28]).

showed the sporadic presence of necrophylactic phellem abutting the impervious boundary. Initiation of phellem formation along the entire boundary usually had occurred about two days after the initial detection of impermeability. It began to show faint staining for suberin, the substance responsible for phellem impermeability.

The short interval of time between development of first and second impermeability, resulting from the phellem, required that the first impermeability be characterized conclusively. Two summers' work led to the conclusion that the phellem in the èarly stages of differentiation did not appear to be suberized. The substances responsible for impermeability failed to dissolve in either ethyl acetate or petroleum ether, or in 50% H_2SO_4, a solvent used in a study of impermeability layer in barley seeds.[29] The possibility that the imperviousness resulted from deposition of substances such as phlobaphenes, phenolic polymers and wound gums, which arise under the stimulus of injury and impregnate cell walls and intercellular spaces, as reviewed by Wood[48] (p. 464, 465), seemed remote because we were not aware of any report implicating the impermeability of such substances to water.

It occurred to us that perhaps the unstained brown tissues (Fig. 16) as a whole were responsible for the impermeability. However, F-F testing through the wound surface toward vascular cambium showed that most of the brown zone was permeable and the test solutions stopped at the narrow brownish concave zone denoted as NIT in Figure 17. This brownish zone remained impervious when the F-F testing was conducted simultaneously through the wound and cambial surfaces, and was observed to consist of only a few of the internal layers of enlarged cells.

A clearcut delineation of the nature of the impervious zone was made possible by the cryofixation technique. The cryofixed section in Figure 18, prepared at the time of impermeability development, shows that the brownish concave zone responsible for the impermeability consists of a strongly fluorescent tissue. Phellem differentiates in the dark zone (double-headed arrow) only after the formation of the impervious tissue. A sudan-stained section from a sample with well-developed NP that was F-F tested through the wound surface is shown in Figure 14. Seven tests for suberin were used to ensure that the tissue was non-suberized; all were negative, hence it was named non-suberized impervious tissue (NIT) to distinguish it from the suberization of injured surfaces prior to wound periderm formation in tuberous crops.[4] NIT was found at all mechanical, freezing and heating injuries before NP formation in the four species.

The differentiation of NP around BWA feeding sites may be complete in some cases (Fig. 15), and partial in others. F-F testing showed that test solutions did not penetrate the feeding zone even where the periderm was absent. In these regions, NIT prevented penetration.

Initial investigations of NIT at rhytidome were unsuccessful because of difficulties in the sectioning of tissues external to the latest NP. We later discovered from work on mechanical injuries that NIT is detected readily only for a few weeks after its formation. Thereafter, NIT is difficult to recognize because its structure becomes highly compressed and distorted from dessication, and the fluorescence of its walls is lost. The injury work had shown that NIT preceeded formation of NP, but there was no method for external detection of recently formed or forming NP abutting the innermost layer of rhytidome (Fig. 5). We developed criteria for

detecting the initiation of rhytidome in young smooth-barked
T. heterophylla (see next section) and were able to detect
NIT (Fig. 19).

The awareness that post-NIT changes can obscure NIT
structures enabled us to detect NIT at various sites of
abscission or disease in the two *Abies* species, *T. heterophylla* and *Pinus monticola*.[28] Recent studies of Soo[44] have
confirmed the occurrence of NIT in representatives from the
four conifer families and several deciduous trees.

The developmental studies of mechanical injuries showed
that NIT invariably preceded the formation of NP, and that
the latter developed specifically from tissues internally
abutting NIT. The same points were confirmed from studies on
the development of rhytidome (see next section). In preliminary developmental studies, the same pattern was observed
at abscission scars. At all other sites of NP, where developmental studies were not undertaken, the periderm was always
found abutting NIT. Never was NP observed without NIT, though
NIT was often seen without the NP.

On the basis of (1) developmental studies; (2) the invariable presence of NIT externally abutting the NP, i.e., NP never
found without NIT; (3) the equivalence of all NP's,[31] and (4)
the existence of NIT without the NP, we concluded that the
specific site of all NP formation was the tissues abutting
NIT, and that NIT was an integral part of the process of periderm formation.[28]

NIT arises as a non-specific response of the bark to
injury and disease regardless of the causal agent. The induction of NIT, like the NP,[37] is thus non-specific. The nonspecificity in general, and its formation at rhytidome in
particular, suggests that there may be an inherent process of
NIT formation in host tissues, and that induction of the process may result from a common cause rather than from the
varied external stimuli which the literature has associated
with the induction of wound and pathological periderms.[2,7,8,19]

Rates of NIT development after wounding vary at different
times of the year.[32] For example, in June, NIT formation may
take about 20 days, and the cells that form NIT are detectable
about five days before complete formation. Thus, beginning
from the reddish-purple pigments which develop after about
three months of injury, we had retraced the process through

periderm and NIT. What remained to be found was what happens
in the first 15 days, i.e., to determine the earlier steps
to NIT formation.

RHYTIDOME DEVELOPMENT: A PROCESS OF NECROPHYLACTIC PERI-
DERM FORMATION AT A PATHOGEN-FREE 'NON-INDUCED' SITE

Attempts at external detection of.rhytidome initiation
in the deeper layers of rough bark would have been a futile
task. The initiation should have been detectable on smooth
bark where the first exophylactic periderm persisted. In
A. *amabilis*, rhytidome appears infrequently and only after
30 to 50 years. In *T. heterophylla*, however, a large number
of small rhytidome patches appear annually after about 12
years. Accordingly, observations on *T. heterophylla* were
initiated in 1969. In 1972, we were able to develop parameters for detecting the initiation of rhytidome on previously
washed bark surfaces. Rhytidome initiation occurs in smooth
bark adjacent to previously formed rhytidome during May and
early June. The first recognizable sign is a slight swelling
in small areas of various shapes, which soon develop off-
white discolorations due to loss of chlorophyll. The off-
white zones become light-brown in about 10 days (Fig. 20).
In another two weeks, the zones become dark-brown, gradually
desicate and adhere as bark scales, the rhytidome.

To understand the interrelationship between morphological and biochemical transformations in cell contents and
walls that occur during the process(es) of NIT, NP and rhytidome formation, we must relate the sequence of biochemical
changes to cellular and morphological alterations. For this,
we need to know the normal chemical states of cell contents
and cell walls and how they are sequentially modified. Conventional histological techniques, although useful in studying cellular transformations during growth and development,
shed little light on the *in vivo* biochemical dynamics, because
of denaturation and loss of chemical contents from cells
caused by chemical fixation, dehydration and embedding. Biochemistry is also limited in studying physiological processes
because of the lack of technology for localizing the site of
isolated chemicals, which is essential for delineating their
function. Cellular fractionation and histochemical techniques
can localize only a relatively few chemicals, while the bulk
are lost in processing. Thus, new technology was necessary.

We felt that our understanding of the physiological basis of swelling and the sequence of color changes in the bark during rhytidome development could be advanced if it were possible to observe by fluorescence microscopy the broad chemical characteristics of cell contents in their original state before, and during, rhytidome development. The idea was similar to that employed in fluorescence analysis of chromatograms. Should such a technique work, its uses would be numerous. For example, we could establish stepwise fluorescent landmarks resulting from cellular alterations in the course of a morphogenetic process. The starting point in delineating a process is the establishment of as many visual landmarks, here fluorescence changes, as possible.

The need for developing such a technique arose originally from our studies on leucoanthocyanins and related polyphenols in normal, mechanically injured and diseased (BWA feeding zones) tissues. We observed dynamic transformations in these compounds and were tempted to pursue them because several phenolics, after oxidative polymerization, inhibit polygalacturonase, an ubiquitous exoenzyme of microbes also found in BWA. We gave up the pursuit because of difficulties in establishing their direct relationship to susceptibility or resistance. We felt biochemical studies should only be undertaken after we were able to establish in the interaction zone some visual chemical landmark associated with resistance or susceptibility.

In the knowledge of the complex mixture and distribution of chemicals within the cell, we felt that, if tissues were frozen appropriately, the contents of a cell could be frozen and *fixed* virtually in their original state. In a frozen state, their spatial distribution would remain almost unaltered and it might be possible to look at the broad chemical nature of cellular contents by fluorescence and other modes of illumination.

Efforts in this direction led to the development of a cryofixation technique,[27] by the use of which we were able to localize the site of reddish-purple pigments, discover differences between the first and the usual sequent periderms and establish the nature of non-periderm impermeability. Improvements in the technique since 1971[27] have revealed many natural characteristics of cells, obtained without staining or any other treatment, which permit physiological interpretation of the anatomical aspects of growth and development.

An example is the delineation of three states of phellogen
through their fluorescence characteristics detailed below.

Using the cryofixation technique, we studied the process
of rhytidome development on samples of newly forming rhytidome which were selected at the earliest detectable stages
in May, 1972 and labeled as the zero-day stage at that time.
Cryofixed sections were examined with various modes of illumination, using a Carl Zeiss Photoscope II, as described by
Mullick and Jensen.[31]   The phellogen may exist in quiescent
or active states. The quiescent state in cryofixed sections
is distinguishable on the basis of the distinctive fluorescence of the cell contents (Fig. 21). Phellogen becomes
active for phellem renewal mainly in May and early June on
the west coast of British Columbia. The renewal is necessary
for providing new phellem to accommodate the annual increase
in circumference. The initiation of the active state can be
readily determined on the basis of fluorescence modifications
undergone by the contents of the quiescent phellogen, as well
as the increase in its size (Fig. 22). Neither the active
nor the quiescent phellogen shows fluorescence in the walls
at any time.

We observed that rhytidome development starts mainly in
May and June, and that phellogen renews phellem at about the
same time.   Cryofixed sections from sites of rhytidome initiation showed that whereas the phellogen layer in areas of
bark surrounding the discolored area was usually active in
phellem renewal, in the discolored whitish zone, it was neither quiescent nor active and possessed fluorescent cell
walls (Fig. 23). The intensity of fluorescence in the walls
increased during rhytidome development. The non-functionality
of the phellogen with fluorescent walls in renewal of phellem
was confirmed by examination of over 100 samples.

The non-functional phellogen is an obvious threat to the
survival of the plant. Without phellogen, a perennial plant
with secondary growth can not renew phellem for accommodating
circumferential growth. How and why certain zones of the
phellogen become non-functional remains to be resolved, but
the observations permit the conclusion that after the development of fluorescent walls, phellogen never divides and eventually dies. Thus the plant initiates the process of phellogen restoration. This is what occurs during rhytidome development.

A cryofixed section from a newly forming rhytidome, three days after detection, with extensively hypertophied cells, as revealed by fluorescence, is seen in Figure 24. On the left side is seen the fluorescence of normal bark tissues. Here the phellogen is either quiescent (left-hand corner) or is becoming active toward the boundary of the hypertophied zone (arrows). The cortical parenchyma (c) have bright yellow fluorescent contents. The phloem parenchyma (b) contents fluoresce green. They are very small near the vascular cambium, but gradually increase in size toward the cortex. The dark zones between the strands of phloem parenchyma consist of sieve elements. Ray parenchyma (r) usually have brighter fluorescence; sclereids fluoresce green.

The phellogen over the entire developing rhytidome zone is non-functional, as indicated by its fluorescent walls. Soon after the initiation of rhytidome, the normal cells had undergone structural and chemical changes (as indicated by fluorescent modifications) and reached the state of hypertrophy seen in Figure 24. This zone arises entirely from pre-existing cells. Normal cortical cells show red fluorescent chloroplasts, which decrease in the rhytidome zone, causing visual discoloration of the bark at the time of rhytidome initiation.

The process by which a mature differentiated cell gives up its original function and undergoes structural and chemical transformations to assume another function is termed dedifferentiation. The dedifferentiated cell enters a meristematic state and redifferentiates to another cell type. These concepts are based on the totipotency of plant cells.

In the beginning, the dedifferentiating cells show cell enlargement and development of varied fluorescent contents, as seen in Figure 23; some of the cells develop fluorescent reticulum (fr, Fig. 23), but cell walls, excepting those of phellogen, usually do not develop fluorescence. A zone of 2 to 3 layers of cells below the phellogen does not develop the fluorescent reticulum (Fig. 23).

Gradually, over the next seven days, the fluorescent reticulum disappears, cells develop varied fluorescent contents and corners of cell walls become fluorescent. The appearance of the dedifferentiating zone after about 18 days of rhytidome initiation is shown in Figure 25. Chloroplasts in the external cells have virtually disappeared.

Dedifferentiation has progressed to a considerable extent in deep areas of the bark but, so far, none of the cells has become embryonic. This entire zone has arisen from pre-existing cells without any cell division. Some of the cells below the outer layers of tissue have developed fluorescent walls. A zone of cells at the internal boundary has become less fluorescent than the surrounding cells. This is the zone where phellem will form. Some of the cells external to this zone show sporadic fluorescence in their walls (arrow). This is the zone of developing NIT.

A 23-day-old sample of rhytidome at the time of NIT formation is shown in Figure 26, with NIT clearly distinguished from other enlarged cells by its brightly fluorescent walls. NIT forms from hypertrophic dedifferentiation and does not involve any meristematic activity. NIT is a complex tissue because it is derived from various cell types, some of which, for example, sieve elements (tailed arrows) and sclereids (tailless arrows, Fig. 26), may become only partially transformed, but they develop substances responsible for impermeability and together act as NIT. Cells in the zone internal to NIT lose all fluorescent contents and become meristematic. NP will develop in this zone.

The appearance of the cryofixed section (Fig. 26) on thawing is shown in Figure 6. The problem of recognizing NIT from other enlarged cells in this section speaks for the value of the cryofixation technique. When such a section is observed under fluorescence after thawing, or as it thaws, the fluorescence of cell contents disappears and all cell walls become fluorescent, resulting in the loss of diagnostic fluorescent signposts.

Although NIT can be recognized with Fl IV/41 (Carl Zeiss fluorescence combination of exciter filter IV and barrier filter 41), it is difficult to distinguish with Fl I/53 (exciter filter I, barrier filter 53) from other enlarged cells after a few days of its differentiation (Fig. 27). Loss of fluorescence from the zone internally abutting NIT is diagnostic for the establishment of meristematic activity leading to the formation of NP. Complete differentiation of phellem usually requires about one month. During differentiation, phellem cells are turgid, and roughly rectangular and they develop yellowish fluorescent compounds[30] and reddish-purple pigments (Fig. 19). As the intensity of reddish pigments increases, the protoplasts of phellem cells disintegrate and phellem

becomes highly compressed and distorted because of the pressures resulting from secondary growth. Rhytidome tissues
dessicate and sectioning becomes difficult, as shown in an
88-day-old sample (Fig. 28). In *T. heterophylla*, *A. amabilis*
and *A. grandis*, additional NP phellem appears to form annually
in order to accommodate the pressures of growth.

The process of rhytidome development thus begins with
the non-functionality of phellogen and, through complex steps
in cellular dedifferentiation, leads to the formation of NIT
and NP, with the attendant restoration of phellogen at a
pathogen-free, "non-induced" site.

PHELLOGEN RESTORATION: THE NON-SPECIFIC COMMON CAUSE OF
NECROPHYLACTIC PERIDERM FORMATION

Detailed studies of the process of NP development around
mechanical injuries in bark of the two *Abies* spp. and *T.
heterophylla* were carried out by cryofixation and conventional
histological techniques. Initiation of hypertrophic dedifferentiation of cells abutting injured zones occurs within a
few hours of injury. A zone of dedifferentiating cells arises
in about three weeks in summer and NIT develops, like rhytidome, at its internal boundary. The extent of tissues involved in hypertrophic dedifferentiation at mechanical injuries
in summer varies, depending upon the physiological state of
the bark. Preliminary analysis of the data indicates that
the process of NIT formation at injuries and rhytidome is
fundamentally the same, differing only in minor details,
depending upon the extent of tissues removed by injury and
the physiological state of bark at the time of injuries. The
invariable formation of NP in the zone internal to NIT and the
fact that NP never forms without NIT is evidence that NIT
provides the specific environment necessary for NP formation.[28]

The function of NIT appears to be analogous to sand-
bagging in a dyking operation. The imperviousness of NIT
cuts off external tissues, which subsequently die. It also
seems to eliminate outside interference with the process of
phellem formation. Therefore, contrary to the literature,
it is NIT, and not periderm, that first segregates injured
and diseased tissues.

We mentioned earlier that death of cells as the cause
of NP formation was justifiable directly or indirectly at

all sites, excepting rhytidome. Studies of rhytidome development showed that although the phellogen was nonfunctional
and was destined to die, death *per se* was not the cause of
rhytidome initiation. Although clusters of cells in the
hypertrophying zone of developing rhytidome may at times die
before development of NIT, the majority remain alive and die
only after its development. What then is the cause if not
death?

Bark swelling is the earliest detectable sign of rhytidome initiation. Observations of cryofixed sections showed
two features; namely, hypertrophy and presence of fluorescence
in the walls of phellogen. Is the common occurrence of hypertrophy at rhytidome, injuries and sites of host-pathogen
interactions the cause of NP development? Why does hypertrophy occur? The literature on wound healing considers that
besides growth hormones, the stimulus of injury and of microbial and insect toxins also induce hypertrophying and phellogen formation. Our studies on injuries show that hypertrophic
dedifferentiation is induced immediately after death. However, death *per se* is not involved in the initiation of rhytidome and, therefore, in hypertrophic response.

Observations by cryofixation showed that at the time of
renewal of the exophylactic phellem, the phellogen enlarges
and its contents become dully fluorescent prior to meristematic activity. Similarly, normal parenchyma is often found
hypertrophied and without fluorescent contents. Hypertrophy
is also observed in phloem parenchyma during sclereid formation. In all these cases, death of cells is not the cause of
cell enlargement or hypertrophy. On the contrary, hypertrophy which, in some instances, is considered an abnormal
response, is the first detectable step in dedifferentiation
leading to formation of new cell types, and thus appears to
be a vehicle for carrying out normal requirements of growth
and adjustment.

Our studies at injuries and rhytidome support this view.
Future work may yield specific landmarks in dedifferentiating
cells that lead to NIT formation.

Non-functionality of phellogen at the earliest stage of
rhytidome initiation, and the process of rhytidome development ending with the restoration of phellogen lead to the
concept that rhytidome formation, in fact, is a process of
phellogen restoration.

The process of NP formation is initiated whenever phellogen is physically removed at mechanical wounds, is damaged or destroyed by agents of disease or is absent, e.g. at abscission zones. Under all these conditions, restoration of the phellogen, a vital tissue necessary for providing continuing protection to plants by seasonal renewal of the phellem, becomes essential for the survival of plants possessing secondary growth. Thus it is the need for restoring phellogen that is the non-specific common causal factor in the formation of necrophylactic periderms, regardless of the site of their origin.

Literature on wound periderm (i.e., NP) is extensive.[7,8,1?] Most of the studies were carried out on agricultural plants, which develop periderm within two to three days. Histological techniques showed only the suberization of wound surfaces, in some cases, and the establishment of mitotic activity associated with wound periderm formation. Biochemical studies yielded the concept that wound hormones, such as traumatin and oxy acids, released under the stimulus of injury, led to the formation of wound periderm. Numerous studies were carried out with chemicals which stimulate mitotic activity, but in spite of extensive investigation of wound periderm formation, the causal factors in its origin and which cells after injury become phellogen remained obscure.

Our studies show that the process of phellogen restoration is a 'non-induced' endogenous process which is triggered whenever the phellogen becomes non-functional. The restorative process is autonomous; once triggered, it requires no external stimulus for its completion. The totipotent cells obviously have the genetic code to dedifferentiate in a stepwise manner for restoring phellogen. Thus, the literature beliefs that cell death, damage to cells, breakdown products of cells, and wound hormones released under the stimulus of injury, growth regulators, agents which stimulate mitotic activity, microbial toxins and salivary secretions of insects, either singly or collectively, induce and/or stimulate the formation of 'wound' and 'pathological' periderms,[2,9,11,17,19,48] failed to clarify the underlying unity of the non-specific common cause in the origin of NP.

NON-SPECIFIC NATURE OF DEFENSE ASSOCIATED WITH PENETRATION OF BARK SURFACE: A PROCESS OF PHELLOGEN RESTORATION

Retracing the phenomenon of reddish-purple pigmentation showed that whenever phellogen becomes non-functional,

regardless of cause, the autonomous, non-specific process of
phellogen restoration involving NIT and NP development is
triggered. Unpublished studies of Soo[44] extend these concepts. Thus, the autonomous process of phellogen restoration
should have a common genetic machinery in all woody plants.
Evidence shows that in trees where the epidermis is wounded
prior to the establishment of exophylactic periderm, there is
no process for epidermal restoration, and phellogen restoration occurs, leading to establishment of NP.

Thus, a deterministic step in establishing the nature
of host response is the initial interaction of the pathogen
with the first layer of living cells. These interactions may
vary from rapid death to virtually no effect at all. If the
effect triggers phellogen restoration either through outright
killing or marked derangement, resistance or susceptibility
should be determined by how successfully the host is able to
complete the restorative process while under a greater or
lesser influence of the pathogen. Pathogens, after triggering
the process, interact not only with the chemicals present in
the normal bark but also with an array of new chemicals produced in the course of attendant ultrastructural and cytological modifications occurring in dedifferentiation. Thus,
whatever the nature of defense, it follows from the foregoing
that defense should be a part of the process of phellogen
restoration.

Susceptibility should also result if penetration is
accomplished without triggering the process because the pathogen would be "unopposed" until possible subsequent derange-
ments triggered the process. Whether or not the process is
triggered initially, pathogenic interactions could modify
cells so that the affected cells would be incapable of carrying out the process.

Trees, being sessile, are prone to considerable damage
by biotic and abiotic agents of injury. Their bark sustains
open wounds of varied origins, where pathogens land at random,
and yet in a majority of cases, wounds heal without disease.
Microbes can break down only a few specific chemical linkages.
The healing of open wounds in bark, usually without disease,
suggests that compounds elaborated in the restorative process
must be toxic or unsuitable substrates for a vast array of
microbes. Thus, relatively few organisms should be able
either to utilize the compounds elaborated in the restorative
process or to avoid triggering the process and, therefore,

only such organisms should be pathogenic. The restorative process could be interfered with at different stages, depending upon the pathogen and the time of elaboration of the compounds it requires. Studies on the specific effect of each pathogen in triggering the process and on the process itself are necessary to understand the basis of disease specificity.

The involvement of classical wound periderms, i.e., NP, in host-pathogen interactions has been extensively investigated. However, they are assigned only a passive role in defense, because of their absence at the time of pathogen inactivation or delimitation. They develop after inactivation, several cells away from the site of pathogen delimitation. They are, therefore, considered only in terms of histological barricades,[48] rather than in terms of the process of their formation, as indicated by our studies. In contrast, our studies with cryofixation enabled us to recognize the development of the yellowish wall fluorescence as an indicator of the non-functional phellogen that gave us the basis for understanding the process of 'wound' periderm, i.e., NP formation. Even if we had isolated the yellow fluorescent compound (Fig. 23) in its native state, it would have been virtually impossible to assign it a role in phellogen disfunction. Our studies show that whereas 'wound' periderm *per se* may not be involved in host-pathogen interactions, the process of its formation should be.

The process of phellogen restoration is non-specific. Similarly, phytoalexin production is non-specific,[47] in particular its production following wounding. This suggested that phytoalexin production was part of the process.

From 1967 to 1970, we carried out studies on the distribution of leucoanthocyanins and other phenolics in normal bark, injured bark and BWA feeding sites. We found only leucocyanidin in the normal bark of *A. amabilis*, but in dead tissues at mechanical injuries and BWA feeding zones, leucodelphinidin and an unknown leucoanthocyanin were also present. Leucoanthocyanins, after oxidative polymerization, are known to be potent inhibitors of polygalacturonase, an enzyme found in BWA.[1] Their induction at injured sites and feeding sites suggested phytoalexin-type activity. At that time, we were not aware of the dynamics of dedifferentiation occurring in "dead" and "dying" tissues surrounding injuries. Subsequently, it became apparent that the leucoanthocyanin transformations

at sites of injury and feeding sites should have occurred in
the process of phellogen restoration. The suspected role of
leucoanthocyanins, as phytoalexins, suggested that phytoalexins in general should develop in the dedifferentiating
cells surrounding injury. Indeed, following this suggestion,
Rahe and Arnold[36] found that the bean phytoalexin, phaseollin,
previously recovered from extracts of the intact stem segments of beans, was localized entirely in tissues abutting
the injured zone. Phaseollin was detected six hours after
injury, reaching a peak at around 24 hours. Studies by cryofixation, using samples provided by Dr. J. H. Rahe, showed
that dedifferentiation leading to the formation of wound
periderm had commenced about four hours after injury. The
appearance of the zone after 23 hours is shown in Figure 31.
Since phaseollin occurred in the dedifferentiating cells,
the question arose if indeed there was a defense process, and
if so, how to isolate it from the process of wound periderm
formation (I have refrained from classifying the wound periderm in beans as necrophylactic periderm because the impervious tissue formed prior to periderm formation in bean stem
is suberized). Viewed in the context of our model, phaseollin
should simply be one of the many compounds formed during periderm formation, and not specifically formed as a defense compound. This view is supported by the numerous reports of the
non-specific induction of phaseollin. The formation of, and
changes in other non-specific compounds, such as ethylene,
cell-wall degrading enzymes, 'early RNase', phenolics, and
terpenoids, may be a consequence of the processes of dedifferentiation leading to NIT and NP formation, rather than
being non-specific defense factors. A similar view is held
of the other dynamic metabolic alterations occurring after
penetration. Biochemical studies on pathogenesis have dealt
with specifics and not the general, obscuring the common base
to a wide, seemingly unrelated number of biochemical interactions and pathways.

The validity of this view, with particular reference to
biochemical studies on injuries, which some have considered
as the ultimate cause of defense, is clear from the following
statement by Allen:[3] "There is a good deal of evidence that
substances released from injured cells are involved in triggering the defense reactions to mechanical injury. These
wound substances, or as they have been called "hormones" initiate processes leading to new cell divisions, and sometimes
to the formation of new types of tissue, particularly cork."

According to our model, the response to injury is not in
defense, but rather a response to the need for phellogen
restoration.

Introduction of foreign chemicals into host cells should
interfere with the stepwise development of this autonomous
process. Evidence of this is presented in Figures 29 and 30.
Here, a BWA, settled for about 10 days on *A. grandis*, was
ringed by needle point injuries. This feeding site, with the
BWA still alive, was examined by cryofixation 60 days after
injury. NP had formed at the sides of injuries but in the
region close to the stylets only NIT was forming (Fig. 30).
When this sample was further sectioned to the zone immediately
adjacent to the stylet, even NIT was absent. The cells in
the feeding zone and processes occurring in them must have
been interfered with by the saliva, because they failed to
differentiate NP even in response to injuries.

Evidence provided so far, together with integration of
literature findings on penetration of pathogens, generate the
view that the process of phellogen restoration should occur
when phellogen is affected as a result of penetration. No
specific process of defense has been so far characterized.
Susceptibility and resistance each should come about under
two different conditions. Susceptibility could result by
either failure to trigger the process or through successful
interference with the process. Resistance would result in
cases where the pathogen was unable to successfully interact
with the process or in cases where the pathogen was inactivated by as yet uncharacterized means prior to, or without
triggering the process. We propose that where the process is
obviously occurring, "defense reactions" are specific interactions of pathogens with this non-specific autonomous process
and not separate processes in themselves.

The process of phellogen restoration thus should constitute a host component in host-pathogen interactions. This is
a measurable process, through monitoring the rate of NIT
formations, against which we can assess the effects of environmental variations on pathogenic interactions. "To determine
the relationship of pathogenesis and NIT formation (phellogen
restoration) an abvious prerequisite is knowledge of the
effect of non-pathogenic conditions on the process."[32] Effect
on environmental fluctuations occasioned by year-to-year
within-season variations generally, and water stress specifically, have been investigated in recent studies by Mullick

and Jensen[32] and Puritch and Mullick,[35] respectively. Since
rhytidome forms only at a specific time of year and pathogenesis occurs at any time of the year, NIT formation at
injury was chosen for the above studies. The rate of NIT
formation was fastest in early summer, slowing in fall to a
virtual halt in winter, with an increase in early spring
back to the peak in early summer (Table 1).[32] Since rhytidome forms on trees too large for controlled laboratory studies, water stress effects were studied at injuries to potted
greenhouse stock.[35] Water stress past a certain level was
found to slow NIT formation in proportion to the level of
stress (Table 2). Thus, conditions known to promote pathogenesis have been shown to adversely affect the process of
phellogen restoration.

Table 1. Rate of NIT formation after wounding at various
times of the year in *Abies amabilis* (abstracted from 32)

Time of injury	Days to impermeability
Jan. 9	147
Feb. 14	109
Mar. 24	70
June 6	14
July 4	16
July 27	23
Sept. 5	35
Nov. 1	214

Table 2. Rate of NIT formation after wounding at different
levels of water stress in *Abies grandis* (abstracted from 35)

Stress level (MPa)*	Days to impermeability
Control	16
1.5	17
2.0	29
3.0	49

*MPa = megapascal = 10^6 Pa = 10 bars.

Recent data (unpublished) have shown that one or more of the components that make up the complex NIT tissue, e.g. cortical and phloem tissues and ray parenchyma, may undergo incomplete transformations as mentioned above, resulting in defects in NIT. Further preliminary findings indicate a strong correlation between susceptibility to BWA and defective NIT formation.

The concept of phellogen restoration suggested that if the functions of other essential tissues were disrupted, there should also exist processes for their restoration. Many pathogens penetrate bark tissues to greater depths than we had studied in establishing phellogen restoration; accordingly, new experiments were initiated. Along with some earlier findings, two additional processes, namely, vascular cambium restoration and blocking of sapwood, were discovered. These are described in the next two sections.

NON-SPECIFIC NATURE OF DEFENSE DURING PENETRATION OF VASCULAR CAMBIUM: A PROCESS OF VASCULAR CAMBIUM RESTORATION

NIT forms from the hypertrophic dedifferentiation of pre-existing bark tissues. It should not form below deep injuries if the remaining tissues are insufficient for NIT formation. This was confirmed with a series of injuries of increasing depth (Fig. 37); the deepest injury caused death of all tissues to the xylem. The bottom sample in Figure 37 shows that the entire bark is capable of forming NIT (arrows) and that NIT was absent directly below the injury.

In further studies, the bark of A. *amabilis* and A. *grandi* was injured to increasing depths to determine the minimum amount of bark needed for NIT formation. In these experiments the conductive integrity of sapwood beneath deep bark injuries was also examined by ascent of 0.5% acid fuchsin solution (see also next section). Acid fuchsin testing showed the presence of a non-conductive light-brown zone directly below the injury (arrow), external to which, in some cases, a strip of conductive xylem had formed (Fig. 35). In deeper injuries, a brown zone abutting the newly formed conductive xylem was found not below, but on either side of the injury, and sapwood below the injury became non-conductive to the dye solution (Fig. 36). In such injuries, a thin brown zone of dead, incompletely differentiated xylem tissue was sometimes present immediately below the injury (Fig. 36).

The origin and the development of the brown zone were
investigated through cryofixation studies of deep bark injuries (*ca.* half-way through bark thickness) made on July 8,
1974, to *Pinus contorta* seedlings. Evaluation of the preliminary data has shown the simultaneous occurrence of two
responses of bark to deep injuries. Besides the initiation
of phellogen restoration, we found that the vascular cambium
beneath such injuries was affected at about the time injuries
were made. While the vascular cambium surrounding the injury
continued to produce xylary and phloic derivatives, that
below the central regions of injury stopped producing xylary
derivatives and, along with the pre-existing phloic zone,
appeared to undergo modifications. There was little enlargement of this cambial and phloic zone until NIT formation,
which took about 10 days. Normally non-fluorescent cell walls
of this zone gradually became fluorescent. After about 16
days, when NP was differentiating below NIT, some cells in
the phloem tier between the periderm and the modified cambial-
phloic zone began enlarging. The role of this enlargement
is under study; it may be associated with the re-establishment
of vascular cambium. By the 21st day, xylem-like cells started
appearing sporadically (Fig. 32) at the external boundary of
the transformed cambial-phloic zone. A cambial zone (tailless white arrow, Fig. 32) began developing between the
sporadic xylem-like cells and the enlarged phloic cells
(tailed white arrow, Fig. 32). The position of the original
vascular cambium at the time of injury is shown by a black
arrow (Fig. 32). The enlarged phloic cells, the sporadic
xylem-like cells and the new vascular cambium zone in the
process of re-establishment are also revealed under polarized
light (Fig. 33). Here, too, the cambial zone appears dark
because of the absence of secondary cellulosic walls in the
newly differentiating cambial zone.

Over the next few weeks, while the regular production of
xylem and phloem by the newly restored vascular cambium
began, the transformed cambial-phloic zone, internal to the
newly developing xylem, underwent dramatic transformations
in fluorescence. Many cells, excepting xylem-like cells sporadically dispersed in the zone, did not conduct acid fuchsin.
This zone, on the basis of its appearance, resembled bark,
especially cortex, and like bark did not conduct acid fuchsin.
Further changes continued to occur in this zone and its appearance, 66 days after injury, is shown in Figure 34.

The cambial-phloic zone is the same brown zone (Figs. 35 and 36) which, in part, led to this investigation. Observations on *Abies* species and *P. contorta* showed that at deep injuries, vascular cambium is restored external to this tissue only after the formation of NIT and NP. Although precise details of the mechanics of vascular cambium restoration are still under scrutiny, the development of the modified cambial-phloic zone appears to be essential for the restoration of vascular cambium. If the vascular cambium at the time of deep injuries had not undergone transformations, the newly-forming NIT could have cracked under pressure resulting from the normal addition of xylem and phloem tissues. In shallow injuries, the modified tissue is not formed.

The observation that normal vascular cambium activity stops without direct injury to the cambium shows that, like the non-specific process of phellogen restoration, there exists a second inherent process, non-specific as to incitant, that comes into play in injuries beyond a certain depth, in the absence of disease agents. Thus, this process should be understood as a control for interpreting interactions of pathogens in this zone.

Preliminary developmental studies using cryofixation indicate that the cambial-phloic zone is derived from modified bark tissues extant at the time of injury. The relationship of this tissue to that of modified xylem tissue formed after injury to xylem[42] remains to be determined. Developmental studies of the latter tissue have not been recorded in the literature.

NON-SPECIFIC NATURE OF DEFENSE DURING PENETRATION OF SAPWOOD: A PROCESS OF BLOCKING SAPWOOD CONDUCTION

The processes of phellogen and vascular cambium restoration occur following deep bark injuries. Knowledge of sapwood response to bark injuries seemed necessary because, should an agent of disease successfully interfere with both processes and enter the sapwood conductive stream, it could lead to further spread of disease. This knowledge would complete our understanding of the continuum of host responses to injuries of varying depths, from exophylactic periderm to sapwood.

Observations on vascular cambium restoration showed that below bark injuries of a certain minimum depth, sapwood did not conduct acid fuchsin solution (Fig. 38). The rate of impermeability development in sapwood was investigated, using *A. grandis* and *A. amabilis* seedlings (about six years old). Injuries were made to three- to four-year-old regions of the stems of the seedlings by slicing off slightly more than half of the bark thickness. The conduction of sapwood below the injured bark zone was tested thrice a week by placing the excised stem of the seedling with intact foliage in 0.5% acid fuchsin. When dye was detectable in the terminal needles, the stem with injured bark was cut open longitudinally through the center of the injury. These studies were carried out with forced flushed stock maintained in a greenhouse in the winter of 1971, and with shadehouse and field-grown trees in the summers of 1972 to 1974. In general, we found that in summer, sapwood directly below injuries remained conductive for at least 16 days and became impervious on about the 19th day (Fig. 38). Studies, using scanning electron microscopy, showed that the imperviousness resulted from encrustation of pits (compare Figs. 7 and 8).

In some instances, even if both phellogen and vascular cambium restoration go to completion, imperviousness may develop in sapwood, but this is not always the case. In injuries that extend to the sapwood or where insufficient tissue remains for both restorative processes to occur, all living tissues in the central area die, phellogen restoration occurs in a roughly cylindrical shape from the exophylactic phellem to the sapwood (Fig. 37) and sapwood impermeability results. The restoration of vascular cambium under these conditions occurs through "rounding off," followed by intrusive growth.

The process in the blockage of sapwood conduction appears to be triggered when an injury reaches beyond a certain depth in the bark. It is initiated in the absence as well as the presence of direct injury to the sapwood and, like the processes of phellogen and vascular cambium restoration, appears to be inherent and non-specific as to incitant.

Rotholz, produced by BWA, is impervious to the conduction of acid fuchsin.[5,22] This suggests the possibility of using formation of impermeability in sapwood as a control with which to compare rotholz formation.

Cryofixation studies show that biochemical transformations as indicated by fluorescence changes also occur in the zone of non-conducting sapwood during its formation. Thus, it is not only the stoppage of dye movement that is significant, but also the potential chemical barriers against pathogens formed during this process. The compartmentalization process[42] occurs over an extensive area and involves plugging of vessels and chemical changes that act as a barrier(s) to decay. Direct demonstration that the compartmentalization tissues are non-conducting is lacking. Similarly, the report[37] of embolisms in the current year's xylem formed adjacent to the galleries of bark beetles on *Pinus contorta* Dougl. did not give direct demonstration of non-conduction. Dessureault and Tattar[12] have demonstrated recently that discolored wood in red maple and a thin zone of normal-appearing wood abutting the discolored wood failed to conduct an acid fuchsin solution. They suggested that this normal-appearing non-conducting wood corresponded to the "reaction" zone of Shain,[40] the "bleached" zone of Shigo and Sharon[43] and the "protection" zone of Shigo and Hillis.[42] However, none of these were developmental studies and direct demonstration of non-conduction in the "reaction," "bleached" or "protection" zone has not been reported. Such demonstration, in conjunction with developmental studies as outlined for the three non-specific processes presented here, is necessary to interrelate the findings

CONCLUSIONS

The studies presented here have led to the discovery of three autonomous processes, initiated during pathogen attack, in the bark of probably all woody plants possessing exophylactic and necrophylactic peridermal habits (Fig. 39). The three processes are non-specific as to incitant. The first process is that of phellogen restoration, which is triggered whenever phellogen becomes non-functional, regardless of cause. This applies to pathogens penetrating through the first layer of living cells on the tree, e.g. phellogen or epidermis. When the extent of penetration of injury to the bark is beyond the first layer of living cells, but is still shallow (Fig. 39a), only the process of phellogen restoration is triggered. The second process, vascular cambium restoration, is triggered when penetration or injury reaches a certain depth of the bark (Fig. 39b). The prior formation of NIT (phellogen restoration) is essential to completion of the second process. The third process is that of blocking sapwood conduction, which is

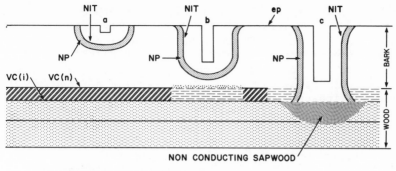

NP – NECROPHYLACTIC PERIDERM

ep – EXOPHYLACTIC PERIDERM

VC(n) – POSITION OF VASCULAR CAMBIUM AT TIME OF PHELLOGEN RESTORATION

VC(i) – POSITION OF VASCULAR CAMBIUM AT TIME OF INJURY

▨ – CONDUCTING SAPWOOD FORMED AFTER INJURY

▨ – CONDUCTING SAPWOOD EXTANT AT TIME OF INJURY

▨ – TRANSFORMED CAMBIAL – PHLOIC ZONE

▨ – ZONE OF NEWLY RESTORED VC

▨ – ZONE OF NON – CONDUCTING SAPWOOD

Figure 39. Diagrammatic views of the non-specific nature of defense during: (a) penetration of bark surface, (b) penetration of vascular cambium, and (c) penetration of sapwood.

triggered when injuries to bark are deeper still (Fig. 39c). NIT forms from the exophylactic periderm to the vascular cambium and the sapwood below the injury becomes impervious. This should permit total isolation of any injurious agent not only from the bark but also from the sapwood.

The three processes could be triggered singly or collectively in pathogenesis, depending upon the site of attack. Pathogens could interact with the chemicals produced in the course of dedifferentiation, or cellular transformations associated with these processes.

A conceptual overview of the non-specific nature of defense is illustrated with the process of phellogen restoration. Whenever the first layer of living cells, i.e., the

phellogen, is rendered non-functional, the process of its
restoration is triggered. The degree of interference with
the process by the pathogen should determine whether the
interaction is one of resistance or susceptibility. From
this view, the restorative process itself is the process of
defense; otherwise, if there is a specific defense, it would
coincide with phellogen restoration and, to date, the two
would be empirically inseparable. Adequate controls would
permit such separation if this were the case. If the pathogen avoids triggering the process, the non-specific defensive
(restorative) process is lacking, potentially leading to susceptibility. A pathogen may be inactivated by an as yet
uncharacterized process (currently under investigation) which
occurs prior to or without triggering the restorative process.
Under this postulation, phellogen restoration would be a process of recovery and a potential defense against secondary
infection. Disease specificity should result from pathogens
specifically manipulating or stopping a certain step in the
process, thus leading to specific disease syndromes. On the
basis of this model, it should be possible to delineate how
various environmental and edaphic factors affect these processes in the predisposition of hosts to attack, and how each
pathogen specifically affects these non-specific responses.

     With respect to specific interactions of the three pro-
cesses in pathogenesis, we have studied in detail only the
process of phellogen restoration, with special reference to
BWA. The specific effects of BWA on this process will be
described elsewhere. Preliminary observations have also
shown the involvement of all three processes during interactions between *Phellinus* (*Poria*) *weirii* and its host *Pseudotsuga menziesii* and between *Cronartium ribicola* and *Pinus
monticola*. It is evident from our studies that by using
these host-processes as controls, against which to compare
modifications resulting from pathogenic interactions, we
should be able to understand the specific nature of each
disease. For example, in biochemical and histological studies
of interactions associated with penetration of phellogen, we
have hitherto compared the results from pathological tissues
with those from normal or senescent tissues where, obviously,
the 'control' process of phellogen restoration, triggered
following penetration, is not occurring.

     The model of host-pathogen interactions proposed here
offers a different perspective on future studies of pathogenesis and in interpreting past studies, and provides the

basis for a unified approach, as opposed to piecemeal and compartmentalized approaches in studying plant diseases.

ACKNOWLEDGMENTS

    The encouragement of Dean J. A. F. Gardner and Dr. K. Graham, Faculty of Forestry, University of British Columbia, Vancouver, Dr. I. C. M. Place, Canadian Forestry Service, Mr. M. H. Drinkwater and Mr. D. R. Macdonald, Pacific Forest Research Center, and the technical assistance of Mr. G. D. Jensen, Mrs. Mary F. Heinreich, Mr. H. Doppelreiter, Mr. B. Schellenberg, Mrs. J. Fraser, and Mrs. O. C'eska are gratefully acknowledged. I especially thank Mr. Benjamin Soo, graduate student, for permission to quote his unpublished work and collaborative assistance in studies on vascular cambium, and numerous colleagues for their interest in this problem, and for reviewing the manuscript. Support of the National Research Council of Canada, HR MacMillan Family Fund, Council of Forest Industries, and Truck Loggers Association is gratefully acknowledged.

REFERENCES

1. Adams, J. B. and J. W. McAllan. 1958. Pectinase in certain insects. *Can. J. Zool. 36:*305–308.
2. Akai, S. 1959. Histology of defense in plants. *In* Plant Pathology Vol. I (J. G. Horsfall and A. E. Dimond, eds.), Academic Press, New York, pp. 392–467.
3. Allen, P. J. 1959. Physiology and biochemistry of defense. *In* Plant Pathology Vol. I. (J. G. Horsfall and A. E. Dimond, eds.), Academic Press, New York, pp. 435–467.
4. Artschwager, E. 1927. Wound periderm formation in the potato as affected by temperature and humidity. *J. Agric. Res. 35:*995–1001.
5. Balch, R. E. 1952. Studies of the balsam woolly aphid, *Adelges piceae* (Ratz.) and its effects on balsam fir, *Abies balsamea* (L.) Mill. Can. Dep. Agric. Publ. 867, 76 pp.
6. Balch, R. E., J. Clark, and J. M. Bonga. 1964. Hormonal action in production of tumours and compression wood by an aphid. *Nature 202:*721–722.
7. Bloch, R. 1941. Wound healing in higher plants. *Bot. Rev. 7:*110–146.

8. Bloch, R.   1952.   Wound healing in higher plants.   II.
    *Bot. Rev. 18:*655-679.
9. Bloch, R.   1953.   Defense reactions of plants to the pre-
    sence of toxins.   *Phytopathology 43:*351-354.
10. Carter, W.   1952.   Injuries to plants caused by insect
    toxins.   II.   *Bot. Rev. 18:*680-721.
11. Carter, W.   1962.   Insects in relation to plant disease.
    Interscience Publishers, New York.
12. Dessureault, M. and T. A. Tattar.   1975.   Dye movement
    associated with discolored wood in red maple.   *Can. J.
    For. Res. 5:*330-333.
13. Doppelreiter, H.   1973.   Preliminary observations on exo-
    phylactic and necrophylactic periderms in western white
    pine.   B.Sc. F. Thesis, University of British Columbia,
    Vancouver, B.C.
14. Esau, K.   1965.   Plant anatomy.   John Wiley, New York.
15. Fahn, A.   1967.   Plant anatomy.   Pergamon Press, New York
16. Forbes, A. R. and D. B. Mullick.   1970.   The stylets of
    the balsam woolly aphid, *Adelges piceae* (Homoptera:
    Adelgidae).   *Can. Entomol. 102:*1074-1082.
17. Hare, R. C.   1966.   Physiology of resistance to fungal
    diseases in plants.   *Bot. Rev. 32:*95-137.
18. Kozlowski, T. T.   1973.   Shedding of plant parts.   Aca-
    demic Press, New York.
19. Lipetz, J.   1970.   Wound healing in higher plants.   *Int.
    Rev. Cytol. 27:*1-28.
20. McMullen, L. H. and J. P. Skovsgaard.   1972.   Seasonal
    history of the balsam woolly aphid in coastal British
    Columbia.   *J. Entomol. Soc. B.C. 69:*33-40.
21. Miles, P. W.   1968.   Insect secretion in plants.   *Annu.
    Rev. Phytopathol. 6:*137-316.
22. Mitchell, R. G.   1967.   Translocation of dye in grand and
    subalpine firs infested by the balsam woolly aphid.
    U. S. For. Serv. Res. Note PNW-46.
23. Mullick, D. B.   1969a.   Reddish-purple pigments in the
    secondary periderm tissues of western North American
    conifers (studies of periderm I).   *Phytochemistry 8:*
    2205-2211.
24. Mullick, D. B.   1969b.   Anthocyanidins in secondary peri-
    derm tissue of amabilis fir, grand fir, western hemlock
    and western red-cedar (studies on periderm II).   *Can.
    J. Bot. 47:*1419-1422.
25. Mullick, D. B.   1969c.   Thin-layer chromatography of anth
    cyanidins, I.   *J. Chromatogr. 39:*291-301.

26. Mullick, D. B. 1969d. New tests of microscale identi-
    fication of anthocyanidins on thin-layer chromatograms.
    Phytochemistry 8:2003–2008.
27. Mullick, D. B. 1971. Natural pigment differences dis-
    tinguish first and sequent periderms through a cryo-
    fixation and chemical techniques (studies of periderm,
    III). *Can. J. Bot.* 49:1703–1711.
28. Mullick, D. B. 1975. A new tissue in the bark of four
    conifers and its relationship to the origin of necro-
    phylactic periderms (studies of periderm, VI). *Can.
    J. Bot.* 53:2443–2457.
29. Mullick, D. B. and V. C. Brink. 1970. A method for
    exposing aleurone tissue of barley for colour classi-
    fication. *Can. J. Plant Sci.* 50:551–558.
30. Mullick, D. B. and G. D. Jensen. 1973a. Cryofixation
    reveals uniqueness of reddish–purple sequent periderm
    and equivalence between brown first and brown sequent
    periderms of three conifers (studies of periderm, IV).
    Can. J. Bot. 51:135–143.
31. Mullick, D. B. and G. D. Jensen. 1973b. New concepts
    and terminology of coniferous periderms: Necrophylac-
    tic and exophylactic periderms (studies of periderm,
    V). *Can. J. Bot.* 51:1459–1470.
32. Mullick, D. B. and G. D. Jensen. 1976. Rates of non-
    suberized impervious tissues development after wounding
    at different times of the year in three conifer spe-
    cies (studies of periderm, VII). *Can. J. Bot.* 54:881–
    892.
33. Oechssler, G. 1962. Studien über die saugchäden mittel-
    europäischer tannenläuse im gewebe einheimischer und
    ausländischer tannen. *Z. Angew. Entomol.* 504:408–454.
34. Puritch, G. S. 1976. Personal communication.
35. Puritch, G. S. and D. B. Mullick. 1975. Effect of water
    stress on the rate of non–suberized impervious tissue
    (NIT) formation following wounding in *Abies grandis*
    (studies of periderm, VIII). *J. Exp. Bot.* 26:903–910.
36. Rahe, J. W. and R. M. Arnold. 1975. Injury-related
    phaseollin accumulation in *Phaseolus vulgaris* and its
    implications with regard to specificity of host-
    parasite interaction. *Can. J. Bot.* 53:921–928.
37. Reid, R. W., H. S. Whitney, and J. A. Watson. 1967.
    Reactions of lodgepole pine to attack by *Dendroctonus
    ponderosae* Hopkins and blue stain fungi. *Can. J. Bot.*
    45:1115–1126.

38. Saigo, R. H. 1969. Anatomical changes in the secondary phloem of grand fir (*Abies grandis* (Doug.) Lindl.), induced by the balsam woolly aphid (*Adelges piceae* Ratz.). Ph.D. Thesis, Oregon State University, Corvallis, Oregon.

39. Schellenberg, B. D. 1974. Extension of necrophylactic and exophylactic periderm terminology to Sitka spruce. B.Sc.F. Thesis, University of British Columbia, Vancouver, B.C.

40. Shain, L. 1971. The response of sapwood of Norway spruce to infection by *Fomes annosus*. *Phytopathology* *61*:301–307.

41. Shaw, M. 1972. Physiology of rust resistance. *In* Biology of Rust Resistance in Forest Trees. Proc. Nato-IUFRO Advanced Study Institute, p. 87–95.

42. Shigo, Alex L. and W. E. Hillis. 1973. Heartwood, discolored wood, and microorganisms in living trees. *Annu. Rev. Phytopathol. 11:*197–222.

43. Shigo, Alex L. and E. M. Sharon. 1970. Mapping columns of discolored and decayed tissues in sugar maple, *Acer saccharum*. *Phytopathology 60:*232–237.

44. Soo, B. 1977. The occurrence of non-suberized impervious tissues in woody plants. Ph.D. Thesis, Faculty of Forestry, University of British Columbia, Vancouver, B.C. (in preparation).

45. Srivastava, L. M. 1964. Anatomy, chemistry and physiology of bark. *Int. Rev. For. Res. 1:*203–277.

46. Treshow, M. 1970. Environment and plant response. McGraw Hill, New York.

47. Van der Plank, J. E. 1975. Principles of plant infection. Academic Press, New York.

48. Wood, R. K. S. 1967. Physiological plant pathology. Blackwell Scientific Publications, Oxford.

COLOR PLATE I

Figure 11.   Settled BWA first instars (arrows) on *A. amabilis* with characteristic dark body color and development of white wax 'wool' fringes.

Figure 12.   Prominent tufts of white wax 'wool' (arrows), characteristic of the adult BWA, completely obscure the underlying insect.

Figure 13.   Settled BWA first instars on a non-host species, *T. heterophylla*. Some insects exhibited dark body color and white wax 'wool' fringes comparable to that on host trees (tailed arrow, c.f. Fig. 12), while others showed poor wool development (tailless arrow).

Figure 14.   A sudan stained section of an injury with well-developed NP, F-F tested from the wound surface. The blue color produced by the test (top) was not present in the layers of enlarged NIT cells or the underlying phellem (ph). Only the NP phellem was colored by the sudan stain (c.f. Fig. 20).

Figure 15.   Cryofixed section through a BWA feeding site completely surrounded by reddish-purple pigmented NP. The pigment is located in the phellem cells (ph). yb, yellow brown interaction zone.

Figure 16.   A sample of injured bark F-F tested through the cambial surface after development of impermeability. Test solutions have colored the entire sample except the brown tissues external to the crescent-shaped impervious boundary. vc, vascular cambium; fep, first exophylactic periderm.

Figure 17.   A sample of injured bark F-F tested through the injury surface after development of impermeability. Most of the outer brown tissue is deeply colored by the test except a narrow brown zone, NIT, at the internal boundary of the injury zone. vc, vascular cambium; fep, first exophylactic periderm.

Figure 18.   Cryofixed section of bark through a freezing injury at the time of impermeability development. The enlarged cells with strongly fluorescent walls (NIT) were found to be responsible for the initial impermeability. NP will form in

the dark zone (double-headed arrows) internally abutting
NIT.   Carl Zeiss fluorescence filter combination IV/41.

COLOR PLATE II

    Figure 19.   A cryofixed section through a rhytidome in
T. heterophylla bark with NP formed, F-F tested through the
wound surface.   Blue color of the test is present only at
the external boundary of NIT.   Reddish-purple pigments are
forming in the NP phellem (ph).

    Figure 20.   Typical appearance of a young *T. hetero-*
phylla stem 10 days after initial detection of rhytidome
formation.   The irregular raised areas are lighter at their
edges than surrounding bark and their central regions show
early stages of browning.

    Figure 21.   A cryofixed section of *T. heterophylla* bark
with phellogen (pg) in the quiescent state.   The fluorescence
of phellogen cell contents is typical of the inactive state
(c.f. Fig. 23), and phellogen cell walls do not fluoresce.
Carl Zeiss fluorescence filter combination I/53.

    Figure 22.   A cryofixed section of *T. heterophylla* bark
with active phellogen (pg).   The dull fluorescence of the
cell content and the increase in cell size are characteristic
of the active state (c.f. Fig. 22).   Like the quiescent state,
the phellogen cell walls do not fluoresce.   Carl Zeiss fluores
cence filter combination I/53.

    Figure 23.   A cryofixed section of *T. heterophylla* bark
from a region of an early stage in rhytidome formation.   The
fluorescence of the phellogen cell contents is dissimilar
from either the quiescent or active states (c.f. Fig. 22 and
23, respectively), and phellogen cell walls now show strong
fluorescence (arrows).   The fluorescent reticulum (fr) appearance of some parenchyma cells is a common feature of early
rhytidome development.   Carl Zeiss fluorescence filter combination I/53.

    Figure 24.   A cryofixed section of *T. heterophylla* bark
through the boundary of a developing rhytidome sampled three
days after detection.   Extensive hypertrophy is clearly seen.
The phellogen and outermost cortical cells do not enlarge
significantly.   The red fluorescence of chlorophyll is marked

diminished in the hypertrophied zone in comparison with the
normal bark on the left side.   In the region of normal bark,
the phellogen is either quiescent (tailless arrows) or active
(tailed arrows).   c, cortical parenchyma; b, phloem parenchyma; r, ray parenchyma; vc, vascular cambium.   Carl Zeiss
fluorescence filter combination I/53.

Figure 25.   A cryofixed section of *T. heterophylla* bark
through the boundary of a developing rhytidome sampled 18 days
after detection.   The red fluorescence of chlorophyll is
virtually gone from the hypertrophied zone.   Cell contents in
this zone have undergone changes in fluorescence characteristics.   Cell walls, beginning at corners, have become fluorescent.   NIT will form from the enlarged cells with sporadic
wall fluorescence (arrows) at the inner boundary of the hypertrophied zone and NP will form in the somewhat dark zone
internal to these cells.   Carl Zeiss fluorescence filter combination I/53.

Figure 26.   A cryofixed section of *T. heterophylla* bark
through the boundary of a developing rhytidome sampled 23 days
after detection.   The brightly fluorescent cell walls of NIT
clearly distinguish it from adjacent tissues.   Incompletely
transformed sieve elements (tailed arrows) and sclereids
(tailless arrows) are part of this complex tissue.   NP will
form in the zone internally abutting NIT.   Carl Zeiss fluorescence filter combination IV/41.

COLOR PLATE III

Figure 27.   A cryofixed section of *T. heterophylla* bark
through the boundary of a developing rhytidome sampled 28 days
after detection.   At this stage in rhytidome development NIT
is difficult to distinguish from adjacent enlarged cells under
this mode of fluorescence.   The zone of NP formation is clearly
seen as a dark zone.   Carl Zeiss fluorescence filter combination I/53.

Figure 28.   A cryofixed section of *T. heterophylla* bark
through the boundary of a developing rhytidome sampled 88 days
after detection.   The major portion of the hypertrophied zone
has been lost (double-headed arrows) because at this stage
the tissues are desiccated and it was not possible to obtain
intact cryofixed sections.   The outermost NP phellem has
developed pigmentation.   NIT at this stage is virtually impossible to detect. Carl Zeiss fluorescence filter combination I/53.

Figure 29. A cryofixed section of *A. grandis* bark wounded by a ring of needle point injuries around a settled BWA first instar about 10 days after the aphid settled and sampled 60 days after injury. Normal NP formation has occurre except in the central region, the zone of BWA interaction. Carl Zeiss fluorescence filter combination I/53.

Figure 30. Higher magnification of a section from the same sample as Figure 29 but closer to the stylets. Hypertrophy has occurred but NIT and NP have not formed in the area affected by the aphid. fr, fluorescent reticulum. Carl Zeiss fluorescence combination I/53.

COLOR PLATE IV

Figure 31. A cryofixed section of an injured *Phaseolus vulgaris* stem 23 hours after injury. The zone of dedifferentiating enlarged cells (arrows) of an analogous process to phellogen restoration in conifers was the site of phaseollin production. This material has very low levels of intrinsic fluorescence necessitating prolonged exposure which resulted in the pronounced background in the photomicrograph. Carl Zeiss fluorescence filter combination I/53.

Figure 32. A cryofixed section of the stem of a *P. contorta* seedling sampled 21 days after a small patch of bark about 1/2 the bark thickness had been removed. The circumferential extent of the modified cambial-phloic zone is marked by white bars. Sporadic xylem-like cells (X') are present at the external boundary of the modified cambial-phloic zone. The newly established dark vascular cambium zone (white tail-less arrows) has formed between the zone of sporadic xylem-like cells and the enlarged phloic cells (white tailed arrows, c.f. Fig. 34). The position of the vascular cambium at the time of injury is marked by a black arrow. Carl Zeiss fluores cence filter combination I/53.

Figure 33. The same section as Figure 33 viewed with crossed polarizing filters. The cell walls of most tissues are more clearly seen with this mode of illumination. Labeled as in Figure 32.

Figure 34. A cryofixed section of a similar injury to that in Figures 42 and 43, sampled 66 days after injury. Normal xylem production has been restored across the entire region, so that the transformed cambial-phloic tissue is completely surrounded by xylem. Most cells in the cambial-phloic zone have fluorescent contents. Note the cells in the 7 to 9 layers of xylem produced externally abutting the cambial-phloic zone have wider lumens, similar to early wood, than do the xylem cells produced at the same time away from the injury. Carl Zeiss fluorescence filter combination I/53.

Figure 35. A deep mechanical injury to the bark of *A. amabilis*. Sufficient tissue for phellogen restoration remained in the central area of the injury. Acid fuchsin testing of sapwood conduction showed that in the xylem a narrow zone of brown tissue (arrow) below the injury failed to conduct the stain.

Figure 36. A deeper injury to *A. amabilis* bark than that shown in Figure 35. Insufficient tissue for phellogen restoration remained below the injury. Tissue is being regenerated across the killed zone by intrusive growth from the sides (tailless arrows). The non-conducting brown tissue appears as two narrow zones, one on either side of the injury (tailed arrows). A zone of non-conducting sapwood (ncs) has formed immediately below the injury. The thin brown zone immediately below the central region is dead xylem tissues.

Figure 37. A series of increasingly deep mechanical injuries to the bark of *A. amabilis* injured in July and sampled in November. The lower most injury left insufficient tissue for phellogen restoration to occur in the central region. All tissues including the vascular cambium died. NIT formed from the first exophylactic periderm (fep) to the vascular cambium (vc).

Figure 38. An *A. grandis* seedling stem which was injured (tailed arrow) by slicing away about 1/2 of the bark tissue. After 33 days, the entire excised plant was tested with 0.5% aqueous acid fuchsin. A small zone of non-conducting sapwood (tailless arrow) has formed.

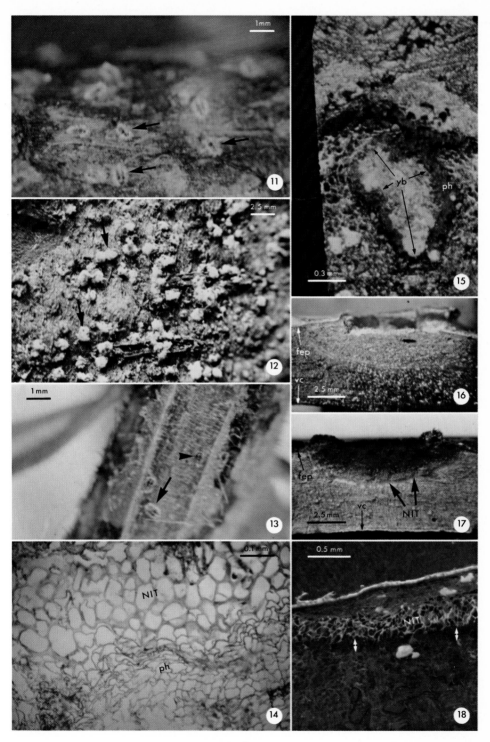

I

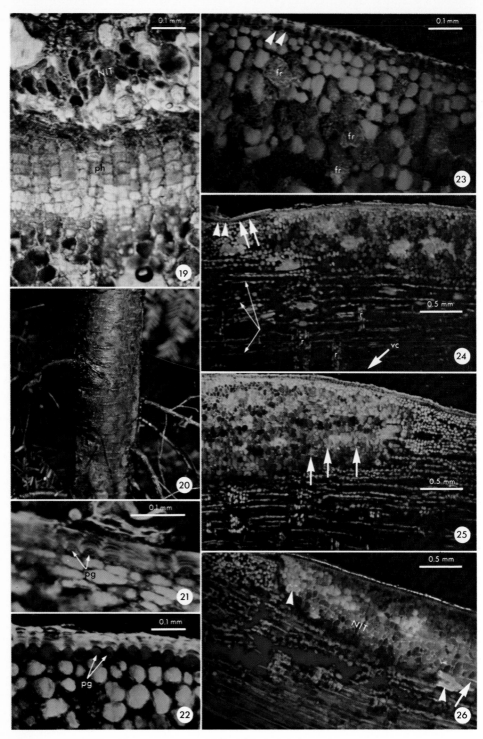

II

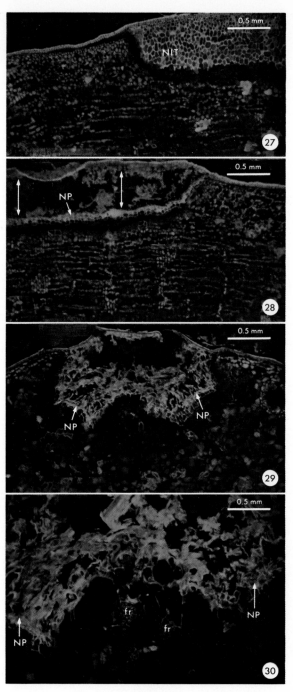

III

IV

Chapter Eleven

UTILIZATION OF CHEMICALS FROM WOOD: RETROSPECT AND PROSPECT*

FRANKLIN W. HERRICK

Olympic Research Division, ITT Rayonier Incorporated, Shelton, Washington 98584

HERBERT L. HERGERT

ITT Rayonier Incorporated, 605 Third Avenue, New York, New York 10016

INTRODUCTION

Extensive research on the chemical structure and utilization of wood constituents has been conducted by the
forest products industry but has remained largely unnoticed
among the chemical fraternity during recent years. This is
a result of the predominant position held by petroleum as
a raw material in the chemical industry. The recent upsurge
in crude oil prices, coupled with a growing public awareness
that petroleum is a finite resource, has prompted serious
review of the prospect for sustaining our economy on

---

*Contribution No. 164 from Olympic Research Division, ITT
 Rayonier Incorporated, Shelton, Washington.

alternative chemical and energy resources. Among these are
the forest since it is capable of sustained yield if
properly managed.

The concept of a forest-based chemical industry has
been the dream of wood chemists for several generations
and, indeed, was a reality in Sweden until the late 1940's.[125]
In spite of a significant research and development (R&D)
investment by some segments of the forest products industry,
the manufacture of chemicals from wood (and bark) represents
only an insignificant fraction of the total chemical production of the developed countries. One of the reasons for
this situation is that the research objective has almost
invariably been byproduct utilization. The principal
products of the industry are lumber, poles, railway ties,
plywood, particleboard, pulp, and paper. Although conversion of the tree to the finished product has vastly improved
over the years, substantial quantities of byproducts remain
to serve as fuel or are disposed of as an environmental
nuisance. These include bark, sawdust, shavings, branches,
roots, and pulping or bleaching effluents. Research,
therefore, has been raw-material-oriented rather than being
based on market demand. Furthermore, the raw material is
a solid, frequently intractable material with an almost
hopelessly complex chemistry.

Because of space limitations, it is not possible to
review all the types of chemicals that have been or could
be made from wood. Recent symposia and reviews have already
dealt with the subject broadly.[64,67,69,154,191,201,203]
Rather, our purpose here will be to review selectively
several categories of chemical compounds which have a production history or have been sufficiently investigated to
the point where the installation of production facilities
has been seriously contemplated. Emphasis will be placed
on North American practice.

Secondly, our aim is to reach some conclusions as to
the direction that wood chemistry research should take in
the future if it is to be ultimately put to commercial use.
Modern management expects the R&D department of a business
to show a payout in the same way as any other function
of the operation. If future R&D projects, to produce
chemicals from wood, are to be funded either at the university or by industry and are to be successful in the sense
of resulting in a profitable wood chemical venture, it is

vital to learn from the past history of such projects. It
will be noted that those which succeeded have involved
substantial process and end-use research. Those which have
not, with few exceptions, were "long" on structural analyses
of the compound in question and "short" on its utility.

Before proceeding with the discussion of the different
types of chemicals obtainable from wood, it is well to
outline briefly the composition of wood and bark from a
typical North American conifer. We have chosen western
hemlock (*Tsuga heterophylla*) for this purpose since it is
a major pulpwood species and its properties and chemistry
are well within the experience of the authors. The composition of bark has been included because this tree component
is traditionally available at saw and pulp mills. Bark is
part of the raw material for which uses have or must be
developed. A summative analysis of western hemlock wood
and bark is given in Table 1. As is well known, the major
components of wood are cellulose, hemicellulose, and lignin.
In addition to these three components, bark contains major
fractions of extractives, alkali soluble phenolic polymers
(termed "phenolic acids"), and natural polyesters such as
cork, some of which have already been discussed in previous
chapters.

Table 1. Summative analysis of western hemlock wood and
bark

	Dry unextracted solids	
	Wood	Bark
	% by weight	
Total extractives	4.5	29.3
Phenolic acids	0.1	18.5
Lignin	29.8	15.2
Hemicellulose	27.5	16.2
Cellulose	38.0	20.5
Ash	0.1	0.3
Total	100.0	100.0

The extractives of western hemlock wood and bark
(Table 2) are similar in type but are appreciably more
concentrated in bark. The cell structure in bark is
encrusted and impregnated with flavonoid polymers in the
categories of tannins, phlobaphenes and phenolic acids,
which are produced from monomeric polyphenols during the
annual bark-stretching phase of growth, possibly as a
defense mechanism against fungal and other attack. The
hemicellulose components of hemlock wood and bark (Table 3)
are similar except that the acetylgalactoglucomannan fraction of bark is less.

Another perspective that may be helpful in understanding the variety and utility of chemicals from wood
is presented in Figures 1 to 3. Figure 1 is a brief outline of the historical pyrolysis or wood "distillation"
route for producing chemicals. Gaseous products were useful
as fuel. By proper control, synthesis gas (CO, H_2) was
generated for the production of ammonia, methanol, formaldehyde, and hydrocarbons. During the 19th century the liquid
products of wood distillation, such as methanol, acetic
acid, cresote, and wood tar, were important locally-produced
commodities. Charcoal and alkali from wood ashes were
essential to the economy of every community, the latter
being used in soap making.

Table 2. Analysis of the extractives in western hemlock
wood and bark

	Dry unextracted solids	
	Wood	Bark
	% by weight	
Wax	0.3	2.8
Flavonoids, lignans	1.7	3.6
Tannin	0.2	12.7
Phlobaphene	1.5	5.3
Water soluble carbohydrates	0.8	4.9
Total	4.5	29.3

Table 3.   Hemicellulose components of western hemlock wood
and bark

	Dry unextracted solids	
	Wood	Bark
	% by weight	
Acetylgalactoglucomannan	18.9	8.4
Glucuronoarabinoxylan	4.6	2.9
Arabinogalactan	1.6	2.0
Pectin	2.4	2.9
Total	27.5	16.2

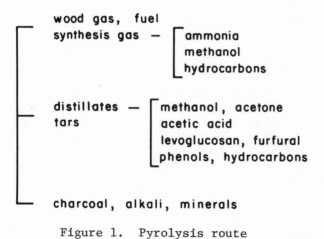

Figure 1.   Pyrolysis route

    A much broader view of wood chemicals is obtained from
Figure 2.   This hydrolysis scheme is pertinent mainly to
the sulfite pulping process where the "spent sulfite liquor"
becomes the collector for sulfonated lignin and the organic

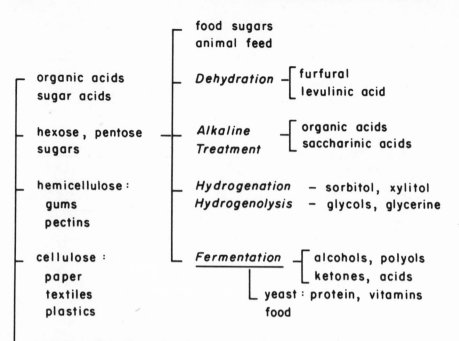

Figure 2.   Sulfite pulping route

acids and sugars resulting from the hydrolysis of hemicellulose. Cellulose fibers remain insoluble and are further
purified prior to conversion into paper or, by chemical
means, into textile filaments, plastics, and a host of
other products. Second to cellulose, the sugar fraction is
potentially useful in a variety of chemical processes
including those involving fermentation. Finally, the
lignin fraction has unique macromolecular properties of its
own and represents an important potential resource for
aromatic and phenolic compounds via alkaline oxidation or
the hydrogenolysis route of Figure 3.

COMPOUNDS OBTAINABLE BY DIRECT PROCESSES

     By "direct" it is implied that a chemical or a product
containing a class of chemicals is collected from the living

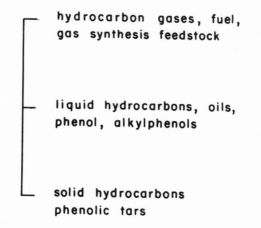

hydrocarbon gases, fuel,
gas synthesis feedstock

liquid hydrocarbons, oils,
phenol, alkylphenols

solid hydrocarbons
phenolic tars

Figure 3. Hydrogenolysis route from wood or lignin

tree or is extracted from some component of the tree. This
category of chemicals is concerned principally with "extractives"; however, we have also included the extractives as
they occur in wood pulping processes and prehydrolysis of
wood chips at moderate pressure and temperature. An
important economic factor of the recovery of many types of
chemicals that are classed as wood extractives is that a
pulping or wood-fiber process serves as the means for
collecting and concentrating an otherwise minor wood component that would be excessively expensive to produce by direct
extraction of the wood.

Terpenes.[79,186,205] The history of turpentine and rosin
is several thousand years old and is related more to species
of pines than other conifers. While the foliage of many
conifers is rich in terpenes, particularly in the season of
new growth, the terpene industry is almost entirely based
on a few pine species. In the United States the traditional
naval stores industry is dependent largely on two southern
pine species: slash (*Pinus elliottii*) and long leaf (*P.
palustris*). However, "sulfate" turpentine recovered during
the kraft pulping process includes the terpenes of loblolly
pine (*P. taeda*). Turpentines are quite variable in

composition depending on production region. Some variability
is attributed to genetic strains or families within the
same pine species. Oleoresin, recovered by tapping the
tree contains 18 to 25% "gum" turpentine, which is similar
in composition to sulfate turpentine and contains 60 to
70% α-pinene [1], 20 to 35% β-pinene [2] and 5 to 12%
other terpenes. Wood or steam-distilled turpentine,
recovered in about 5% yield (dry wood) from stumps and
roots, contains 75 to 80% α-pinene and 20 to 25% other
terpenes.

During the past 25 years the expansion of the kraft
pulping industry in the Southeast has been the major factor
in sustaining annual turpentine production in the range
30 to 35 million U.S. gallons[79] (114 to 132 million liters).
About 80% of current production is sulfate turpentine
from the kraft pulping of southern pine. The advantage of
the latter process is that it provides an excellent means
for collecting a rather minor fraction of volatile oil in
a yield on the order of 0.2 to 0.3% of (dry) pine wood.

The traditional large market for turpentine was as a
solvent in oil- or resin-based paints. With the advent of
water-based latex paints this market declined but was
replaced by a growing number of other industrial uses in
adhesives, disinfectants, insecticides, and essential oil
or perfume components. Since the turpentine industry is
well established and chemical transformations and product
uses are extensively reviewed elsewhere,[2,5,42,79,186,205]
this discussion will deal with other aspects and prospects
for terpene production.

α-pinene   β-pinene

[1]   [2]

There has been some concern that the turpentine resource in the southern pine region has reached its maximum capability. Most kraft pulp mills recover about half the turpentine as compared to that available in the freshly harvested wood.[127] Major evaporation losses are caused by long storage of logs, as needed to maintain wood supplies at the mill, and the practice of storing wood chips in outside chip piles.[190] In the case of mills which operate continuous digesters, turpentine recovery is, at best, only 50% of that obtained in batch digesters, presumably because of resinification. In batch digesters, turpentine is steam distilled and removed continuously during the process, whereas in the continuous digester, the turpentine is mostly trapped in the system.

On the positive side of turpentine production, there are efforts to develop genetic types of slash and longleaf pine that have high yields in terms of turpentine and rosin.[55] More recently, chemical treatments, specifically with "Paraquat" herbicide, have been applied to the growing tree to promote saturation of the wood above the wound with turpentine and rosin. The content of turpentine and rosin in the "lightwood" log segment is 7 to 8 times that of normal wood.[49] The saturated wood contains 5 to 8% turpentine, based on dry wood weight, or about 20 times the content of average normal wood. Assuming that such treatment becomes acceptable and general, the turpentine resource and production could be increased two-fold over present levels.

Finally, the recovery of terpenes and other valuable components of conifer foliage is of renewed interest.[13,83,84,126] The development of whole-tree harvest equipment and central yarding procedures has increased the volume of foliage present at a given collection site and has made the economics of the utilization of foliage more attractive. Leaf oils are present to the extent of 0.2 to 2.0% in many conifers and are recovered by steam distillation or solvent extraction. The term "technical foliage" is now being applied to mixtures of leaf and twig material obtainable at harvest or collection sites. A yield of 1 to 1.2% of volatile oil has been obtained from European pine (*P. sylvestris*). This oil contains 48% monoterpenes (including 20% α-pinene) and 46% sesquiterpenes.[83,84]

Resin Acids.[200,205] Like turpentine, the resin acids
or rosins have been articles of commerce from the dawn of
civilization. Resin acids are diterpene derivatives and
are generally found in nature in association with monoterpenes, as in the case of the exudates and pitches of the
pine species and other conifers. The resin acids of pine
include some eight closely related compounds and structural
isomers, abietic acid [3] being representative of the predominant type. Pine woods are quite variable in their
content of extractives and may contain less than 1% resin
acids in average (dry) wood, while heartwoods or resin-
saturated segments may contain as much as 20% resin acids.
The nonvolatile fraction of southern pine oleoresin contains
25 to 40% resin acids, 40 to 60% fatty acid esters, and
7 to 11% neutral or unsaponifiable material. Stump and
root woods (aged) contain 16 to 20% resin acids on a dry
weight basis.

As with turpentine the resin acid or rosin industry in
the United States is located principally in the Southeast
and is based on slash, longleaf and loblolly pines. The
traditional rosin industry, involving the collection and
processing of oleoresin or the extraction of stumpwood has
been converted during the past 25 years to strong dependence
on "tall oil" obtained from the kraft pulping process. Tall
oil or "sulfate" rosin now accounts for more than 50% of
total production, while extraction of stumpwood or rosin-

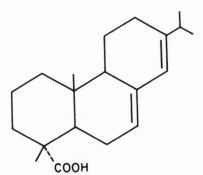

Abietic acid
[3]

saturated wood accounts for nearly all of the remainder.
Total United States production of rosin has declined from
about 500,000 tonnes in the 1950's to 330,000 tonnes in
1970.

The story of tall oil (Swedish for pine oil) and its
development as a large scale industrial commodity during
the past 25 years is a hopeful sign that other wood chemical
resources will become important in the future. Present
United States capacity for tall oil fractional distillation
is about one million tons/year (907,000 tonnes/year). One
tonne of crude tall oil yields 300 kg of fatty acids,
350 kg of rosin, and 350 kg of head, intermediate and pitch
fractions.

Rosin uses are well established in paper sizing,
adhesives and chemical intermediate markets. Because composition varies according to source and processing, rosin
products are classified for end-use purposes and are often
chemically modified to improve utility and stability.

Concern has been expressed because of the diminishing
resource for wood rosin, obtained from pine stumpwood.
Such stumpwood is usually recovered after tree harvest from
natural forests and includes the remainder of stumps and
roots from previous wood-cutting cycles. Plantation forestry
is not expected to yield stumpwood of similar volume or
value. On the other hand, other factors favor an increase
in overall rosin production. In a long range program, the
breeding of pine trees, selected on the basis of high yields
of oleoresin or wood extractives, will provide forests that
are twice as productive as present natural or plantation
types.[55] The use of Paraquat herbicide treatment of the
living tree, as already mentioned above, holds even greater
and near-future gains in production[49] that could greatly
support the wood rosin extraction industry, since the rosin-
and turpentine-saturated portion of the tree bole could be
easily diverted to this market. The resin acid content of
the lightwood log segment is five times greater than present
in normal wood. The lightwood or saturated wood portion
contains 22 to 28% resin acids, based on dry wood weight.
Routine, preharvest treatment of trees with Paraquat in
the accessible southern pine plantations could increase the
annual production of rosin by two-fold over present levels.

Finally, in the case of tall oil, there is some room for improved production and recovery efficiency. The practice of storing wood chips in outside chip piles results in significant biological degradation of resin acids. For example, the storage of chips for four weeks has been found to result in 50% loss in tall oil yield and corresponding losses in resin and fatty acid fractions.[170] Tall oil soap recovery in most kraft pulp mills is 80 to 85% of the potential amount available in black liquor system. Process improvements aimed at recovering such losses have been developed.[21]

Fatty Acids.[200,205] Solvent extraction of pine wood yields about 1% of fatty acids and their esters. Conifer woods vary considerably in total extractives within a tree species and within a given tree, heartwoods generally being richer in extractives. Hardwoods usually contain less solvent-extractable material than conifers, but the fatty acid components may be predominant. Since seed-oil crops are widely grown and vegetable oils are obtained as coproducts in such industries as cotton and corn products, it is surprising that of the production of more than 450,000 tonnes/year of fatty acids in the United States, 35% comes from wood *via* the recovery of tall oil. The kraft pulping process thus serves as a collector for three major classes of wood chemicals, all three of which could not be extracted economically from normal wood.

The components of fatty acids derived from tall oil include about 50% oleic [4], 35% linoleic [5], and 8% conjugated linoleic [6]. Other fatty acids are present in small amounts. Advances in the fractional distillation of fatty acids permit the production of oleic acid in grades that are 99.5% pure. Markets are well established for tall oil fatty acids. The production of conjugated linoleic acid, dicarboxylic acids and "dimer acids" is important in the synthesis of polyamide resins and a variety of other polymer products, detergents, adhesives, and printing inks.

Any major increase in the resource of fatty acids from wood will probably depend on the expansion of the kraft pulping industry and improvements in tall oil recovery. Paraquat herbicide treatment of pine tree species does not enhance fatty acid production as in the case of turpentine or rosin. Genetic studies for increasing fatty acid extractives in pine wood have not progressed far enough

$$CH_3-(CH_2)_7-CH=CH-(CH_2)_7-COOH$$

Oleic acid
[4]

$$CH_3-(CH_2)_4-CH=CH-CH_2-CH=CH_2-(CH_2)_7-COOH$$

Linoleic acid
[5]

$$CH_3-(CH_2)_5-CH=CH-CH=CH-(CH_2)_7-COOH$$

Conjugated linoleic acid
[6]

for judgement. Conifer foliage contains a larger proportion of fatty acid esters (C_{16}, C_{18}, C_{20} derivatives) than resin acids. Triglycerides obtained by solvent extraction of pine technical foliage have amounted to 2.5% of dry weight, while resin acid yield was 1.1%.[83]

Steroids. The form in which such compounds are usually found in plants is as sterols or generally, "phytosterols." Although many sterols have been identified as minor components of wood or tree foliage, β-sitosterol [7] is relatively abundant. Sitosterol is associated with fatty acids or esters and, being neutral, is efficiently extracted from soaps by the use of organic solvents. As is the case of the resin and fatty acid components of wood, sitosterol is concentrated in the kraft pulping process in the tall oil soap product. The sitosterol content of crude tall oil falls in the range of 2 to 3% and is the main component of the neutral or "unsaponifiable" material associated with tall oil. In a recent study, some 80 chemical compounds were identified in the neutral fraction of tall oil soap from southern pine wood.[38] Total steroid content was 32.4% by weight, including 25.1% sitosterol.

Recovery of sitosterol at a kraft pulp mill could best be done by solvent extraction of concentrated soaps prior to acidification of the soap to recover crude tall oil. It is estimated that 100-160 tonnes/year of sitosterol could be recovered at a single kraft pulp mill, depending on pulping capacity.

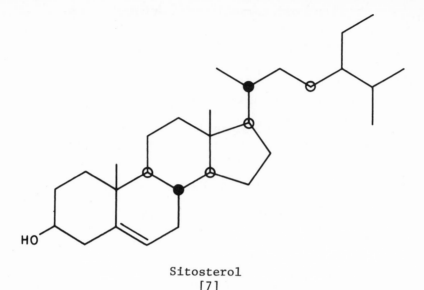

Sitosterol
[7]

Another pulp mill fraction that contains sitosterol is that called "fines", corresponding to the ray cells or similar structures that are too small to be useful as fiber. The fines from sulfite pulp mills are rich in extractives, yielding about 1% sitosterol of the unextracted fines dry weight. Pine foliage has also been found to yield 1% sitosterol on a dry weight basis.[83]

Over the years, much of the interest in sitosterol has been related to its potential use in the synthesis of cortisone derivatives and other steriods or hormones. Some progress has been made in biochemical routes (fermentation) to transform sitosterol to more desirable products.[39] Another area of developing interest is the use of sitosterol as a dietary additive in the prevention or treatment of atherosclerosis. It has been reported that about 500 tonnes/year are required in Russia for this purpose.[183] The presence of β-sitosterol in corn oil has been associated with the beneficial effects of this oil in the diet, for the purpose of reducing blood cholesterol.[115]

Waxes. The bark of most conifers and hardwoods contains
a mixture of long chain fatty acids, alcohols and esters
with wax-like properties. Usually the content of this
fraction is not sufficiently large for commercial exploitation, or the isolation of a saleable product is made difficult by the co-occurrence of resin acids, low melting esters
and sticky sesquiterpenes and diterpenes. Sequential
analyses indicate that the latter group of compounds are
biosynthesized in the phloem (inner bark) while the aliphatic
wax esters primarily originate in the cork cells.[96,114]
Thus, species that have bark with a substantial volume of
cork are most likely to be suitable for wax production,
provided that the particular mixture of compounds has a
suitably high melting point, hardness and absence of color,
to qualify as a commercial wax.

Douglas fir is the most valuable timber species in
western United States and Canada. There are many small
sawmills cutting this species, and bark disposal represents
a significant environmental problem. Douglas fir (*Pseudotsuga menziesii*) contains the highest volume of bark of any
of the common North American conifers and correspondingly,
contains the largest quantity of wax (up to 7%).[99] E. F.
Kurth was among the first to recognize the potential of
bark for the production of useful products and chemicals.
He devised a sequential extraction procedure with solvents
of increasing polarity. Pulverized Douglas fir bark was
extracted with an aliphatic (C_5 to C_9) hydrocarbon to give
a pale yellow-brown wax in 3 to 5% yield. Subsequent extraction with benzene gave 2% of a dark brown wax.[129,130,132,134]
Ether yielded 0.5 to 3% dihydroquercetin, and ethanol or
hot water gave a mixture of polymeric polyflavonoids which
could tan leather.[131,133]

A small Douglas fir wax extraction plant was constructed
in 1948 at the Oregon Forest Products Laboratory (now part
of the School of Forestry, Oregon State University) at
Corvallis, Oregon. The aliphatic hydrocarbon-soluble wax
fraction was sampled rather widely and considered to be a
reasonably good substitute for imported carnauba or candel-
illa wax. In the middle 1950's, the M. W. Kellog Company
licensed the Kurth process and built a larger pilot plant
which incorporated petrochemical separation technology.[202]
Plans were made to build a full-scale wax extraction plant
in Oregon but were subsequently terminated as a result of

a major drop in the price of carnauba wax and a shift to
the use of acrylics in place of wax in liquid floor polishes.

The Weyerhaeuser Company also had a major research and
pilot plant program to produce Douglas fir wax and polymeric
phenolic fractions. Instead of using organic solvents
with their attendant problems of recovery from residual
bark, they used aqueous alkaline extraction of bark at
elevated temperature and pressure.[24,43,102] While wax is
not soluble in dilute, aqueous sodium hydroxide solution,
the co-occurring polyphenols act as dispersants and aid in
extraction from the bark. Furthermore, the yield of wax
is increased by the addition of the ω-hydroxy acids resulting from the saponification of the polyestolides (suberin)
in the cork fraction which cannot be isolated by direct
organic solvent extraction of bark. Partial acidification
of the alkaline extract yields the "wax" fraction. Wax
obtained by this method is darker in color and has a higher
acid number (undesirable in many liquid polish formulations)
than wax obtained directly from Douglas fir bark by organic
solvent extraction. Presumably insufficient markets could
be developed because this development program was terminated
in the middle 1960's.

Wax production from Douglas fir bark was resurrected
in 1974 with the announcement of a combination solvent
extraction and physical separation process which would yield
wax, a powdered filler and a phenolic extender for adhesives.
[70,192,193] A full scale plant was built by the Bohemia
Lumber Company in Oregon and came on stream in 1975.[7] The
wax is isolated by a mixed aliphatic-aromatic solvent,
b.p. 95 to 101 C and is reported to be useful for shoe
polish, floor wax, automobile wax, internal lubricants for
plastics, carbon paper, and as a wax for use in investment
casting.[194]

A number of studies have been conducted on the composition of Douglas fir wax. Predominant components are
mixed esters based on ferulic acid [8],[130] behenic (C22),
lignoceric (C24), and cerotic (C26) acids [9], C16 and
C18 dicarboxylic and ω-hydroxy acids [10], behenyl (C22),
and lignoceryl alcohols and the usual mixture of plant
sterols (predominantly β-sitosterol).[137,138,139] The
polyfunctionality of some of the components results in
molecular weights up to a thousand and is partly responsible
for good hardness (3 to 4, standard ASTM penetration test).[194]

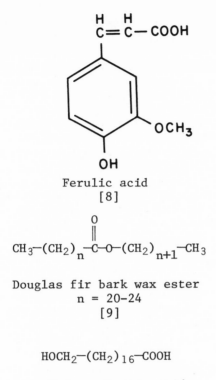

Ferulic acid
[8]

$$CH_3-(CH_2)_n-\overset{\overset{\displaystyle O}{\|}}{C}-O-(CH_2)_{n+1}-CH_3$$

Douglas fir bark wax ester
n = 20-24
[9]

$$HOCH_2-(CH_2)_{16}-COOH$$

ω-Hydroxystearic acid
[10]

A similar wax exists in the bark of the white fir (*Abies concolor*).[100] This species is of much narrower geographic distribution and does not present the same problems of disposal. However, among all North American conifers, it is the second most attractive raw material from the standpoint of yield and wax properties.

Lignans.[41,89,92] The term lignan was introduced by Haworth[90] in 1936 as a category for those plant or wood extractives of a phenolic nature, having two phenylpropane structural components linked at the β-position. The structural relationship to lignin was also noted. Lignans are extractable from wood with hot water, polar solvents, ether, aromatic solvents, and alkaline solutions. Yield may vary from a fraction of 1% to as high as 30% of dry heartwood. These substances are generally only slightly soluble in

cold water and are crystalline when purified and separated
from associated materials. It is probable that the present
list of lignans[41,89,189] will be expanded in years to come
as the extractives of different tree species are examined
and classified. As in the case of phenolic extractives in
the flavonoid category, there is good evidence that these
compounds are formed and deposited at the heartwood-sapwood
boundary and are not produced via the lignin biosynthesis
route.[113,189] Many lignans have fungicidal, insecticidal
and antioxidant properties and may be associated with
natural plant defense mechanisms.

   Potentially commercial extraction or recovery processes
have been developed for two types of lignans occurring
in conifer woods. Lignans related to conidendrin [12] are
present in the wood of hemlock and spruce species, while
lignans related to plicatic acid [14] are present in western
red cedar (*Thuja plicata*) wood. Historically, conidendrin
has been called "sulfite liquor lactone"[91] because it was
observed as an organic residue in spent sulfite liquor
storage or handling equipment as used in the sulfite pulping
process on hemlock or spruce woods. Treatment of warm spent
liquors (10% solids) with 1 or 2% of an immiscible solvent
such as toluene or trichloroethylene hastens the direct
crystallization and nearly quantitative recovery of conidendrin during storage for 18 to 24 hours.[135] Yields as
high as 0.6 to 0.8% of spent liquor solids, or about 0.35%
of dry wood are obtainable. If a pulp mill producing
500 tonnes/day of cellulose and an equivalent quantity of
spent sulfite liquor solids were to recover conidendrin in
0.5% yield based on spent liquor solids, the potential
annual (350 day) production of conidendrin would be 875
tonnes.

   It has been shown that the lignans of hemlock wood
include hydroxymatairesinol [11] and conidentrin [12] in a
ratio of about 5:1, corresponding to yields of about 0.25
and 0.05% of dry wood. The acidic condition of the sulfite
pulping reaction promote condensation and dehydration of
hydroxymatairesinol to form conidendrin.[66] Extensive
research has been conducted on conidendrin, particularly
in the laboratory of Crown Zellerbach Corporation.[35,93]
Demethylation to conidendrol (norconidendrin) [13], which
has a substituted catechol structure, has been demonstrated.
Many uses for conidendrin and its derivatives as antioxidants, chemical intermediates and medicinals have been

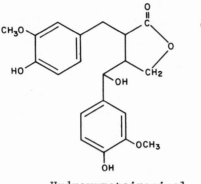

Hydroxymatairesinol
[11]

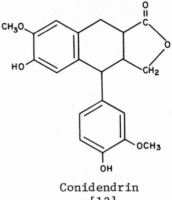

Conidendrin
[12]

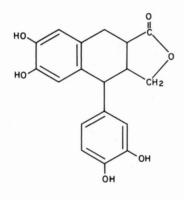

Conidendrol
[13]

proposed. Research samples were offered by Crown Zellerbach for many years. To date, these lignans have not been
utilized on a commercial scale.

The lignans present in western red cedar heartwood
have been investigated extensively at the Western Forest
Products Laboratory, Canada Department of Environment,
Vancouver, B.C.[59,60,61,189] The heartwood of western red
cedar was found to be a rich though variable source of
hot water extractives, obtained in yields from 3 to 23% of

the dry wood.[141]   The highest yields were obtained from
the darkly stained heartwood segments of old trees.   The
chemical structure of seven related lignans, including
plicatic acid [14] and its lactone plicatin [15] were elucidated during the 1960's.[189]

The development of commercial processes for these
lignans was of interest to two Canadian companies: Mac-
Millan Bloedel Limited and Raonier Canada Limited, both of
Vancouver, B.C.   The former company devised a hot-water
extraction process that was applied to cedar sawdust or
wood chips, derived from sawmill residues, to obtain an
extract in 8 to 10% yield.   The hot extract was treated with
calcium hydroxide to precipitate the basic calcium salts
(pH 10.5) of the "polyphenols" which were recovered in
about 4% yield of dry wood.   This separation process was
claimed to yield the active ingredient of the extract with
respect to an end-use in the electrorefining of lead.[40,195]

Studies at the laboratory of Rayonier Canada Limited
were directed to processes for recovering pure or crystalline plicatic acid salts or derivatives.[118,136]   Crude,
hot-water extract of cedar sawdust (8% yield) was solvent-
extracted to remove neutral (lactone) lignans.   Crude plicatic
acid was recovered in 4% yield of dry wood.   The purity of
this product was 75%.   In order to prepare crystalline
plicatic acid tetrahydrate it was advantageous to treat the

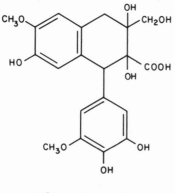

Plicatic acid
[14]

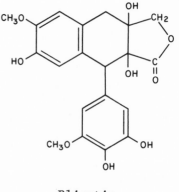

Plicatin
[15]

dry crude cedar extract with methanol and mineral acid to
form methyl plicatate, which is readily crystallized from
the crude ester mixture. Methyl plicatate is then treated
with sodium hydroxide to form crystalline sodium plicatate
which is acidified to yield pure plicatic acid tetrahydrate
crystals.[117] Purified plicatic acid was readily converted
to crystalline derivatives including plicatin[116] and a
number of esters[119] and amides[26] having valuable properties
as antioxidants for fats and oils and other food products.
This process and end-use technology is described in eleven
U.S. patents in addition to the references cited. The
investigation of plicatic acid derivatives also produced some
unusual products. Acetylation of dry cedar extract or
plicatic acid in ethyl acetate with a perchloric acid catalyst
yielded a crystalline, fully acetylated product [16], where
nuclear acetylation followed by condensation resulted in
the formation of an added pyrone ring. Hydrolysis of the
peracetate (deacetylation) yielded a bright-yellow crystalline compound for which the conjugated, zwitterionic structure [17] is proposed.[118] Although research samples of
many plicatic acid derivatives were offered and the economics
for producing such derivatives from cedar sawdust were
favorable, markets for these products have yet to be established.

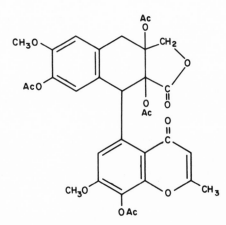

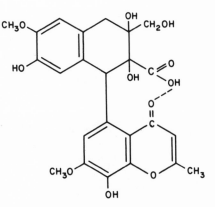

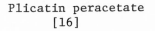

Plicatin peracetate
[16]

Hydrolysis product
[17]

Polyphenols and Their Polymers. [89,98,171] The most
important chemicals in this category of extractives are
those having flavonoid structures. This large class of
compounds and polymers is widely distributed in woody plants;
the compounds are often highly concentrated in heartwoods,
barks, leaves, fruits, and roots. Monomeric flavonoids,
as typified by dihydroquercetin [18], may be present in
wood to the extent of a fraction of 1% and yet be responsible
for the durability or decay resistance of Douglas fir.
This and other flavonoid chemicals act as inhibitors in the
sulfite pulping process, preventing satisfactory delignification. Polymeric flavonoids of the condensed or "catechin"
tannin type are important commercial products having a long
history in the manufacture of leather, and other uses such
as in dyes, preservatives and medicinal products. At the
peak of vegetable tannin usage and production during the
1950's, world production of the three major tannin extracts
(quebracho and chestnut wood and wattle bark) was estimated
as 300,000 tonnes/year. [120]

Our knowledge of the chemical relationships between
monomeric flavonoids and different types of tannins has
been progressing for nearly 100 years; however, many of the
fine structural details have only been determined in recent
decades. In general, such monomers as catechin [19], which
is detectable in the leaves and bark cambium of most conifers,
undergo enzymatic oxidative coupling to form tannin polymers
[20]. Further oxidative modification of tannins leads to

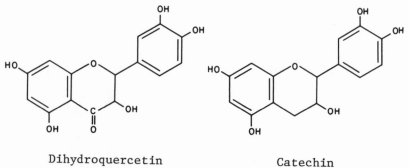

Dihydroquercetin   Catechin
     [18]   [19]

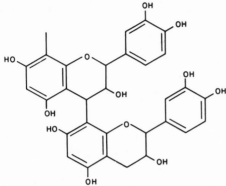

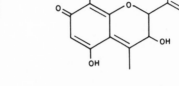

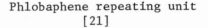

Tannin terminating unit   Phlobaphene repeating unit
        [20]   [21]

the formation of water-insoluble polymers that are more
intensely colored. These "phlobaphenes" and related higher
molecular weight polymers termed "phenolic acids" are major
constituents of western hemlock bark. Their chemical
structures include conjugated flavonoid units [21] where
some phenolic groups are present in the quinone form. The
large number of functional and structural isomers and
closely related flavonoid molecules account for the diversity
and variety in the chemical compositions of the condensed
tannins. Thus, the water-soluble tannin of western hemlock
bark is considered to be a polymer derived from catechin and
leucocyanidin.[97]

Industrial interest in chemical products from conifer
bark evolved during the 1950's in the Pacific Northwest of
the United States and Canada. Supplies of bark at the large
lumber and pulp mills were adequate for the commercial
production of bark polyphenols or tannins on a year-round
basis. Work at the Oregon Forest Products Laboratory
(1947) resulted in a process for extracting dihydroquercetin
[18] from Douglas fir bark.[80,97,98] Initial tests on
heavy, corky bark indicated yields of 5% dihydroquercetin
based on whole dry bark or 22% yield from the cork fraction.
Extraction with hot water, polar solvents or mixed solvents
was effective for the recovery of crude dihydroquercetin.
Early estimates were that 68,000 tonnes/year (150 million
pounds) of dihydroquercetin were potentially available from

Douglas fir bark available at lumber and pulp mills in the
states of Oregon and Washington.

Subsequent development of dihydroquercetin processes
at the Research Division of the Weyerhaeuser Company[50,77],
[78,168] were somewhat less optimistic. Mill run bark contained about 2.5% dihydroquercetin, but recovery in a hot-
water countercurrent extraction process was 1.8% based on
dry bark. Hot water extracts containing 3% total solids and
0.54% dihydroquercetin did not yield a readily crystalliz-
able product because of the presence of associated tannins
and carbohydrates. An air oxidation process, applied to
the hot extract solution was devised to convert dihydroquercetin to quercetin, which is much less soluble in water
and was obtained as a crystalline product in about 70% yield
based on dihydroquercetin analysis.[77] Later, a process was
worked out for treating the hot water extract with amyl
acetate, a good selective solvent for dihydroquercetin, from
which a crystalline product could be obtained.[50] Work at
Weyerhaeuser included the operation of a pilot plant for
the production of both dihydroquercetin and quercetin to
provide adequate samples for testing. A complete review
of the chemistry and literature on quercetin was made
available.[80] Markets for these products as antioxidants,
dyes and medicinals were too small to warrant further
commercial developments.

Aqueous extraction processes for bark, aided by chemicals such as sodium sulfite and bisulfite, sodium hydroxide
or ammonia, were developed by several companies. Perhaps
the initial intent was to increase the yield of water-
soluble extract to make production of a domestic vegetable
tannin economically attractive. Investigations at the
Olympic Research Division of ITT Rayonier Incorporated were
broadly oriented to include the development of high yield
bark extracts (20 to 50% of dry bark weight) for use as oil
well drilling mud additives (dispersants), water treatment
chemicals, phenolic adhesive components, carriers for agricultural micronutrients, and chemical grouting reagents.[103]
Western hemlock bark was preferred as a raw material because
of its high content of tannin, phlobaphene and phenolic
acids (Tables 1 and 2), from which net yields of sulfonated
products as high as 36% of dry bark weight were obtained,
or gross yields, including sulfite reagent weight, as high
as 50%.[101] From a study of the sulfonation of catechin[177]
it was concluded that the monomeric or repeating unit in the

polymer structure of sulfonated bark polyphenols and tannins is that of a sulfonated polychalcone [22]. Commercial production of sulfonated extracts of hemlock bark was begun by Rayonier in 1954 at Hoquiam, Washington and expanded in 1956 at Vancouver, B.C. A similar type of bark extract was produced commercially for a number of years from redwood (*Sequoia sempervirens*) bark by Pacific Lumber Company, Scotia, California. The principal use of this product was as a dispersant in oil well drilling fluids.[151]

The extraction of bark with alkaline reagents such as aqueous sodium hydroxide or ammonia yields products that are quite different from the sulfonated polyflavonoids. Crude extracts composed of the modified tannin, phlobaphene and phenolic acid fractions of bark contain a high proportion of acid (water) insoluble polyphenolic polymers. These polymers are only soluble in water at high pH (9-12) and are intensely colored (black appearance) indicating a high content of chromophoric structures. The term "phenolic acids" was applied generally to alkaline extracts of bark because of apparent titratable acidity and evidence for carbonyl content in infrared spectra.[80,98] More recently, a study of the base rearrangement of catechin to form "catechinic acid" has revealed that the acidity is attributable to a conjugated *enol* structure [23] and the formation of a new cyclohexanone ring.[179] It is probable that similar alkaline rearrangement accounts for chemical changes in alkaline bark extracts.

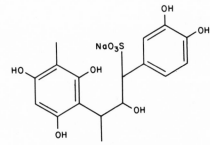

Sulfonated tannin
repeating unit
[22]

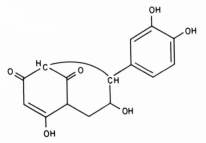

Catechinic acid
[23]

Selective extraction of hemlock and other conifer barks with aqueous ammonia or sodium hydroxide was carried out on a commercial scale by Rayonier[103] for the production of bark extracts that were reactive to formaldehyde and were useful as components of phenolic resin adhesives for exterior type plywood[105,106,107] and particleboard. Extracts of this type were obtained in about 25% yield based on dry bark. Higher extract yield was achieved under more drastic conditions, but such products were less reactive to formaldehyde and relatively less tolerated in adhesive formulations. Cold-setting adhesives based on combinations of bark extracts with resorcinol- or phenol-resorcinol resins and paraformaldehyde were used successfully in the manufacture of weatherproof laminated timbers from Douglas fir and other conifer lumber.[37,108,109] Commercial use of adhesives containing significant proportions (50% of solids) of bark extracts was not realized because of the low price of phenol and synthetic resins during the 1960's.

The most recent development by ITT Rayonier, in terms of uses for alkaline bark extracts, was that involving "chemical grouting". In this application a water solution of the bark extract is mixed with a solution of formaldehyde and a complexing metal salt such as sodium dichromate or ferrous sulfate. This reaction mixture is injected into porous soils, gravel or rock and forms a gel or immobile solid grout which adds strength, stability, and water impermeability to the grouted structure. The gelation time of the chemical grout can be varied from a few seconds to several hours, according to the engineering requirements to stop water flow or to provide stabilized soils for tunneling, excavation and foundation needs.[104]

Other companies such as Pacific Lumber Company and Weyerhaeuser were engaged for some time in the production of alkaline extracts of redwood and Douglas fir bark, respectively. The Weyerhaeuser Company also sold a Douglas fir bark fraction as a filler for phenolic resin adhesives for plywood. Extraction of formaldehyde reactive components of bark *in situ* was assumed in this case, even though it was difficult to demonstrate any advantage for the bark filler formulation over others that contained different acceptable fillers. Parallel work in the use of bark or wood polyphenols as phenolic resin components has been conducted in Australia,[158] New Zealand[81] and South Africa.[171]

Carbohydrates. When wood is subjected to extraction with water at temperatures in the range of 100 to 170 C for 30 minutes or longer, as much as 10% of the dry weight of conifer wood is solubilized. Somewhat larger proportions of common hardwoods are extracted. As the temperature and time of extraction are increased, the acids present in the wood, or those released by hydrolysis promote extensive depolymerization of hemicelluloses. The water-soluble carbohydrate fraction obtained in industrial water or steam "prehydrolysis" processes is composed largely of low molecular weight oligo- and polysaccharides and smaller amounts of monomeric sugars. Carbohydrate polymers containing pentose sugar units are more readily hydrolyzable and appear earlier or at lower temperatures in the prehydrolysis process. Hemicelluloses containing hexose units are more resistant to hydrolysis and require longer or more drastic treatment. Extensive research has been conducted on the course of the prehydrolysis reaction and on the composition of soluble fractions.[31,32,82,178] A typical composition for a prehydrolysate from southern pine wood is given in Tables 4 and 5.

From the above carbohydrate composition it is seen that about 75% of the water-soluble carbohydrate is present in the form of polymers.[178] Mannose is the principal wood sugar component of conifer wood prehydrolysates, while

Table 4. Composition of prehydrolysate from southern pine wood

	Dry solids
	% by weight
Carbohydrates: monomeric	19
polymeric	57
Lignin (derived)	9
Aldonic acids (as gluconic)	7
Furfural	3
Formic and acetic acids	1
Nitrogen	1
Ash	2

Table 5. Carbohydrates in prehydrolysate from southern
pine wood

| | Dry solids | |
	Monomeric	Total*
	% by weight	
Galactose	4	10
Glucose	2	14
Mannose	5	40
Arabinose	3	3
Xylose	5	9
Total	19	76

*After hydrolysis.

polymers of xylose (30%) and glucose (25%) are predominant
in hardwood extracts.

    A commercial product, classed as a "hemicellulose
extract" has been marketed since 1965 under the trade
name of "Masonex".[57] This material is a concentrate resembling molasses and is representative of carbohydrates
extracted during the wet process manufacture of fiberboard
products (hardboards). Such concentrates of wood sugars and
low molecular weight polysaccharides are valuable as ingred-
ients in feeds for ruminant animals, in the same manner as
conventional molasses from sugar refining.

    Growth potential in the prehydrolysis industry is
predicated on the basis of removal of carbohydrate fractions
(especially from hardwoods) that adversely affect the properties of particle- or fiberboard products. In the case
of two- or multiple-stage pulping processes, partial removal
of hemicellulose not only reduces chemical usage but improves
pulp quality in other ways.

    Certain woods contain specific or desirable polysac-
charide gums. The content of arabinogalactans in the larches
is of commercial interest, particularly when combined with
further pulping or with the conversion of the wood to board

products. Larch gum is extracted at 100 C and is obtained
in variable yield according to age and location of wood in
the tree. Average yields of 10% are obtainable.[154,201]

CHEMICALS DERIVED FROM
CARBOHYDRATE COMPONENTS

   Wood Sugars. Wood and bark contain pentose sugars in
the form of d-glucuronoarabinoxylan, a polymer composed
predominantly of xylose with side units of 4-*O*-methylglu-
curonic acid and arabinose, and hexose sugars in the form
of cellulose (glucan) and galactoglucomannans, polymers
derived predominantly from mannose with lesser amounts of
galactose and glucose. From the above discussion of the
mild aqueous hydrolysis of the hemicellulose components of
wood, it follows that more drastic acidic conditions will
ultimately result in the hydrolysis of cellulose. Mild
acidic prehydrolysis of wood using 0.5 to 1% sulfuric acid
at 120 to 130 C was used in Germany during the 1940's in
two-stage acid-alkaline pulping processes.[201] Such hydrolysates are rich in monomeric wood sugars and were thus useful
for aerobic fermentations by *Torula* or fodder yeast. Prehydrolysis with dilute sulfurous acid is also practiced
and, as in the case of the sulfite pulping process, the
spent liquor contain the five common wood sugars derived
from hemicelluloses, in varying proportions depending on
the wood raw material, while glucose content is increased
as the hydrolysis of cellulose becomes more significant.
The three common hexose sugar monomers associated with wood
carbohydrates are: D-galactose [24], D-glucose [25] and
D-mannose [26]. Two pentose sugar monomers are present and
one uronic acid: L-arabinose [27], D-xylose [28] and
glucuronic acid [29].

   The composition of the acid hydrolysate of the hemicelluloses of spent sulfite liquor from western hemlock
wood chips treated with the acid sulfite process is shown
in Table 6.

   A process for recovering mannose from conifer spent
sulfite liquors via the sodium bisulfite adduct has been
developed.[33,110] The intermediate "sodium mannose bisulfite"
[30] was crystallized selectively from the crude wood sugar
bisulfite adducts and lignosulfonate mixture. Mannose was
then regenerated along with sodium sulfite for recycle to
the process.

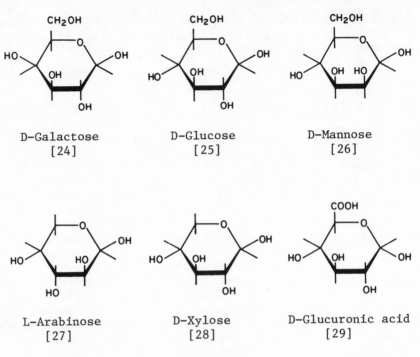

D-Galactose   D-Glucose   D-Mannose
   [24]   [25]   [26]

L-Arabinose   D-Xylose   D-Glucuronic acid
   [27]   [28]   [29]

Processes for recovering xylose from hardwood hydrolysates are usually based on the selective fermentation of
hexose sugars to ethanol, which leaves xylose and arabinose
unchanged. Such crude solutions are then concentrated and
purified and xylose is recovered by crystallization. A
process for preparing crystalline xylose and xylitol from
hardwood spent sulfite liquor (dry solids) via methanol
extraction and purification is available.[20] The xylose
remaining in fermentation broths may also be hydrogenated
directly to produce crystalline xylitol.[123] Xylitol is
primarily of interest as a sweetener for use by diabetics.
Recent studies have demonstrated physiological acceptance
in the human diet.[95]

Table 6. Composition of spent sulfite liquors from western hemlock wood

	Dry solids	
Base:	Sodium	Ammonium
	% by weight	
Wood sugars – total	29.5	24.9
Galactose	4.0	3.4
Glucose	4.4	3.2
Mannose	15.5	13.4
Arabinose	1.5	1.2
Xylose	4.1	3.7
Carbohydrate polymers	0.0	3.5
Lignosulfonates (as sodium)	67.0	66.0
Aldonic acids (as gluconic)	1.6	3.5

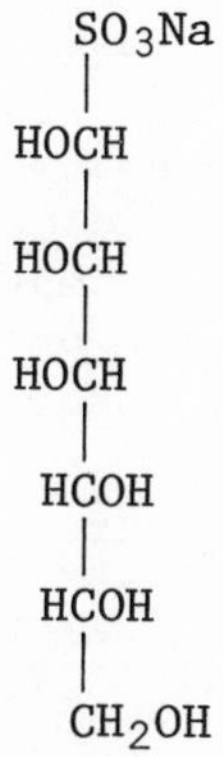

$$SO_3Na$$
$$|$$
$$HOCH$$
$$|$$
$$HOCH$$
$$|$$
$$HOCH$$
$$|$$
$$HCOH$$
$$|$$
$$HCOH$$
$$|$$
$$CH_2OH$$

Sodium mannose bisulfite
[30]

Acid hydrolysis of wood[154,162,201] for the specific purpose of producing sugar (glucose) has been of interest for more than 100 years. Though not presently practical in North America, such processes have been a part of the sugar industry in European countries, the Soviet Union and Japan. Hydrochloric acid is the most efficient hydrolyzing

agent but dilute sulfuric acid (0.4-0.6%) is generally used
in industry such as in the Scholler-Tornesch or comparable
processes.   In order to preserve the hydrolysate or monomeric sugar solution from the destructive effects of hot
acid, most processes are based on continuous hydraulic flow
of reagent, while hydrolysate is removed to a cool zone.
Total hydrolysis time to process a charge of wood chips
may last several hours while hydrolysis temperature is
gradually increased from 140 to 180 C.   Hydrolysate solutions,
containing 4 to 6% reducing sugar, are neutralized with
calcium carbonate, filtered and concentrated by evaporation.
The yield of sugar varies with the wood species used.
Average yield from dry conifer wood falls in the range of
50 to 55% based on reducing sugar analyses or about 40%
fermentable sugars, while deciduous woods yield 35% fermentable sugars and 20% unfermentable pentoses.   These fermentable sugar values are pertinent to alcoholic fermentation.
If the objective is to produce food or fodder yeast, both
hexose and pentose sugars and other constituents such as
acetic acid are utilized.   Crystalline glucose has been
manufactured in Japan by wood saccharification processes.[162]

Renewed interest in the enzymatic hydrolysis of cellulose to produce glucose is the result of the development of
powerful new cellulase enzymes from the tropical fungus,
Tricoderma viride, described in Chapter 8.   The U.S. Army
process[56,203] continues to be improved and holds promise
as a noncorrosive, atmospheric pressure process for converting cellulosic wastes to glucose.

Furfural and Hydroxymethylfurfural.   The dehydration
of pentose sugars results in the formation of furfural
[31] while the corresponding reaction on hexose sugars
yields hydroxymethylfurfural [32].

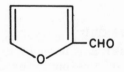

Furfural   Hydroxymethylfurfural
[31]   [32]

These reactions are acid-catalyzed and are favored by high temperature and pressure. High conversion efficiency and yield are contingent on rapid conversion during very short reaction times, such as less than one minute, at high temperature (200 to 300 C) and the immediate removal of the product from the reaction zone to prevent degradation or resinification.

Since furfural is steam volatile, removal from a reaction mixture and purification are relatively simple. The furfural industry[44] has been long established on the basis of pentosan-rich agricultural residues such as corn cobs, grain hulls and sugar cane bagasse. Hydrolysis to xylose and conversion to furfural are concurrent in a digestion with dilute mineral acid that may last several hours. The use of hardwoods[86,169,198,206] is also advantageous in furfural production. Wood from many species contains hemicellulose components, that on hydrolysis yield 20 to 25% pentose sugars on a dry wood basis. Such "pentosan" yield is comparable to that obtained from all but the best agricultural residues. A long established practice in Europe has been to recover furfural from the sulfite pulping of hardwoods. Partial conversion of xylose to furfural occurs during the acidic sulfite digestion and furfural is recovered from relief, blowdown, stripping, and evaporation condensates. Hardwood spent sulfite liquors contain unconverted xylose. In a recent study,[206] a hardwood spent sulfite liquor containing 23.3% xylose on a dry weight basis was tested for furfural yield under optimum high temperature, fast reaction conditions. Conversion to furfural was nearly quantitative yielding 9.25% furfural based on dry starting material weight. Prehydrolysis of hardwoods, especially in the presence of dilute sulfuric acid (0.5%), provides a convenient way to remove and hydrolyze the pentose-rich hemicellulose fraction for conversion to furfural, while the remainder of the wood is pulped by the kraft process to produce a high grade cellulose.

In the formation of hydroxymethylfurfural[58,87,149] it is advantageous to have the raw material converted to a hexose sugar by conventional saccharification. Since hydroxymethylfurfural is not steam-volatile to any extent, the acidic reaction mixture is exposed to high temperature for a very short time and is then cooled rapidly. Hydroxymethylfurfural is recovered by solvent extraction and purified by distillation.[124]

World production of furfural was estimated as 90,000 tonnes in 1969.[188] Furfural is a well established commercial chemical having wide utility as a unique solvent and as a chemical intermediate for a variety of valuable furan derivatives.[44] Although hydroxymethylfurfural has only been produced in pilot plant quantities, its uses as a chemical intermediate are promising.

Levulinic Acid. Levulinic acid [33] is a γ-keto-carboxylic acid. Kinetic studies have shown that hydroxymethyl furfural most likely is an intermediate in the formation of levulinic acid from hexose sugars.[88,148] Equimolar amounts of formic acid are co-produced during the formation of levulinic acid, cleavage being predominantly between the first and second carbon atoms of the original hexose sugar.[184] Further additional dehydration of levulinic acid occurs under the same condition as its initial formation to form the cyclic compound α-angelica lactone [34] which is readily convertible to the β-isomer [35].

The theoretical yield of levulinic acid from glucose is 64.5%. Optimization experiments in our laboratory indicated the best practical yield was 42% from glucose or about 33%, based on cellulose.[28,155] These yields were obtained when the reactions were carried out in 6% aqueous hydrochloric acid for times as long as 150 minutes at 140 C or as short as 15 minutes at 160 C. Raw wood, aqueous prehydrolysate from coniferous dissolving pulp manufacture, or screen rejects from sulfite pulp manufacture could be used equally well as the carbohydrate resource, but the

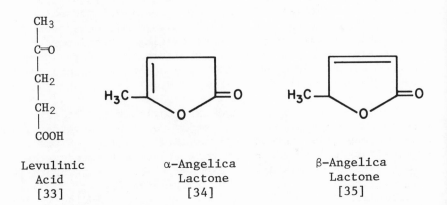

| Levulinic
Acid
[33] | α–Angelica
Lactone
[34] | β–Angelica
Lactone
[35] |

product contained colored impurities (probably derived from lignin) which required extensive decolorization with ion exchange resin or charcoal. The colored impurities were not obtained when alpha cellulose or glucose were used as raw materials.

The Quaker Oats Company was the industrial pioneer in the commercial production of levulinic acid. A plant at Omaha, Nebraska, was put into operation in 1958 to utilize hexose-containing corncob residues remaining from furfural production. An extensive review of uses and chemical reactions was published to encourage market development.[163]

In 1962 the Crown Zellerbach Corporation began commercial production of levulinic acid in a five million pound per year (2270 tonnes) plant located adjacent to the kraft pulp mill at Port Townsend, Washington.[174,184] Douglas fir sawdust, a byproduct of the pulp mill and local lumber mills, was used as the raw material. Wood and dilute hydrochloric acid were heated under pressure in a batch digester. The filtrate from the reaction was heated to remove water and hydrochloric acid, the latter being recycled to the digester. Crude levulinic acid was decolorized and fractionated to yield a product of 95 to 98% purity. Severe corrosion problems were encountered, and the plant was closed down in 1965.

The main point of market interest that was encouraging in the above production ventures was the contemporary development of "diphenolic acid" [36] or 4,4'-bis-(p-hydroxyphenyl pentanoic acid)[11] by the S. C. Johnson Company (waxes) as a new intermediate for the synthesis of protective and decorative finishes.[10,74,75,76,122] Unfortunately, the market for levulinic acid in this and other uses, such as in foods as an acidulent and a carrier for calcium in the diet, was not strong enough to gain permanence.

The lactones of levulinic acid are interesting monomers in their own right; however, they represent a serious nuisance when the preparation of pure levulinit acid is the objective. A Crown Zellerbach process sought to minimize this problem and maintain overall yield of levulinic acid by recycling angelica lactone distillate fractions to the initial hydrolysis-dehydration reaction step.[174]

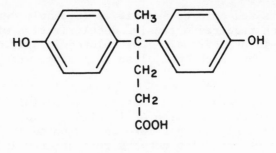

Diphenolic acid
[36]

        Research in the Soviet Union has focused on a two-
stage process for the production of furfural and levulinic
acid.[152] Such a process is similar to that used by the
Quaker Oats Company. The raw material is selectively
hydrolyzed to pentoses with 5% sulfuric acid, dehydrated to
furfural at 150 C, and steam distilled to remove furfural.
The residue is heated for 120 minutes at 180 C, filtered to
remove lignin, and extracted with diethyl ether to selectively remove a levulinic acid fraction which is subsequently
purified by fractional distillation. Subsequent work has
been devoted to the special metallurgy required for the
process[185] and to increasing the yield through the use of
ferric ion[153] or a mixture of bromate and bromide ions as
catalysts[197] and by the addition of concentrated hydrochloric
acid to the woody raw material to facilitate dissolution
prior to subsequent hydrolysis at 170 to 250 C.[53] A study
of the most suitable acid showed that hydrochloric acid
gave the best yields[52] presumably because the acid anion
is involved as an intermediary in the dehydration step.

        Recent two-stage process studies on sugar cane bagasse
are undoubtedly equally applicable to hardwood residues.
When sulfuric acid is used as a catalyst, concentration of
the residue following steam distillation to remove furfural
results in a better yield of levulinic acid during the
subsequent high temperature (185-210 C) hydrolysis-dehydration reaction.[164,166] The calcium or barium salt is then
used as a method of obtaining color stability.[165]

        Levoglucosan. The dehydration of sugars leads to the
formation of a large number of anhydrosugar derivatives

depending on the conditions used and the number of molecules of water that are abstracted. Furfural and hydroxymethyl-furfural are formed by the removal of three moles of water per molecule of starting sugar. Levoglucosan [37] is representative of anhydrosugars where only one molecule of water has been removed from a hexose sugar.

Levoglucosan, or more properly 1,6-anhydro-β-D-gluco-pyranose, is a crystalline solid having a long chemical history.[180,201] It was first characterized in 1894 and prepared from cellulose in 1918. It is formed in yields as high as 40% by the pyrolysis of dry cellulose at 300 C, under conditions of high vacuum and removal of products from the reaction zone. The direct formation of levogluco-san from glucan polymers such as cellulose or starch is advantageous because the polymer structures are already partial anhydrides. The yield of levoglucosan from glucose is only 6%. The use of partially hydrolyzed cellulose, having a degree of polymerization of about 150, results in an increased yield of 60% of the starting material weight.

Small amounts of levoglucosan and other anhydrosugars are present in the tars associated with the burning of wood or carbohydrate materials. Dry distillation of wood, as in the preparation of charcoal, also results in the formation of low yields of levoglucosan, which may be recovered from the tar or liquid condensate.

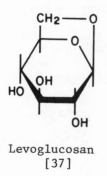

Levoglucosan
[37]

There has been some interest in the commercial production of levoglucosan from waste wood and cellulose pulp. Pyrolysis of hydrolyzed wood cellulose at 350 to 400 C in a stream of superheated steam has resulted in a yield of 62% levoglucosan.[29]

Recent research[180,181] using levoglucosan as a model compound in pyrolysis reactions, such as occur in the combustion of wood, paper and cellulose textiles, has the objective of providing control over the reaction, to develop flameproofing for textiles or wood products, or to direct the course of decomposition to increase the yield of valuable chemical products such as furan derivatives and 1- to 4-carbon aliphatic derivatives: hydrocarbons, alcohols, aldehydes, ketones, and acids. Levoglucosan has also been used as a monomer in polymerization reactions that yield "synthetic" high molecular weight, branched-chain polysaccharides.[30,204]

Sugar Acids.[73] This chemical category is represented by a wide variety of compounds having carboxyl groups at one or the other end of the molecule or at both ends.

The uronic acids [38], as exemplified by glucuronic acid, are ubiquitous in plant materials and are especially abundant in plant gums, pectins and many seaweeds. Glucuronic acid is very important in animal physiology, serving as a detoxifying agent for otherwise poisonous substances eliminated in the urine. The 4-O-methyl ether of D-glucuronic acid is a common building unit in wood hemicellulose and is probably the source of most of the methanol produced in the sulfite pulping process. Uronic acids are chemically less stable than other sugar acids and are decarboxylated under acid or thermal conditions. Hexuronic acids yield furfural on decarboxylation and dehydration.

$$
\begin{array}{ccc}
\text{CHO} & \text{COOH} & \text{COOH} \\
| & | & | \\
\text{(CHOH)}_n & \text{(CHOH)}_n & \text{(CHOH)}_n \\
| & | & | \\
\text{COOH} & \text{CH}_2\text{OH} & \text{COOH}
\end{array}
$$

Uronic acids	Aldonic acids	Saccharic acids
[38]	[39]	[40]

Aldonic acids [39] such as gluconic acid [41], where the aldehyde group has been oxidized to carboxyl, are very stable under alkaline conditions and are the basis for a growing number of commercial applications. Gluconic acid and sodium gluconate are manufactured from glucose by fermentation processes[161] and may also be produced by chemical oxidation. Spent sulfite liquors contain the aldonic acids derived from the common wood sugars, from which they are formed by partial chemical oxidation during the process.[173] Although total aldonic acid content only amounts to about 4% of dry solids in spent sulfite liquors (Table 2) or 7% in hemicellulose prehydrolysate (Table 1), the presence and analysis of these sugar acids is of interest. An analysis of total "lactonizible hydroxy acids" calculated and calibrated on the basis of gluconic acid is useful.[112]

Because of the industrial importance of aldonic acids it has now become common to convert glucose to glucoheptonic acid [42] via the Kiliani cyanohydrin synthesis.[121] This reaction has also been applied to spent sulfite liquor to convert all sugars to stable aldonic acids. A process for recovering crystalline sodium mannoheptonate from a crude mixture of sodium aldonates and lignosulfonates has been developed.[19]

	COOH	
	$\vert$	
COOH	HCOH	COOH
$\vert$	$\vert$	$\vert$
HCOH	HCOH	HCOH
$\vert$	$\vert$	$\vert$
HOCH	HOCH	HOCH
$\vert$	$\vert$	$\vert$
HCOH	HCOH	HCOH
$\vert$	$\vert$	$\vert$
HCOH	HCOH	HCOH
$\vert$	$\vert$	$\vert$
CH_2OH	CH_2OH	COOH
D-Gluconic acid [41]	D-Glucoheptonic acid (two isomers) [42]	D-Glucaric acid [43]

The saccharic acids [40] are also of commercial interest and are prepared by the oxidation of sugars or carbohydrate polymers with warm concentrated nitric acid. Glucaric acid [43] is obtained in 65% yield from starch. Galactaric (mucic) acid may be prepared from galactans such as the arabinogalactan from larch wood.

Uses for sugar acids are well established in foods, medicinals, and in cleaning and metal ion sequestering or complexing applications. Aldonic acids are used in ferrous metal derusting baths and as retarders for Portland cement concrete.

A final group of monocarboxylic sugar acids known as the "saccharinic" acids[72,150] are of importance because they are major constituents of kraft black liquor and alkaline process bleaching or refining effluents in the pulp and paper industry. These acids are produced from cellulose and hemicellulose by the so-called "peeling degradation, whereby the sugar units are rearranged and fragmented. Saccharinic acids are classified as meta- or iso- according to their chain structure. Examples of these are: D,L-2,4-dihydroxybutyric acid [44], a metasaccharinic acid, and β-D-glucoisosaccharinic acid [45], usually called isosaccharinic acid.

On a solids basis, kraft black liquors contain 30 to 39% organic acids, where major constituents are the saccharinic acids and lactic, glycolic, acetic, and formic acids. The alkaline refining of wood cellulose is practiced in

$$
\begin{array}{cc}
 & \text{COOH} \\
 & | \\
\text{COOH} & \text{HOC}\!-\!\text{CH}_2\text{OH} \\
| & | \\
\text{HCOH} & \text{CH}_2 \\
| & | \\
\text{CH}_2 & \text{HCOH} \\
| & | \\
\text{CH}_2\text{OH} & \text{CH}_2\text{OH}
\end{array}
$$

2,4-Dihydroxybutyric acid (two isomers) [44]	Glucoisosaccharinic acid (two isomers) [45]

the production of very pure chemical cellulose or "dissolving pulp". Hot sodium hydroxide extraction liquors from
such processing contain as much as 54% organic hydroxy
acids on a solids basis, of which major components are the
glucoisosaccharinic acids (24%) and dihydroxybutyric, lactic
and glycolic acids. The resource of hydroxy or sugar acids
in the pulp and paper industry is enormous. Thus far, no
commercial processes or economic procedures to recover
these sugar derivatives have been put into practice.

   Formic and Acetic Acids. As has already been mentioned
above, many sugar fragments or degradation residues are
formed by the processing of wood or wood pulps under alkaline
conditions. Formic acid is stabilized by salt formation
and is thus prevented from decomposing to carbon dioxide
as occurs in acid systems. The alkaline refining of cellulose results in liquor containing as much as 12% formic and
2.5% acetic acids, present as sodium salts, based on total
dry solids content.

   Kraft black liquor also contains appreciable quantities
of formic and acetic acid. The production of acetic acid
is greater from hardwoods than conifer wood. The acetyl
content of hardwoods, present as acetylated sugar components and some free acetic acid, is about 3.9% of dry wood,
whereas this analysis for conifer wood yields a value of
about 1.4%.

   The commercial production of acetic and formic acid
from a hardwood black liquor of a neutral sulfite semi-
chemical pulping process was initiated in 1958 by the Sonoco
Products Company at Hartsville, South Carolina.[15] The
sodium acetate and formate content of this liquor is typically 25.0 and 3.0%, respectively, of dry solids, equivalent
to 18.3 and 2.3% acetic and formic acids. The recovery
process is based on concentration of dilute black liquor
to 40% solids, acidification with sulfuric acid, and extraction of acetic and formic acids with 2-butanone. Acetic
acid is further refined to a specification glacial grade of
99.5% purity. Formic acid production has been discontinued.
Sodium sulfate (salt cake) recovered from the process is
sold to kraft pulping mills for the regeneration of pulping
chemicals.

   A process to recover acetic acid from spent sulfite
liquor evaporation condensates has been developed.[8] Acetic

acid is first adsorbed on activated carbon and is desorbed as ethyl acetate. This process is being demonstrated at Flambeau Paper Company, Park Falls, Wisconsin.

The Sonoco plant is representative of the innovative and pragmatic process development required to compete with synthetic acetic acid. Direct distillation of acetic and formic acids from water solution is not practical, nor is solvent extraction from dilute water solutions. The advantageous extraction coefficient obtained by extracting from a strong electrolyte solution, which also suppresses solvent solubility in the aqueous phase, and the use of the same solvent for azeotropic distillation to remove water from the crude acid mixture, are reasons why this process works. Similar advantageous conditions need to be searched for in other segments of the wood products industry.

Finally, it should be mentioned that acetic acid has a long history of production by the dry distillation (pyrolysis, 270–400 C) of hardwoods. It is still produced in this manner in many countries and is obtained along with formic and propionic acids, methanol, furfural, tars, and charcoal. The aerobic fermentation of sugars to acetic acid solutions (vinegar) has been practiced for several thousand years.

Alcohols, Polyols. It has long been noted that steam-volatile fractions collected as water condensates from pressure relief or "blow-down" gases of wood pulping processes contain a variety of low-boiling organic compounds, including methanol. Such minor components were not considered to be economically recoverable. Methanol is formed by acid or alkaline hydrolysis from 4-*O*-methyl glucuronoxylan components of hemicellulose. New or modified pulp mill systems, particularly in sulfite pulping, are now designed to capture even minor components which may then be concentrated or accumulated in spent liquor evaporating systems. The evaporator condensate may be processed by means of fractionation towers to recover low-boiling or steam-volatile components. The recovery of methanol, furfural and *p*-cymene by such means has been proposed[12] and a demonstration unit for the recovery of methanol and furfural is under construction at Flambeau Paper Company, Park Falls, Wisconsin.[8] Even though recovery of only 1.5 to 4 tonnes/day of methanol, depending on the pulp mill size, does not represent much in terms of the commercial capacity for methanol, its recovery along with similar amounts of furfural is of interest in terms of local markets.

The fermentation of sugars in spent sulfite liquors[146] to produce ethyl alcohol has been practiced in Europe for many decades. In 1945, it was reported that alcohol (95%) production in Sweden amounted to 31.6 million gallons or 96,700 tonnes and was conducted in 33 plants. Only two North American sulfite mills are engaged in this type of fermentation: Georgia Pacific at Bellingham, Washington and Ontario Paper Company, Ltd. at Thorold, Ontario. Annual production is estimated[22,140] as 3.2 and 1 million gallons, respectively, or 9790 and 3060 tonnes. The anaerobic fermentation is carried out with a *Saccharomyces* yeast which convert hexose sugars to alcohol in a theoretical yield of about 51% by weight. Practical yields are at best about 40% based on fermentable sugar weight and fermentation efficiency. Alcohol is steam-distilled from the residual lignosulfonate and pentose sugar material and is fractionally distilled to 95% purity. Pentose sugars are not fermented and thus hardwood spent sulfite liquors are a poor resource for alcohol production, fermentation being used primarily as a route to the recovery of xylose.

The large scale production of sugar alcohols or polyols, ethylene glycol [49], glycerol [50] and other glycol derivatives from sugars via hydrogenation and hydrogenolysis has been commercial for several decades.[14] Primary reduction to the sugar alcohol can be carried out by electrolytic or low-pressure catalytic hydrogenation processes. The latter route is favored and is quantitative in yield. Typical processes utilize continuous reactors. Sugar solutions containing as much as 40% solids are mixed with Raney nickel or other catalysts and reacted with hydrogen at 50 to 100 atms and 80 to 100 C. Under more drastic conditions of temperature and pressure, 200 to 300 C and 150 atms, hydrogenolysis of 5- and 6-carbon chains results in the formation of good yields of glycerol, ethylene glycol, and other glycols, depending on reaction conditions. It is best to conduct hydrogenolysis as a second step, following hydrogenation to the sugar alcohol, in order to avoid sugar decomposition at the higher reaction temperatures. Hydrogenolysis yields a mixture of products that is not easily separated.

In recent years there has been interest in two wood sugar polyols: mannitol [47] and xylitol [48]. Mannitol[17,110]

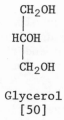

```
        CH2OH                CH2OH
         |                    |
        HCOH                 HOCH                CH2OH
         |                    |                   |
        HOCH                 HOCH                HCOH
         |                    |                   |
        HCOH                 HCOH                HOCH
         |                    |                   |
        HCOH                 HCOH                HCOH
         |                    |                   |
        CH2OH                CH2OH               CH2OH

      Sorbitol            D-Mannitol            Xylitol
     (D-Glucitol)            [47]                [48]
        [46]
```

```
                   CH2OH                  CH2OH
                    |                      |
                   CH2OH                  HCOH
                                           |
                                          CH2OH

             Ethylene glycol            Glycerol
                  [49]                    [50]
```

is potentially crystallizible directly from a mixture of
wood sugar polyols. The commercial production of mannitol
is as a coproduct with sorbitol [46] from the mild alkaline
hydrogenation of invert sugar or glucose. Mannitol is
formed partly from the hydrogenation of fructose but also
from mannose resulting from the alkaline epimerization of
glucose. As much as 20% by weight of mannitol may be
obtained in the sorbitol processes from which it is readily
separated by crystallization. If mannitol were sold at
the same price as sorbitol it would have a much larger
market because of its crystalline, nonhygroscopic and
dietetic sweetener properties.

The production of xylitol from hardwood prehydrolysates is now commercial in Finland.[9] Purification of
xylose by alkaline fermentation of the hexose sugar components of the wood sugar hydrolysate yields a solution that
is then concentrated and hydrogenated. Pure xylitol is
recovered by crystallization. The reported capacity of the

xylitol plant in Finland is 2722 tonnes/year. Markets for xylitol were previously considered as being in the area of dietetic sweeteners,[95] sweetness being comparable to that of sucrose. However, recent study has revealed that xylitol in the diet may reduce dental cavities by as much as 90%.[9] This property of xylitol could result in high demand and rapid expansion in production capacity.

CHEMICAL PRODUCTS DERIVED
FROM LIGNIN

   Vanillin and Related Compounds. Oxidation of lignin in alkaline media, using various oxidizing agents, results in the degradation of the polymer and the formation of a multiplicity of monomeric aromatic compounds, most of which may be classed as substituted phenols.[23,156] Vanillin [51], the most important, is produced commercially from the crude lignosulfonate liquors obtained by fermentation or by the fractionation of spent sulfite liquor to remove or utilize sugar components. Vanillin and to a lesser extent vanillic acid [52] and acetovanillone [53] are the major products derived from conifer wood, while syringaldehyde [54] and *p*-hydroxybenzaldehyde [55] and closely related derivatives are also formed in the oxidation of hardwood lignin and the lignins derived from grasses, canes and bamboos.

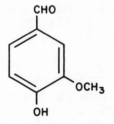

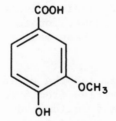

Vanillin
[51]

Vanillic Acid
[52]

Acetovanillone
[53]

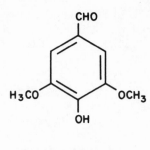

Syringaldehyde   *p*-Hydroxybenzaldehyde
    [54]   [55]

      The formation of vanillin in the alkaline degradation
of spent sulfite liquor has been known for about 50 years.
Gradual development of the vanillin production process,
from a simple alkaline treatment to a high temperature and
pressure oxidation, including injection of air or oxygen
and the use of catalysts, took place from about 1935 to
1955. Early process yields of vanillin of 2 to 3% based on
lignin weight were increased to 14.4% by using a copper
hydroxide-oxide catalized, pressure oxidation with air or
oxygen for 3 hours at 170 C, in the presence of optimum
ratios of sodium hydroxide to lignin.[142] By use of dilute
aqueous reaction media (3.5% lignin, 10% NaOH) and oxidation
at 200 C, the yield of vanillin was further increased to
21%, of which 80% was recoverable.[175]

      Nitrobenzene is a unique oxidizing agent in alkaline
systems and has been utilized in research on the structure
of lignin under conditions of mild alkaline oxidative
degradation. The yield of vanillin in nitrobenzene oxidation from a variety of lignin materials is much higher than
obtained otherwise. Lignosulfonate oxidation yields 25%
vanillin and claims have been made for yields as high as
32.4%.[187]

      In commercial practice the yield of vanillin from
spent sulfite liquor varies from 5 to 10% depending on the
type of raw material and sodium hydroxide and catalyst
usage. Vanillin is extracted from oxidized alkaline liquors
with butanol and is purified by crystallization or by
vacuum co-distillation with an inert hydrocarbon.[25]

Vanillic acid is useful as a monomer for the synthesis of polyesters having good fiber-forming characteristics as needed for textiles.[16,18,48] Ethyl vanillate[156,157] [56] is less toxic to humans than sodium benzoate but very toxic to specific microorganisms, and has thus found commercial use as a food preservative and medicinal in the treatment of human and animal diseases. Vanillic acid diethylamide [57] is widely used in Europe as an analeptic medicinal for the control of respiration and blood pressure. More recently vanillin has been used in the synthesis of the medicinal "L-dopa" (L-dihydroxyphenylalanine), which is effective in the treatment of Parkinson's disease.[6]

Guaiacol, Catechol, Protocatechuic Acid, and Related Phenols.[1,68,201] Continuing from the above discussion, if wood or lignin are subjected to degradation by pyrolysis, it is possible to isolate small yields of a large number of monomeric phenols and low-molecular weight "lignols". The presence of salt-forming reagents such as potassium or sodium hydroxide is advantageous since the stability of the phenolic fragments is increased. The dry distillation of wood (400-600 C), concurrent with charcoal manufacture (Fig. 1) has a history that extends into ancient times, because of the uses of wood tar as a valuable preservative, sanitizing agent and medicinal product. The yield of wood

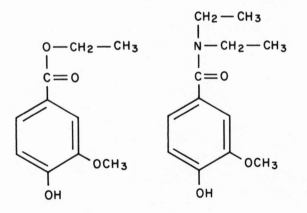

Ethyl Vanillate   N-Diethyl-vanilloylamide
     [56]   [57]

tar is about 10% of dry wood in efficient batch processes,
but may be increased substantially by continuous pyrolysis
at high temperature.[201] About half of the tar is composed
of steam-volatile phenols boiling in the range of 180 to
225 C. Guaiacol [58] and 4-methylguaiacol are major constituents, representing as much as 30% each of the distillable
phenols.[147] Crude technical guaiacol(s) are obtainable in
5% yield based on dry wood and are marketed as modifiers
for paint and varnish resins. Pyrolyses of lignins under
reduced pressure and in the presence of inert carrier oils
have been reported to yield 7% to 10% phenolic products.[1,201]

Alkali fusion (KOH) is a classical procedure in organic
chemistry and was applied to wood and lignin in structural
studies as early as 1866. Catechol [59] and protocatechuic
acid [60] may be obtained from isolated lignin in yields of
9 and 19%, respectively.[111] Catechol yields (23%) were
increased, at the expense of decarboxylating the protocatechuic acid, when alkali fusion was carried out in the
presence of iron. Catechol can also be prepared from
guaiacol by demethylation or from vanillin via vanillic and
protocatechuic acid intermediates. Interest in the conversion of vanillin to protocatechuic acid by an alkali fusion
method was based on potential uses for the latter in the
production of polyesters having fiber-forming properties.[17]

The search for low-cost means for the production of
phenols by the alkaline degradation of commercial lignins
has continued to the present time. Processes developed in
Finland[36,47] during the 1960's were based on high temperature
(250-290 C) treatment of kraft pulping black liquors

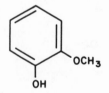

Guaiacol
[58]

Catechol
[59]

Protocatechuic acid
[60]

containing 16 to 20% organic solids, in the presence of
additional sodium hydroxide and sodium sulfide, for time
periods in the range of 10 to 20 minutes. This type of
reaction promoted demethylation of the lignin and the formation of methylmercaptan and dimethylsulfide as will be
discussed later. About 50% of the organic material was
converted into an ether-soluble product composed largely
of phenols. Catechol and 4-methylcatechol were obtained
in about 5 and 2% yield, respectively, based on lignin
weight. A crude phenol(s) fraction and a crude protocatechuic acid (derivatives) fraction, each in 27% yield
of lignin, were also obtained. Residual lignin (18%)
recovered from this process is claimed to have high phenolic
hydroxyl content and utility in the formulation of phenolic
resins.

The most recent development in this field represents
a return to dry pyrolysis.[46] Concentrated softwood kraft
black liquor was treated with 18% sodium hydroxide based
on solids, and spray dried. The lignin material, containing
5% moisture, was then subjected to continuous pyrolysis at
310 C for about seven minutes in an atmosphere of nitrogen.
Catechol and its 4-methyl and 4-ethyl derivatives were
obtained in a yield of 48 kg/tonne of 90% pulp, calculated
as equivalent to about 8.7% yield based on lignin weight.
Part of this yield was obtained by "cracking" ether-soluble
derivatives of guaiacols and protocatechuic acids.

Other approaches to the production of phenols from
lignin have been based on hydrogenolysis or degradation
under various conditions of hydrogenation. The most
practical or nearly commercial process in this area is
discussed below. Hydrogenation of acid hydrolysis lignin
(sulfur-free) at 250 C for one hour, in a benzene solvent
medium using special nickel catalysts, yielded 30 to 36%
monomeric phenols.[176] The single compound obtained in the
highest yield was 4-propylguaiacol (20% of lignin). Hydrogenolysis of the same type of lignin at 400 C for one hour,
in a cyclohexane medium using a nickel carbonyl catalyst,
resulted in a yield of 31.6% of catechol.[172] It has also
been proposed that "phenol lignin", resulting from the
pulping of wood in aqueous acidic phenol, is well suited
for the production of the pulping chemical itself or a
commercial mixture of monomeric phenols.[199] A continuous
pyrolysis process at 500 to 600 C, applied to dry sodium

"lignophenolate", gave optimum yields of 30% phenols based
on sodium-free lignin weight.

   Phenols from Hydrogenolysis. Work on the hydrogenation
of wood, lignin, lignite, and coals dates back to about
1925.[23,156] Exhaustive hydrogenation of lignin under
extreme conditions results in conversion of all aromatic
structures into cyclic hydrocarbons and alcohol derivatives.
Under less severe conditions a large number of products are
formed, the aromatic structure is generally retained and
phenolic products are predominant. Because of high catalyst
usage or cost and the need for sulfur-free lignin raw
materials to avoid catalyst poisoning, the "hydrogenolysis"
of lignin has been applied primarily in the study of lignin
structure.

   The use of metal sulfides as inexpensive hydrogenation
catalysts was developed in Japan at the Noguchi Institute.[63]
By 1952, in a study of the liquefaction of lignin, it was
found that such catalysts were not deactivated by sulfur
compounds and also substantially suppressed hydrogenation
of aromatic structures, thus increasing the yield of monomeric phenolic products. In 1961, the Crown Zellerbach
Corporation began extensive testing of the Noguchi process
with the hope of entering into commercial production.
Lignin powder was mixed with about three times its weight
of "pasting oil" or inert carrier, and 1 to 10% catalyst
based on lignin weight, and subjected to hydrogenation at
temperatures in the range of 370 to 430 C, initial pressures
of 100 atms and reaction times of 0.5 to 4 hours. Under
optimal economical conditions of raw material type, catalyst
usage and conservation of pasting oil for recycle, the yield
of monophenols was 21 to 23% based on lignin. The mono-
phenol fraction was composed of about 3% phenol [61], 10%
cresols [62], 4.2% ethyl phenols, 2% propyl phenols, 1.2%
2,4-xylenol [63], and other minor components.

   The best lignin starting material for the process was
prepared by pretreating a lignosulfonate raw material with
lime for 1 hour at 200 C. The desulfonated, de-ashed dry
lignin was recovered in 40% yield from spent sulfite liquor.
Kraft lignin was not suitable as such, yielding only 13%
monophenols. It was concluded that the overall process
for phenols from lignin would require considerable improvement to compete with synthetic phenol. The Crown Zellerbach Corporation has also developed a catalytic process

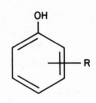

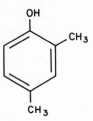

Phenol
[61]

Monoalkylphenols
R=methyl, ethyl,
propyl (three
isomers each)
[62]

2,4-Xylenol
[63]

for dehydroalkylation of alkyl phenols that yields 40 to 50% phenol and 30% benzene.[162]

Dimethyl Sulfide, Dimethyl Sulfoxide. The characteristic odor of volatile sulfur compounds produced during the kraft pulping process led to early identification of methyl mercaptan [64] and dimethyl sulfide [65]. The amount of these compounds formed during pulping is small, of the order of 0.1% of lignin weight. Nucleophilic demethylation of lignin under alkaline conditions by sulfide and mercaptide ions was studied extensively at the research laboratory of Crown Zellerbach Corporation.[34,62,94] Concentrated kraft black liquors were treated with additional sulfur (1 to 2%), to provide total sulfur in proportion to methoxyl content, and heated for 10 to 60 minutes at 200 to 250 C. Dimethyl sulfide yields as high as 3.25% of black liquor solids were obtained in 10 minute reactions at 250 C. Recovery of pulping chemicals and heat by combustion of black liquors were not affected by the process.

Chemical production of dimethyl sulfide and a smaller amount of methyl mercaptan was initiated in 1961 at the Bogalusa, Louisiana plant of Crown Zellerbach. In this process, black liquor is concentrated to 45% solids and heated with a small amount of sulfur at 232 C in a continuous reactor. Dimethyl sulfide is recovered by flashing the product liquor and is obtained in 2.5% yield based on black liquor solids.[167] Yield expressed in terms of pulp production is 60 lbs/ton or 30 kg/tonne. The rated annual capacity

of this plant is currently 20 million lbs or about 9 million
kg. Crude dimethyl sulfide is purified to remove methyl
mercaptan and is converted to dimethyl sulfoxide [66] by
cold, liquid phase oxidation with nitrogen tetroxide (N_2O_4)
in the presence of oxygen.[34,94,167] If a stronger oxidizing
agent is used the end product is dimethyl sulfone [67]. In
practice it is more advantageous to oxidize dimethyl sulfoxide.[65] A process has been developed where the latter is
mixed with concentrated nitric acid at a mole ratio of
1:1.5. High yields of dimethyl sulfone (86%) are obtained
when the intermediate conjugate acid is heated for about
two hours at 120 to 150 C.

   Methyl mercaptan and dimethyl sulfide are used principally as odorants for natural gas but also have important
solvent properties for inorganic salts and uses in chemical
synthesis. Because of the unique solvent properties of
dimethyl sulfoxide,[3] it is the choice solvent in manufacture
of many synthetic fibers and for many difficult reactions
where reaction rate is increased greatly. It is also used
extensively in the synthesis of other chemicals.

   Lignin Polymers.[23,71,143,144,145,156] Simplistically,
lignin is an aromatic heteropolymer made up of crosslinked
and highly branched phenylpropane units. Various aspects
of lignin structure have been reviewed in other chapters.
The polymer in its native and isolated forms might be viewed
as a mixture of sponges of widely varying sizes, representing amorphous molecules that differ 1000-fold in size but
are able to interact physically in a "polydisperse" system
of mutual properties.

$CH_3{-}S{-}H$ $CH_3{-}S{-}CH_3$

Methyl Mercaptan   Dimethyl Sulfide
      [64]   [65]

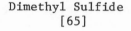

Dimethyl Sulfoxide   Dimethyl Sulfone
      [66]   [67]

The lignin component of wood may be solubilized by
a number of chemical treatments. The common wood pulping
systems, generically typed as "sulfite" or "kraft", yield
distinctly different lignin polymer products. The separation of lignin from wood by sulfite processes (acid sulfite,
bisulfite, neutral sulfite) is based on the sulfonation of
lignin, along with partial hydrolysis and depolymerization,
yielding a mixture of water soluble "lignosulfonates" and
sugars resulting from the hydrolysis of hemicellulose. In
the kraft or "sulfate" process, lignin is extensively
depolymerized or degraded by alkaline hydrolysis and oxidation and is water-soluble only under the prevailing strongly
alkaline conditions, which also degrade hemicelluloses to
yield saccharinic acid derivatives as discussed previously.

Lignosulfonates are water-soluble or dispersible across
a wide range of pH and are relatively stable under neutral
or mild acidic conditions. Molecular weight will vary
widely within a given lignosulfonate preparation, from
1,000 to 500,000 D for conifer wood lignosulfonates. Hardwoods yield products having lower ranges in molecular weight.
Because of the wide range of molecular weights represented
in lignosulfonate preparations they are classed as polydisperse systems.[71]

Kraft process lignin is usually referred to as "alkali
lignin" but has also been termed "thiolignin", because of the
presence of sulfide derivative substituents, or "sulfate
lignin" because of the use of sodium sulfate in the preparation of pulping chemicals. Alkali lignin is insoluble in
water under neutral or acidic conditions and is recovered
by carbon dioxide or acid precipitation processes. The
molecular weight range for alkali lignin lies below
10,000 D. The average molecular weight of kraft lignins
prepared from hardwood and pinewood has been determined as
2,900 and 3,500 D, respectively.[144] Alkali lignins have
softening points as low as 140 C and usually less than 200 C.
These lignins are generally quite soluble in polar solvents
such as methanol.

The history of the utilization of lignin products,[45]
and particularly lignosulfonates, is as old as the European
development of the calcium bisulfite pulping (1866) process.
By the 1880's claims were being made regarding the use of
lignosulfonates as tanning agents, dye-bath additives and
water-soluble adhesives. Evidently, then as now, the

cellulose, paper and fiber products were more easily
recovered, useful and marketable than the water-soluble
lignin derivatives. Starting in 1905, a United States
developer, J. S. Robeson, was successful in producing and
marketing substantial tonnages of concentrated crude and
modified dry calcium lignosulfonates for use as tanning
aids, and as mineral and foundry core binders. Processes
were developed in the 1930's for the fractional precipitation and purification of calcium lignosulfonate. Meanwhile
research on the mineral- and dye-dispersing properties of
lignosulfonates led to broader and more significant markets
as additives for concrete, oil-well drilling fluids,
ceramic clays, gypsum, mineral pigments, and organic dyes.
The refining of lignosulfonates via the fermentation of
sugars in the crude liquors to produce ethyl alcohol and
food yeast (*Torula*) contributed greatly to the utility of
lignin products.

From about the 1950's to the present, efforts to reduce
and eliminate pulp mill pollution led to conversion of
sulfite pulping bases from calcium to ammonium, sodium or
magnesium, so that concentrated spent liquors could be
burned with the recovery of heat and chemicals. Major
industrial research effort was also placed on the development
of new uses and markets for lignin products. The total
volume of crude and refined lignosulfonates marketed in
the United States rose from 70,400 tonnes in 1956 to about
200,000 tonnes in the mid-1960's. Large-scale uses in oil-
well drilling muds and as binders in animal food pellets
led to this market expansion.[45,140] The 1973 market volume
of 313,000 tonnes[196] includes further utilization in animal
feeding, this time as a component of feeds for ruminant
animals, and increases in mineral treatment including dirt
road soil stabilization and dust abatement.

The primary utility of lignosulfonates may be assigned
to properties involving dispersion and adhesion or "binding".
Crude lignosulfonates are better in the animal feed applications because sugar content, particularly in ruminant
feeds, can be assigned energy value. Otherwise, adhesion
in consolidated mineral or soil systems appears to be
improved as the result of dispersion properties. In the
case of Portland cement concrete, a lignosulfonate additive
will be absorbed on the mineral surface, reducing the amount
of hydraulic water necessary to provide suitable fluidity
or plasticity for handling. The set concrete is thus

denser, less permeable and much stronger. The same princi-
ple applied to mineral or clay slurries or drilling muds
permits the control of slurry viscosity at higher solids
content or the maintenance of a standard solids content
while reducing viscosity and pumping energy. In soil
stabilization and mineral binding, the adsorbed lignosulfonate reduces the water required to provide a plastic or
compactable mixture, yielding denser and stronger pellets
or soil structures.

    The evolution of alkali or kraft lignin products
followed a somewhat different route. Early in the development of the kraft pulping process it became practical to
recover heat and chemicals by burning concentrated black
liquors. This branch of the wood pulping industry has
grown rapidly during the past 40 years, since a much wider
variety of trees are pulpable under kraft (alkaline) conditions than sulfite. Chemical recovery systems evolved
earlier than in the sulfite process and were advantageous
in eliminating major pollution problems. The 1975 capacity
of the United States kraft pulping industry was of the order
of ten times that of the sulfite, while in Canada it was
about three times larger. Early interest in kraft lignin
centered around its properties of solubility in polar
organic solvents, its fusion point and compatibility with
phenolic resins. Alkali lignin has also been rendered
soluble in water by sulfonation and has been chemically
modified to add phosphate ester and other substituents for
various applications in water treatment and dye and pigment
dispersion. United States production of kraft lignin
products is estimated at about 9,000 tonnes/year.[22]

    Lignin residues resulting from the alkaline oxidation
of lignosulfonates to produce vanillin are also recovered
in commercial quantities. These lignins are more like alkali
lignin than their lignosulfonate starting materials, having
been oxidized and degraded to molecular weights of less
than 10,000 D. They are recovered by acid precipitation
or fractionation from the alkaline process liquor. These
products have been found to be excellent dye and pigment
dispersants. In Canada, this type of lignin has been used
as a component of thermosetting resins for decorative paper
laminates of the "Arborite" trade name.[4]

UTILIZATION OF WOOD RAW MATERIALS
AND PULP MILL BYPRODUCTS

Fermentation. The wood sugar fraction of spend sulfite liquor, together with minor components such as acetic and formic acids, constitute a large resource of readily fermentable material. Since this fraction accounts for the high biological oxygen demand of sulfite liquors, it has been almost standard practice in Europe to engage in spent liquor fermentation at or near the pulp mill site, and to produce ethyl alcohol or *Torula* yeast. Fermentation to alcohol has already been discussed. The production of *Torula* yeast or other fungal biomasses or "single-cell proteins" is more advantageous from the overall carbohydrate utilization standpoint. Only hexose sugars are fermented to alcohol, but both hexose and pentose sugars, and certain sugar acids and carbohydrate fragments, are converted to *Torula* (*Candida utilis*) yeast in an average yield of 50%, based on the dry weight of wood sugar raw materials.

Torula yeast is currently produced at two sites in the United States: Lake States Division of St. Regis Paper Company, Rhinelander, Wisconsin and Boise Cascade in Salem, Oregon. Annual production capacity is estimated at 3,600 and 5,400 tonnes, respectively, or a total of 9,000 tonnes/year. World production of *Torula* yeast was estimated as 113,400 tonnes in 1951.[45] North American sulfite pulp capacity is estimated as 5,700,000 tonnes in 1975.[51,182] Assuming production of 1.2 tonnes of spent sulfite liquor solids per tonne of dry pulp, the annual North American potential for *Torula* yeast production is outlined as follows:

Spent sulfite liquor solids:	6,840,000 tonnes
Wood sugar fraction (25% of above):	1,710,000 tonnes
Torula yeast (50% of sugar weight):	855,000 tonnes
Yeast protein (50% of yeast weight):	427,500 tonnes

The concept of producing a variety of single-cell proteins (SCP) from the large volume of soluble carbohydrates available at pulp mills should provide a high challenge for research in future years. The development of the "Pekilo" process in Finland,[54] involving the production of the fibrous fungus, *Paecilomyces varioti*, is a good example of the selection of microorganisms for improved processing.

In this case, the fibrous product is recovered by simple filtration, avoiding the expense of centrifuging as required for yeast cell recovery.

Secondary treatment of pulp mill effluents, as required for pollution abatement purposes and reduction in biological oxygen demand, is another example of fermentation. As presently conceived, these processes are aerobic treatments with mixed bacterial microorganisms, collectively termed "activated sludge". Freshly produced sludge biomass from a pulp mill secondary treatment plant, if properly dewatered and dried, yields a product that is similar in composition (50% protein), color and taste to other single-cell proteins. Though not presently approved, these products could be acceptable as animal feed components. Many secondary treatment plants produce in excess of 20 tonnes of dry sludge solids per day or 8,000 tonnes/year. The development of specific fermentation processes to utilize the carbohydrate components of pulp mill bleaching effluents will require further research. For example, isosaccharinic acids, representing rearranged sugars, are not fermented by yeast organisms but are efficiently utilized by the unclassified bacteria present in activated sludge.

The conversion of waste wood or cellulose to fungal or bacterial protein has also attracted attention in recent years. The ensilage processes for treating straws, vines and other agricultural residues represent primitive applications in this area. The product is rendered more digestible, via partial delignification and cellulose hydrolysis, more nutritious by the growth of SCP, and is preserved by acids derived from the fermentation. Treatment of wood fiber, sawdust or chips with specific white rot fungi holds promise for greatly increasing the digestibility and value of wood in the feeding of ruminant animals.[128] Finally, it has been proposed to grow certain fungi or bacteria on waste wood or agricultural residues and harvest the product for use as animal[27,85] or human food.

Wood Prehydrolysis Extracts. This topic has already been discussed in general in the second section of this chapter. A typical composition for conifer prehydrolysate is given in Tables 4 and 5. The Masonite Corporation has developed markets for the entire available supply of water-soluble carbohydrate fraction derived from their manufacture of hardboard. The hot water extract is concentrated to 65%

solids by evaporation. Production of "Masonex" amounted to
over 53,000 tonnes, dry basis, in 1974, or 90,000 tons[57]
of liquid concentrate. The addition of Masonex or similar
wood extracts, at 10% by weight of cattle (ruminant animal)
feed is permitted by government regulation. Market value
on a solids basis is comparable to that for cane or beet
sugar molasses.

Expansion in the use of prehydrolysis processes is
predicted on the above present markets and on future development of the integrated production of furfural and hardwood
pulps. Since most of the pentose components of hardwoods
are solubilized by mild acid hydrolysis, this fraction,
which contains as much as 25% xylose on a solids basis,
could become the prime domestic resource for furfural
production. Principal raw material competition would come
from sugar cane bagasse.

Uses for Crude Spent Sulfite Liquor (Concentrated).
This topic was discussed in a preliminary manner under the
subtitle "lignin polymers". The 1975 North American
capacity for producing sulfite liquor solids is estimated
at 6,840,000 tonnes, or twice this weight as a 50% concentrate. This raw material represents a major annual resource
that has a variety of uses, all of relatively low value.
The principal uses or markets are listed below in order of
decending approximate size:

1. Fuel for heat and power generation, with inorganic
     chemical recovery.
2. Manufacture of dispersants and oil-well drilling
     mud additives.
3. Additive in animal feeds, at 4% by weight.
4. Soil stabilization, binder; dust abatement.
5. Extender for phenolid and urea resin adhesives
     for plywood and particleboard.
6. Binder for coal, ores and dry minerals.
7. Mineral treatment, processing and refining.
8. Raw material for vanillin production.

Within a few years, if not already so, all existing
sulfite pulp mills will be equipped to concentrate and burn
their liquors, or alternatively market their entire output.
This dramatic conversion in the sulfite pulping industry,
though required for its plants to stay in business and meet
environmental regulations, means that large volumes of

concentrated sulfite liquor will be available, transported, stored, and traded. Value as fuel will continue to dominate all other uses, but should lend stability to the gradual development of higher-value markets. It is interesting that markets in animal feeding, if properly developed, are capable of utilizing the entire annual North American production of sulfite liquor concentrate; however, this would require distribution, storage and transportation far beyond present facilities. Market areas 4 to 7 also have very large potential growth.

Other factors that will become important in the expansion of markets in areas 2 and 4 to 8, include the installation of fermentation or membrane ultrafiltration process to separate sugar components and provide purified lignosulfonates having improved properties in specific uses. Separation of the sugar and low molecular weight lignosulfonate component, and concentration of this fraction, would hasten the development of the animal feed market, which is based largely on the nutritive (energy) value of the wood sugars in ruminant feeding.

The future for marketing large volumes of concentrated spend sulfite liquor, or relatively crude fractions of this commodity, appears favorable. The 1973 estimated nonfuel market of 313,000 tonnes, solids basis, is capable of growth to 1,000,000 tonnes by 1980, perhaps doubling again in volume by 1990. If the price of energy continues to rise, or rises faster than the average market value of concentrated sulfite liquor, then the fuel value of the product will tend to inhibit market growth in nonfuel applications.

Uses for Kraft Black Liquor Components. As in the case of spent sulfite liquor, it is well to consider the total volume of this material and its relative marketability in different applications. Precise production figures are not available except in terms of the pulp products. Total 1975 kraft pulp production in the United States and Canada was calculated as 38,880,000 tonnes.[54,182] Assuming that, on an average, every tonne of dry, finished pulp is equivalent to the production of 0.6 tonnes of lignin and 0.6 tonnes of saccharinic acids, the potentially available supply of each of these materials is 23,330,000 tonnes/ year. Considering these very large resource numbers, it is of some comfort to remember that the fuel value of these

materials is being utilized as standard practice in the
recovery of energy and inorganic chemicals for recycle to
the pulping process.

Only 9,000 tonnes/year of lignin or about 0.02% of the
total organic material resource in kraft black liquor is
currently being used in nonfuel applications. Processing
of black liquor to produce dimethylsulfide at (also) 9,000
tonnes/year requires about 360,000 tonnes of black liquor
solids. Kraft or alkali lignin has a bright future as a
component of resin and adhesive systems. Its recovery
from black liquor involves considerable expense and it cannot
be priced in competition with spent sulfite liquor. The
challenge for research involves development of lower-cost,
combination recovery processes, where inorganic chemicals,
lignin and isosaccharinic acids are separated into useful
fractions. As in the case of spent sulfite liquor products,
market development in well justified uses has been very
slow. No present market demand exists for isosaccharinic
acids, crude or purified. Under these circumstances, it is
not surprising that industrial management considers "fuel
value" as the best available use for all pulping liquors.

Pyrolysis, Combustion. The pyrolysis route to chemicals
from wood (Fig. 1) is also the oldest. The main problem
with this route is competition. Coal raw materials are
readily available, cheaper to produce and higher yielding
of marketable products. Under present economic rules it
is not practical to produce pyrolysis products, or even
synthesis gas, from wood, in view of the total on-site cost
of dry wood as compared with that for coal. Even though
much valuable recent research[159,160] has been performed on
the controlled pyrolysis of wood, bark and spent sulfite
liquor solids, under modern processing conditions, it is
doubtful that any such process is competitive with a similar
one based on coal.

In integrated operations involving the production of
solid wood products, plywood, wood fiber- or particleboards,
and at pulp mills, it is still very practical to burn wood
residues and bark to produce heat or power. The initial
justification for construction of a hog-fuel-burning power
generation plant may have included disposal of the bark and
wood waste. Alternate disposal of these wastes becomes
impractical or expensive. Under present conditions of gas
and oil fuel pricing, the value, collection and storage of

hog fuel have taken on new significance. Most combustion engineers will readily quote the equivalence of a hog fuel unit in terms of barrels of oil, and they will purchase hog fuel as long as the price is competitive and capacity to handle this material is available.

CONCLUSION

   The Real World of Chemical Economics and Markets. It is quite evident that gas and oil and then coal will continue to dominate as raw materials for chemical production for many decades. The "golden age" of wood chemistry has not yet arrived and it is a misplaced hope to believe that wood or wood waste can compete broadly with the more convenient and concentrated resources presently adapted for the masive production of organic chemicals, synthetic polymers and plastics. Areas where natural products and wood derivatives are able to compete are outlined below.

   A simple comparison of raw material prices will support the above view:

Raw Material	Price Range ¢/kg	¢/lb
Crude oil, 1970	2	1
Crude oil, 1976	9	4 ($10/bbl)
Crude oil, 1980	15	7 ($15-20/bbl)
Coal	7-9	3-4
Waste paper	9-13	4-6
Spent sulfite liquor solids	7	3
Alkali lignin	22-33	10-15
Cellulose pulp, 1970	22	10
Cellulose pulp, 1976	44	20
Paper pulps	31-66	14-30

   In the above comparison, crude oil would still be the raw material of choice for organic synthesis and polymer production, even if present prices were to double. At some point, strong interest in conversion to coal could develop; however, the idea of reserving gas and oil materials for chemical production is becoming popular and will undoubtedly receive government action.

Another perspective that is disturbing with regard to resource utilization, is to compare the relative per capita consumption of key manufactured goods and energy in North America and the developing nations.

Markets for Chemicals: Annual per Capita Weight
Usage of Major Commodities

	United States and Canada		Developing Nations	
	kg	lbs	kg	lbs
Textiles	25	55	1.5–3	3–6
Plastics	115	230	1–2	2–5
Paper	320+	700+	2–11	4–25
Energy (as oil)	3,600–5,500	8–12,000	23–180	50–400

In this perspective it is noted that energy uses for transportation, heat and light in North America are quite excessive, even in relation to the estimated annual per capita carbon fixation of about 2,700 kg or 6,000 lbs. Oil raw material used for textile and plastics production is only on the order of 5% of that consumed as energy. Therefore, it is likely that energy uses and markets will require dramatic changes during the coming decades, while the use of oil as a prime raw material for chemicals will continue to be well justified economically.

One of the major conclusions in our intensive search for new energy resources is that there is *no free* raw material. Collection and transportation of so-called wastes or residues, to provide a continuous supply at a production site, is often more expensive than the chemical processing cost. This is why waste wood is principally useful at a wood products or pulp mill, but not when it is still out in the woods, saturated with water, dirty, and possibly partly decayed. Another false hope has been that waste wood could be used to make "good" paper or chemical cellulose pulp. The quality of raw materials invariably affects the quality of products, or the cost involved in achieving high quality.

A factor that will continue to affect adversely the use of wood raw materials in chemical processes involves "environment protection". Using pulp mills as the example, the doubling in the manufacturing cost of cellulose between 1970 and 1976 has a great deal to do with the enormous investment in equipment to meet environmental standards, while pulp production was not increased materially. It is implicit in the development and legislation of environment protection standards that the unit cost to produce a given product will increase, sometimes more than the market will bear. Therefore, many older or smaller pulp mills cannot afford to stay in business.

The Case for Wood Chemicals. Throughout this review we have tried to present a brief, but realistic picture of various facets of the subject. Several attempts to produce chemical products from wood have failed. Markets for these products have often been very small or exotic, or have been very slow in developing to significant size. On the positive side, many successful processes have been described. A brief summary of these and some of the better prospects are presented below.

Traditional naval stores products, such as turpentine and tall oil derived from the kraft pulping of southern pines, are well established in their markets and have good prospects for future growth. The pulping of wood serves as an extraction and collection process that strongly supports the economics of producing these products.

Steam and mild acidic prehydrolysis of hardwood chips is attractive as a process for extracting pentose sugar components, prior to conventional kraft pulping of the remaining wood substance to produce cellulose and paper grade pulps. Such pentose-rich prehydrolysates are advantageously used for the production of furfural or alternatively, xylitol. Pretreatment of hardwood chips also yields superior wood fiber for the manufacture of fiberboard or hardboard products. Concentrated prehydrolysis extracts from both hardwoods or softwoods have a ready market in the animal feeding industry.

The sulfite pulping industry has been virtually rebuilt during the last decade, for recovery and combustion of concentrated spent sulfite liquor. The large wood sugar resource available in spent sulfite liquors could best be

used in fermentation processes, to produce *Torula* yeast or
similar single-cell proteins. The sizable market for
concentrated spent sulfite liquor in the animal feeding
industry, based largely on the value of the sugar fraction,
should continue to grow. This market could become very
large if an economic means for producing a concentrated
wood sugar fraction were developed.

Uses for lignin products will continue to grow in a
steady, but slow manner. Transportation and storage
facilities will need to be expanded greatly in the marketing
of bulk, low-value crude products. Any one of several
markets for these products could grow rapidly within the
next decade. Meanwhile, the steady major use for kraft
black- and spent sulfite liquors will be as fuel. Except
for the production of vanillin from lignosulfonates and
dimethyl sulfide from kraft black liquor, other uses for
lignin as a raw material for the production of chemicals
(phenols) are still economically unattractive.

Problems Requiring Research. Nearly half a century
has been spent unraveling the chemistry of wood and assign-
ing formal structures to various components. The frequent
reference to lignin as mysterious or "unknown" in structure
has become a popular myth. The time has come to admit
that we know a great deal about the chemistry of every
major component of the commonly available woods. Our pre-
occupation with structural details has not been balanced
by an equal zeal for practical utilization. An appeal is
therefore extended: to catch up in end-use research, to
develop information that will help to sell the product and
to develop test methods that will demonstrate the utility
of a given product.

One of the most pressing needs in the pulp and paper
industry is for separating or purifying processes. Ultrafiltration and electrodialysis may provide suitable routes
for recovering sugars or sugar derivatives, while purifying
the lignin fraction. Fermentation routes also need to be
explored. Emphasis in research needs to be placed on
lignin as a distinct type of polymer and on defining the
advantageous properties of this polymer in combinations
with many other materials. As with cellulose, the highest
value of lignin may result from the proper application of
its polymeric properties.

In our view, research should be market-oriented from beginning to end. The proliferation of products, in competition with each other for small markets needs to be minimized. Major efforts should be directed to designing standard products that fit massive markets.

REFERENCES

1. Allan, G. G. and T. Mattila. 1971. *In* Lignins. K. V. Sarkanen and C. H. Ludwig, eds., Chap. 14, 916 pp. Wiley-Interscience, New York.
2. Anon. 1973. *Naval Stores Rev. 83*:7-10.
3. Anon. 1963. "Dimethyl Sulfoxide" Technical Bulletin, 23 pp. Crown Zellerbach Corp., Camas, Washington.
4. Anon. 1964. *Can. Chem. Process 48(12)*:55-60.
5. Anon. 1968. *Oil, Paint & Drug Reporter 193(6)*:3, February 5.
6. Anon. 1970. *Chem. Eng. News 48(3)*:14, January 19.
7. Anon. 1975. *Business Week No. 2389*:110 b,d, July 14.
8. Anon. 1975. *Chem. Eng. 83(19)*:75-6, September 15.
9. Anon. 1976. *Chem. Week 119(2)*:37.
10. Bader, A. R. 1960. U.S. Patent 2,933,472, to S. C. Johnson & Son, Inc.
11. Bader, A. R. and A. D. Kontowicz. 1954. *J. Am. Chem. Soc. 76*:4465-6.
12. Baierl, K. W. 1973. U.S. Patent 3,764,462 to Scott Paper Co.
13. Barton, G. M. 1975. *J. Appl. Polymer Sci., Appl. Polymer Symposia No. 28*:465-484.
14. Benson, F. R. 1963. *In* Kirk-Othmer Ency. Chem. Tech. 2nd Ed. *1*:569-88. Interscience, New York.
15. Biggs, W. A., J. T. Wise, W. R. Cook, W. H. Baxley, J. D. Robertson, and J. E. Copenhaver. 1961. *Tappi 44*:385-91.
16. Bock, L. H. 1954. U.S. 2,686,198 to Rayonier Inc.
17. Bock, L. H. and J. K. Anderson. 1955. U.S. Patent 2,699,438 to Rayonier Inc.
18. Bock, L. H. and J. K. Anderson. 1955. *J. Polymer Sci. 17*:553-8.
19. Boggs, L. A. 1967. *Tappi 50(6)*:133-5.
20. Boggs, L. A. 1968. Preprint, 55th National ACS Meeting, San Francisco (March 31).
21. Bolger, J. C., D. C. Tate, and H. B. Hopfenberg. 1967. *Tappi 50(5)*:231-6; 247-52.
22. Bratt, L. C. 1965. *Paper Trade J. 149(10)*:57-60.

23. Brauns, I. E. 1952. The Chemistry of Lignin, Chap. 19, 807 pp. Academic Press, New York. *Also* Brauns, F. E. and D. A. Brauns. 1960. Suppl. Vol. 804 pp.

24. Brink, D. L., L. E. Dowd, and D. F. Root. 1966. U.S. Patent 3,234,202, to Weyerhaeuser Co.

25. Bryan, C. C. 1950. U.S. Patent 2,506,540 to Monsanto Chemical Co.

26. Buchholz, R. F. and M. Reintjes. 1974. U.S. Patent 3,810,941, to ITT.

27. Callihan, C. D., G. H. Irwin, J. E. Clemmer, and O. W. Hargrove. 1975. *J. Appl. Polymer Sci., Appl. Polymer Symposium No. 28*:189-96.

28. Carlson, L. J. 1964. Canadian Patent 679,330, to Rayonier Inc.

29. Carlson, L. J. 1966. U.S. Patent 3,235,541 to Crown Zellerbach Corp.

30. Carvaldo, J. D., W. Prins, and C. Shuerch. 1959. *J. Am. Chem. Soc. 81*:4054-8.

31. Casebier, R. L., J. K. Hamilton, and H. L. Hergert. 1969. *Tappi 52*:2371-7.

32. Caebier, R. L., J. K. Hamilton, and H. L. Hergert. 1973. *Tappi 56*:135-9; 150-2.

33. Casebier, R. L., F. W. Herrick, K. R. Gray, and F. A. Johnston. 1972. U.S. Patent 3,677,818, to ITT.

34. Cisney, M. E. and J. D. Wethern. 1957. U.S. Patent 2,816,832 to Crown Zellerbach Corp.

35. Cisney, M. E., W. L. Shilling, W. M. Hearon, and D. W. Goheen. 1954. *J. Am. Chem. Soc. 76*:5083-7.

36. Clark, I. T. and J. Green. 1968. *Tappi 51(1)*:44-8.

37. Conca, R. J., A. Beelik, and F. W. Herrick. 1966. U.S. Patent 3,238,158, to Rayonier Inc.

38. Conner, A. H. and J. W. Rowe. 1975. *J. Am. Oil Chem. Soc. 52(9)*:334-8.

39. Conner, A. H., M. Nagaoka, J. W. Rowe, and D. Perlman. 1976. *Appl. Environ. Microbiol. 32(2)*:310-11.

40. Creighton, R. H. J. 1961. U.S. Patent 3,007,971, to MacMillan and Bloedel Ltd.

41. Dean, F. M. 1963. Naturally Occurring Oxygen Ring Compounds. 661 pp. p. 39, Butterworths, London.

42. Derfer, J. M. 1963. *Tappi 46(9)*:513-17.

43. Dowd, L. E., D. L. Brink, A. S. Gregory, and A. K. Esterer. 1966. U.S. Patent 3,255,221, to Weyerhaeuser Co.

44. Dunlop, A. P. and F. N. Peters. 1953. The Furans. ACS Monograph No. 119. 867 pp. Rheinhold Publishing Corp., New York.

45. Elgee, H. 1973. *AIChE Chem. Eng. Prog. Symp. No. 133,* *69*:6-10.
46. Enkvist, T. 1975. *J. Appl. Polymer Sci., Appl. Polymer Symposium No. 28*:285-95.
47. Enkvist, T., J. Turunen, and T. Ashorn. 1962. *Tappi 45(2)*:128-35.
48. Era, V. and J. Hannula. 1974. *Paperi ja Puu 5*:489-95.
49. Esser, M. H. ed. 1976. Proc. Lightwood Research Coord. Council, 154 pp.
50. Esterer, A. K. and L. E. Dowd. 1964. U.S. Patent 3,131,198, to Weyerhaeuser Timber Company.
51. Evans, J. C. W. 1975. *Pulp & Paper 50(7)*:23-7.
52. Evshov, B. N. 1970. *Izv. VVZ, Pishch, Teknol. No. 6*:32.
53. Fomichev, A. V., et al. 1968. U.S.S.R. Patent 249,364.
54. Forss, K. and K. Passinen. 1976. *Pap. ja Puu 58(9)*: 608-18.
55. Franklin, E. C., M. A. Taras, and D. A. Volkman. 1970. *Tappi 53(12)*:2302-4.
56. Gaden, E. L., M. H. Mandels, E. T. Reese, and L. A. Spano. 1976. Enzymatic Conversion of Cellulosic Materials: Technology and Application. Biotech. and Bioeng. Symposium No. 6, 316 pp. Interscience, New York.
57. Galloway, D. 1975. *Pulp and Paper 49(9)*:104-5.
58. Garber, J. D., C. Jones, and R. E. Jones. 1960. U.S. Patent 2,929,823, to Merck and Co.
59. Gardner, J. A. F., G. M. Barton, and H. MacLean. 1959. *Can. J. Chem. 37*:1703-9.
60. Gardner, J. A. F., B. F. MacDonald, and H. MacLean. 1960. *Can. J. Chem. 38*:2387-94.
61. Gardner, J. A. F., E. P. Swan, S. A. Sutherland, and H. MacLean. 1966. *Can. J. Chem. 44*:52-8.
62. Goheen, D. W. 1964. *Tappi 47(6)*:14A-26A.
63. Goheen, D. W. 1965. *In* Lignin Structure and Reactions. R. F. Gould, ed. pp. 205-25. Advan, Chem. Ser. No. 59, ACS, Washington, D.C.
64. Goheen, D. W. 1973. *AIChE, Chem. Eng. Prog. Symposia No. 133, 69*:20-24.
65. Goheen, D. W. and C. F. Bennett. 1961. *J. Org. Chem. 26*:1331-2.
66. Goldschmid, O. and H. L. Hergert. 1961. *Tappi 44(12)*: 858-70.
67. Goldstein, I. S. 1975. Am. Chem. Soc., Div. Chem. Marketing & Econ., Preprint:25 pp.
68. Goldstein, I. S. 1975. *J. Appl. Polymer Sci., Appl. Polymer Symposium No. 28*:259-67.

69. Goldstein, I. S. 1976. *Am. Chem. Soc. Centennial Mtg. Abstr*:MACR 4.

70. Good, R. D. and F. S. Trocino. 1974. *Chem. Eng. 81 (11)*:70-1.

71. Goring, D. A. I. 1962. *Pure Appl. Chem. 5(1,2)*:233-54.

72. Green, J. W. 1956. *Tappi 39(7)*:472-7.

73. Green, J. W. 1957. *In* The Carbohydrates. W. W. Pigman, ed., pp. 902, Chap. VI, Academic Press, New York.

74. Greenlee, S. O. 1960. U.S. Patent 2,933,471; U.S. Patent 2,933,528, to S. C. Johnson & Son Inc.

75. Greenlee, S. O. 1962. U.S. Patent 3,047,627, to S. C. Johnson & Son Inc.

76. Greenlee, S. O. 1967. U.S. Patent 3,300,444, to S. C. Johnson & Son Inc.

77. Gregory, A. S. 1959. U.S. Patent 2,890,225, to Weyerhaeuser Timber Co.

78. Gregory, A. S., D. L. Brink, L. E. Dowd, and A. S. Ryan. 1957. *Forest Prod. J. 7(4)*:135-40.

79. Guenther, E. 1948. The Essential Oils. Six Volumes, D. Van Nostrand Co., New York.

80. Hall, J. A. 1971. Utilization of Douglas Fir Bark. 138 pp. Forest Service, USDA, Portland, Oregon.

81. Hall, R. B., J. H. Leonard, and G. A. Nicholls. 1960. *Forest Prod. J. 10(5)*:263-72.

82. Hamilton, J. K., E. V. Partlow, and N. S. Thompson. 1958. *Tappi 41*:803-16.

83. Hannus, K. 1975. *J. Appl. Polymer Sci., Appl. Polymer Symposia No. 28*:485-501.

84. Hannus, K. and G. Pensar. 1973. *Paperi ja Puu 55(7)*: 509-17.

85. Harkin, J. M., D. L. Crawford, and E. McCoy. 1974. *Tappi 57(3)*:131-4.

86. Harris, J. F. and J. M. Smith. 1961. *Forest Prod. J. 11*:303-9.

87. Harris, J. F., J. F. Saeman, and L. L. Zoch. 1960. *Forest Prod. J. 10*:125-8.

88. Harris, J. F. 1975. *J. Appl. Polymer Sci., Appl. Polymer Symposia No. 28*:121-144.

89. Hathway, D. E. 1961. *In* Wood Extractives. W. E. Hillis, ed. pp. 159-190. Academic Press, New York.

90. Haworth, R. D. 1942. *J. Chem. Soc.* 448.

91. Haworth, R. D. and G. Sheldrick. 1935. *J. Chem. Soc.* 636.

92. Hearon, W. M. and W. S. MacGregor. 1955. *Chem. Rev. 55*:957-1068.

93. Hearon, W. M., H. B. Lackey, and W. W. Moyer. 1951. *J. Am. Chem. Soc.* *73*:4005-7.

94. Hearon, W. M., W. S. MacGregor, and D. W. Goheen. 1962. *Tappi 45(1)*:28A-36A.

95. Hennecke, H. 1970. *Deut. Lebensm. Rundsch. 66*:329-31.

96. Hergert, H. L. Unpublished work.

97. Hergert, H. L. 1960. *Forest Prod. J. 10(11)*:610-17.

98. Hergert, H. L. 1962. *In* The Chemistry of Flavonoid Compounds. T. A. Geissman, ed. pp. 553-92. Mac-Millan Co., New York.

99. Hergert, H. L. and E. F. Kurth. 1952. *Tappi 35(2)*: 59-66.

100. Hergert, H. L. and E. F. Kurth. 1953. *Tappi 36(3)*: 137-44.

101. Hergert, H. L., L. E. Van Blaricom, J. C. Steinberg, and K. R. Gray. 1965. *Forest Prod. J. 15(11)*:485-91.

102. Heritage, C. C. and L. E. Dowd. 1959. U.S. Patent 2,890,231, to Weyerhaeuser Co.

103. Herrick, F. W. 1971. Proceedings of Conference: Converting Bark into Opportunities. pp. 101-4, Oregon State University School of Forestry, Corvallis, Oregon.

104. Herrick, F. W. and R. K. Brandstrom. 1968. U.S. Patent 3,391,542, to Rayonier Inc.

105. Herrick, F. W. and L. H. Bock. 1958. *Forest Prod. J. 8(10)*:269-74.

106. Herrick, F. W. and L. H. Bock. 1958. U.S. Patent 2,819,295, to Rayonier Inc.

107. Herrick, F. W. and L. H. Bock. 1962. U.S. Patent 3,053,784, to Rayonier Inc.

108. Herrick, F. W. and L. H. Bock. 1966. U.S. Patent 3,232,897, to Rayonier Inc.

109. Herrick, F. W. and R. J. Conca. 1960. *Forest Prod. J. 10(7)*:361-8.

110. Herrick, F. W., R. L. Casebier, J. K. Hamilton, and J. D. Wilson. 1975. *J. Appl. Polymer Sci., Appl. Polymer Symposia No. 28*:93-108.

111. Heuser, E. and A. Winsvold. 1923. *Ber. 56B*:902-9.

112. Hilf, R. and F. F. Catano. 1958. *Anal. Chem. 30*:1538-40.

113. Hillis, W. E. 1977. Recent Advances in Phytochemistry *11*:(in press).

114. Holmes, G. W. and E. F. Kurth. 1961. *Tappi 44(12)*: 893-8.

115. Houst and Beveridge. 1963. *J. Nutr. 81(1)*:13-16.

116. Howard, J. 1972. U.S. Patent 3,689,505, to ITT.

117. Howard, J. and T. D. McIntosh. 1973. U.S. Patent
     3,716,574, to Rayonier Inc.
118. Howard, J. and T. D. McIntosh. 1969. Paper presented
     at the Canadian Chemical Conference, Chem. Inst.
     Can., Montreal, Quebec, Canada. (May 28).
119. Howard, J. and T. D. McIntosh. 1972. U.S. Patent
     3,644,481, to ITT Rayonier Inc.
120. Howes, F. N. 1953. Vegetable Tanning Materials.
     325 pp. Butterworths Sci. Pub. London.
121. Hudson, C. S. 1945. *Advan. Carbohyd. Chem. 1*:1-36.
122. Jacobsen, N. A. and E. A. Wilder. 1967. U.S. Patent
     3,297,746, to S. C. Johnson & Son Inc.
123. Jaffe, G. M., W. Szkrybals, and P. H. Weinert. 1974.
     Canadian Patent 945,090.
124. Jones, R. A. and H. B. Lange. 1961. Canadian Patent
     629,710, to Merck and Co.
125. Jullander, I. 1975. *J. Appl. Polymer Sci., Appl.
     Polymer Symposia No. 28*:55-60.
126. Keays, J. L. 1975. *J. Appl. Polymer Sci., Appl.
     Polymer Symposia No. 28*:445-464.
127. Keyes, W. E. 1972. *Am. Pap. Ind. 54(3)*:18-22.
128. Kirk, T. K. and W. E. Moore. 1972. *Wood & Fiber 4*:72-9.
129. Kurth, E. F. 1950. U.S. Patent 2,526,607, to State
     of Oregon.
130. Kurth, E. F. 1950. *J. Am. Chem. Soc. 72*:1685-6.
131. Kurth, E. F. 1953. U.S. Patent 2,662,893, to State
     of Oregon.
132. Kurth, E. F. 1954. U.S. Patent 2,697,717, to State
     of Oregon.
133. Kurth, E. F. and F. L. Chan. 1953. *J. Am. Leather
     Chem. Assoc. 48(1)*:21-32.
134. Kurth, E. F. and H. J. Kiefer. 1950. *Tappi 33(4)*:
     183-6.
135. Lackey, H. B., W. W. Moyer, and W. M. Hearon. 1949.
     Tappi 32(1):469.
136. Langille, D. W., J. Howard, and L. H. Bock. 1969.
     Paper presented at the Canadian Chemical Conference,
     Chem. Inst. Can., Montreal, Quebec, Canada. (May 28).
137. Laver, M. L., H. H. L. Fang, and H. Aft. 1971.
     Phytochemistry 10:3294-5.
138. Loveland, P. M. and M. L. Laver. 1972. *Phytochemistry
     11*:430-2.
139. Loveland, P. M. and M. L. Laver. 1972. *Phytochemistry
     11*:3080-1.
140. Lowe, K. E. 1974. *Pulp & Paper 48(2)*:80-83.

141. MacLean, H. and J. A. F. Gardner. 1956. *Forest Prod. J. 6(10)*:510-16.

142. Marshall, H. B. and C. A. Sankey. 1950, 1951. U.S. Patents 2,516,827, 2,544,999 to Ontario Paper Co.

143. Marton, J. 1964. *Tappi 47(11)*:713-19.

144. Marton, J. and T. Marton. 1964. *Tappi 47(8)*:471-76.

145. McCaleb, K. E. 1968. Lignosulfonates. Chem. Econ. Handbook 596, 5020, 28 pp. Stanford Research Inst., Menlo Park, California.

146. McCarthy, J. L. 1954. *In* Industrial Fermentation. L. A. Underkofler and R. J. Hickey, eds. Vol. 1, Chap. 4. Chem. Publ. Co., New York.

147. McGinness, J. D., J. A. Whittenbaugh, and C. A. Lucchesi. 1960. *Tappi 43(12)*:1027-9.

148. McKibbins, S. W., J. F. Harris, J. F. Saeman, and W. K. Neill. 1962. *Forest Prod. J. 12(1)*:17-23.

149. Mednick, M. L. 1962. *J. Org. Chem. 27*:398-403.

150. Meller, A. 1960. *Holzforschung 14*:78-89, 129-39.

151. Miller, R. W. and W. G. Van Beckum. 1960. *Forest Prod. J. 10(3)*:193-5.

152. Minina, V. S., A. E. Sarukhanova, and K. V. Usmanor. 1962. Fiz i Khim. Prirod. i Sintetich. Polimerov, Akad. Nauk Ua. SSR, Inst. Khim. Polimerov No. *1*:72, 87.

153. Minina, V. S., A. E. Sarukhanova, and K. U. Usmanov. 1962. U.S.S.R. Patent 211,724.

154. Nikitin, V. M. 1974. The Chemical Processing of Wood and Its Perspective. 83 pp., Moscow Press, Moscow, Idaho.

155. Olympic Research Division, ITT Rayonier Inc., unpublished work.

156. Pearl, I. A. 1967. The Chemistry of Lignin. 339 pp. Marcel Dekker Inc., New York.

157. Pearl, I. A. and J. F. McCoy. 1945. *Food Ind. 17*:1458.

158. Plomely, K. F. 1966. Tannin-Formaldehyde Adhesives. pp. 16-19. CSIRO Div. of Forest Prod. Tech. Paper No. 46, Melbourne, Australia.

159. Prahacs, S. 1967. *In* Fuel Gasification. F. C. Schora, ed. pp. 230-52. Advan. Chem. Series No. 69, ACS, Washington, D.C.

160. Prahacs, S., H. G. Barclay, and S. P. Bhatia. 1971. *Pulp Paper Mag. Can. 72(6)*:69-83.

161. Prescott, F. J., J. K. Shaw, J. P. Billello, and G. D. Cragwall. 1953. *Ind. Eng. Chem. 45*:338-42.

162. Pulp-Paper Research Institute of Canada. 1975. Feasibility Study of Production of Chemical Feedstock from Wood Waste. File No. 1354 K L020-4-0002.

163. Quaker Oats Company. 1958. Bulletin 301A, 16 pp.
      Chicago, Illinois.
164. Ramos, E., L. A. Carlow, and R. Vasquez-Romero. 1968.
      Proc. Int. Soc. Sugar Cane Technol. 13:1900.
165. Ramos-Rodriguez, E. 1972. U.S. Patent 3,663,612 to
      Government of Puerto Rico.
166. Ramos-Rodriguez, E. 1972. U.S. Patent 3,701,789 to
      Government of Puerto Rico.
167. Robbins, M. D. 1961. *Chem. Eng. 68(13)*:100-2.
168. Roberts, J. R. and A. S. Gregory. 1958. U.S. Patent
      2,832,765, to Weyerhaeuser Timber Co.
169. Root, D. F., J. F. Saeman, and J. F. Harris. 1959.
      Forest Prod. J. 9:158-65.
170. Rothrock, C. W., W. R. Smith, and R. M. Lindgren.
      1961. *Tappi 44(1)*:65-73.
171. Roux, D. G., D. Ferreira, H. K. L. Hundt, and E. Malan.
      1975. *J. Appl. Polymer Sci., Appl. Polymer Symposia
      No. 28*:335-353.
172. Sakakibara, A. 1963. Japan Patent 26,668, December 23.
173. Samuelson, O. and R. Simonson. 1962. *Svensk. Papper-
      stid. 65*:363-9, 685-9.
174. Sassenrath, C. P. and W. L. Shilling. 1966. U.S.
      Patent 3,258,481, to Crown Zellerbach Corp.
175. Schoeffel, E. W. 1952. U.S. Patent 2,598,311 to
      Salvo Chemical Corp.
176. Schweers, W. H. M. 1969. *Holzforshung 23(1)*:5-9.
177. Sears, K. D. 1972. *J. Org. Chem. 37*:3546-7.
178. Sears, K. D., A. Beelik, R. L. Casebier, R. J. Engen,
      J. K. Hamilton, and H. L. Hergert. 1971. *J.
      Polymer Sci.: Part C, Polymer Symposia No. 36*:425-43.
179. Sears, K. D., R. L. Casebier, H. L. Hergert, G. H.
      Stout, and L. E. McCandlish. 1974. *J. Org. Chem.
      39*:3244-7.
180. Shafizadeh, F. 1968. *Advan. Carbohyd. Chem. 23*:419-74.
181. Shafizadeh, F. 1975. *J. Appl. Polymer Sci., Appl.
      Polymer Symposia No. 28*:153-74.
182. Shaw, C. L. 1975. *Pulp & Paper 50(7)*:65-69.
183. Shevehenke and Klyukvina. 1967. *Bumazh Prom. (5)*:13-14.
184. Shilling, W. L. 1965. *Tappi 48(10)*:105-8A.
185. Shil'nikova, L. L., V. I. Sharkob, and E. F. Bakhmut.
      1966. *Sb. Tr., Vses. Nauch-Issled. Inst. Gidroliz.
      i. Sul'filno-Spirt. Prom. 15*:263.
186. Simonsen, J. L. and L. N. Owen. 1949. The Terpenes.
      2nd ed., Two volumes. University Press, Cambridge.
187. Sorensen, N. A. and J. Mehlum. 1956. U.S. Patent
      2,752,394 to Mechlum.

188. Strohlman, L. E. 1969. *Paper Trade J. 153(19)*:55.
189. Swan, E. P., K. S. Jiang, and J. A. F. Gardner. 1969.
   Phytochemistry 8:345–51.
190. Thornburg, W. L. 1963. *Tappi 46(8)*:453–455.
191. Timell, T. E., ed. 1975. Proceed. 8th Cellulose
   Conf. I. Wood Chemicals – A Future Challenge.
   J. Applied Polymer Sci., Applied Polymer Symposia
   No. 28, 380 pp., John Wiley & Sons, New York.
192. Trocino, F. S. 1971. U.S. Patent 3,616,201, to
   Cascade Fiber Co.
193. Trocino, F. S. 1973. U.S. Patent 3,781,187, to
   Bohemia Lumber Co., Inc.
194. Trocino, F. S. 1976. Canadian Patent 984,382, to
   Bohemia Lumber Co., Inc.
195. Turner, G. H. 1959. U.S. Patent 2,877,165, to
   Consolidated Mining and Smelting Co. of Canada Ltd.
196. 1973. U.S. Tariff Commission Reports.
197. Vdovenko, N. N., et al. 1966. U.S.S.R. Patent
   227,318.
198. Voci, J. J., R. C. Canty, K. Ginvala, and F. G. Perry.
   1958. Paper presented at the 12th National Meeting,
   Forest Products Research Society, Madison, Wisconsin.
   29 pp. (June 22–28).
199. Vorher, W. and W. H. M. Schweers. 1975. *J. Appl.*
   Polymer Sci., Appl. Polymer Symposium No. 28:277–84.
200. Ward, F. B. 1975. *J. Appl. Polymer Sci., Appl. Polymer*
   Symposia No. 28:329–34.
201. Wenzl, H. F. J. 1970. The Chemical Technology of
   Wood. 692 pp., Academic Press, New York.
202. West, F. W. et al. 1961. U.S. Patent 3,009,938 to
   M. W. Kellogg Co.
203. Wilke, C. R., ed. 1975. Cellulose as a Chemical and
   Energy Resource. Symposium No. 5, Biotech. Bioeng.
   361 pp., Interscience, New York.
204. Wolfrom, M. L., A. Thompson, and R. D. Ward. 1959.
   J. Am. Chem. Soc. 81:4623–5.
205. Zinkel, D. R. 1975. *J. Appl. Polymer Sci., Appl.*
   Polymer Symposia No. 28:309–327.
206. Zoch, L. L., J. F. Harris, and E. L. Springer. 1959.
   Tappi 52:486–8.